T0406764

Ngoc Thanh Nguyen and Lakhmi C. Jain (Eds.)

Intelligent Agents in the Evolution of Web and Applications

Studies in Computational Intelligence, Volume 167

Editor-in-Chief
Prof. Janusz Kacprzyk
Systems Research Institute
Polish Academy of Sciences
ul. Newelska 6
01-447 Warsaw
Poland
E-mail: kacprzyk@ibspan.waw.pl

Further volumes of this series can be found on our homepage: springer.com

Vol. 145. Mikhail Ju. Moshkov, Marcin Piliszczuk and Beata Zielosko
Partial Covers, Reducts and Decision Rules in Rough Sets, 2008
ISBN 978-3-540-69027-6

Vol. 146. Fatos Xhafa and Ajith Abraham (Eds.)
Metaheuristics for Scheduling in Distributed Computing Environments, 2008
ISBN 978-3-540-69260-7

Vol. 147. Oliver Kramer
Self-Adaptive Heuristics for Evolutionary Computation, 2008
ISBN 978-3-540-69280-5

Vol. 148. Philipp Limbourg
Dependability Modelling under Uncertainty, 2008
ISBN 978-3-540-69286-7

Vol. 149. Roger Lee (Ed.)
Software Engineering, Artificial Intelligence, Networking and Parallel/Distributed Computing, 2008
ISBN 978-3-540-70559-8

Vol. 150. Roger Lee (Ed.)
Software Engineering Research, Management and Applications, 2008
ISBN 978-3-540-70774-5

Vol. 151. Tomasz G. Smolinski, Mariofanna G. Milanova and Aboul-Ella Hassanien (Eds.)
Computational Intelligence in Biomedicine and Bioinformatics, 2008
ISBN 978-3-540-70776-9

Vol. 152. Jarosław Stepaniuk
Rough – Granular Computing in Knowledge Discovery and Data Mining, 2008
ISBN 978-3-540-70800-1

Vol. 153. Carlos Cotta and Jano van Hemert (Eds.)
Recent Advances in Evolutionary Computation for Combinatorial Optimization, 2008
ISBN 978-3-540-70806-3

Vol. 154. Oscar Castillo, Patricia Melin, Janusz Kacprzyk and Witold Pedrycz (Eds.)
Soft Computing for Hybrid Intelligent Systems, 2008
ISBN 978-3-540-70811-7

Vol. 155. Hamid R. Tizhoosh and M. Ventresca (Eds.)
Oppositional Concepts in Computational Intelligence, 2008
ISBN 978-3-540-70826-1

Vol. 156. Dawn E. Holmes and Lakhmi C. Jain (Eds.)
Innovations in Bayesian Networks, 2008
ISBN 978-3-540-85065-6

Vol. 157. Ying-ping Chen and Meng-Hiot Lim (Eds.)
Linkage in Evolutionary Computation, 2008
ISBN 978-3-540-85067-0

Vol. 158. Marina Gavrilova (Ed.)
Generalized Voronoi Diagram: A Geometry-Based Approach to Computational Intelligence, 2009
ISBN 978-3-540-85125-7

Vol. 159. Dimitri Plemenos and Georgios Miaoulis (Eds.)
Artificial Intelligence Techniques for Computer Graphics, 2009
ISBN 978-3-540-85127-1

Vol. 160. P. Rajasekaran and Vasantha Kalyani David
Pattern Recognition using Neural and Functional Networks, 2009
ISBN 978-3-540-85129-5

Vol. 161. Francisco Baptista Pereira and Jorge Tavares (Eds.)
Bio-inspired Algorithms for the Vehicle Routing Problem, 2009
ISBN 978-3-540-85151-6

Vol. 162. Costin Badica, Giuseppe Mangioni, Vincenza Carchiolo and Dumitru Dan Burdescu (Eds.)
Intelligent Distributed Computing, Systems and Applications, 2008
ISBN 978-3-540-85256-8

Vol. 163. Pawel Delimata, Mikhail Ju. Moshkov, Andrzej Skowron and Zbigniew Suraj
Inhibitory Rules in Data Analysis, 2009
ISBN 978-3-540-85637-5

Vol. 164. Nadia Nedjah, Luiza de Macedo Mourelle, Janusz Kacprzyk, Felipe M.G. França and Alberto Ferreira de Souza (Eds.)
Intelligent Text Categorization and Clustering, 2009
ISBN 978-3-540-85643-6

Vol. 165. Djamel A. Zighed, Shusaku Tsumoto, Zbigniew W. Ras and Hakim Hacid (Eds.)
Mining Complex Data, 2009
ISBN 978-3-540-88066-0

Vol. 166. Constantinos Koutsojannis and Spiros Sirmakessis (Eds.)
Tools and Applications with Artificial Intelligence, 2009
ISBN 978-3-540-88068-4

Vol. 167. Ngoc Thanh Nguyen and Lakhmi C. Jain (Eds.)
Intelligent Agents in the Evolution of Web and Applications, 2009
ISBN 978-3-540-88070-7

Ngoc Thanh Nguyen
Lakhmi C. Jain
(Eds.)

Intelligent Agents in the Evolution of Web and Applications

Ngoc Thanh Nguyen
Institute of Control and Systems Engineering
Wroclaw University of Technology
ul. Janiszewskiego 11/17
50-370 Wroclaw
Poland
Email: Ngoc-Thanh.Nguyen@pwr.wroc.pl

Lakhmi C. Jain
SCT-Building,
University of South Australia,
Adelaide City,
Mawson Lakes Campus,
South Australia SA 5095
Australia
Email: Lakhmi.jain@unisa.edu.au

ISBN 978-3-540-88070-7 e-ISBN 978-3-540-88071-4

DOI 10.1007/978-3-540-88071-4

Studies in Computational Intelligence ISSN 1860949X

Library of Congress Control Number: 2008935500

Typeset & Cover Design: Scientific Publishing Services Pvt. Ltd., Chennai, India.

Printed in acid-free paper

9 8 7 6 5 4 3 2 1

springer.com

Preface

Intelligent agents have revolutionised the way we do business, we teach, we learn, design systems, and so on. Agent applications are increasingly being developed in domains as diverse as meteorology, manufacturing, war gaming, UAV mission management and the evolution of Web [1]. The Web has also has the same effect on our daily life as the intelligent agents. We use Web for information search, shopping, news, communication and so on. We wonder how we lived without Web in the past [2].

The book presents a sample of some of the most innovative research on the use of intelligent agents in the evolution of Web. There are thirteen chapters in the book. Chapters are on theoretical foundations as well as practical applications.

We are grateful to the contributors and reviewers for their contribution. We believe that the research reported in the book will encourage researchers to develop the robust human-like intelligent machines for the service of humans.

We sincerely thank Springer-Verlag for their editorial support during the preparation of the manuscript. The editors appreciate the resources provided by Wroclaw University of Technology and the University of South Australia to edit this volume.

N.T. Nguyen
L.C. Jain

References

[1] Tweedale, J., Ichalkaranje, N., Sioutis, C., Jarvis, B., Consoli, A., Phillips-Wren, G.: Innovations in multi-agent systems. Journal of Network Computing Applications 30(3), 1089–1115 (2007)

[2] Nayak, R., Ichalkaranje, N., Jain, L.C. (eds.): Evolution of the Web in Artificial Intelligence Environments. Springer, Heidelberg (2008)

Contents

1

The Evolution of Intelligent Agents within the World Wide Web

Jeffrey Tweedale and Lakhmi Jain

School of Electrical and Information Engineering,
Knowledge Based Intelligent Engineering Systems Centre,
University of South Australia, Mawson Lakes, SA 5095, Australia
{Jeff.Tweedale, Lakhmi.Jain}@unisa.edu.au

Abstract. The definition of an agent still needs to be agreed upon [1] and the use of multiple agents to form a team is being examined by many researchers [2]. The study of Artificial Intelligence (AI) is diverse because of each domain has encountered a bottleneck or some impasse has forced research to look further a field to find solutions [3]. Agent teaming was one of those choices. Each team consists of one or more agent which form a Multi-Agent System (MAS) [4]. Currently these have a fixed hierarchy and predetermined functionality to achieve specified goals [5]. Ideally that teams should seamlessly interoperate within its environment, autonomously adapt to new tasks and rapidly switch context as required. Learning, cooperation, collaboration and trust are other characteristics that deserve discussion and development, however, the above challenge would represent a significant leap in the natural progression to agent oriented programming.

Keywords: Computational Intelligence, Artificial Intelligence, Agents, Multi-Agent Systems, Teaming.

1.1 Introduction

Researchers have been presented with many Artificial Intelligence (AI) challenges as they attempt to increase the level of personification in intelligent systems. These challenges are both technical and psychological in nature. Agent technologies, and in particular agent teaming, are increasingly being used to aid in the design of "intelligent" systems [6]. In the majority of the agent-based software currently being produced, the structure of agent teams have been reliant on that defined by the programmer or software engineer.

Over the last decades, Artificial Intelligence (AI) has made a great deal of progress in many fields, such as knowledge representation, inference, machine learning, vision and robotics [7, 8]. Minsky poses that Artificial Intelligence is the science of making machines do things that would require intelligence if done by man [9]. Many researchers regard Artificial Intelligence as more than engineering, demanding the study of science about human and animal intelligence be included. Intelligence considers cognitive aspects of human behaviour, such

N.T. Nguyen, L.C. Jain (Eds.): Intel. Agents in the Evol. of Web & Appl., SCI 167, pp. 1–9.
springerlink.com

as perceiving, reasoning, planning, learning and communication. AI was initially conceived by Newell and Simon using production systems [10]; however, the study quickly divided into two streams with John McCarthy and Nil Nillson considered the *Neats* (using formal logic as a central tool to achieving Artificial Intelligence), while Marvin Minsky and Roger Schanks where considered the *scrufs* (using a psychological approach to Artificial Intelligence). Russel and Norvig entered the argument by describing an *environment* as something that provides input and receives output, using *sensors* as inputs to a program, producing outputs as a result of *acting* on something within that program. The Artificial Intelligence community now uses this notion as the basis of definition of an agent [1].

Intelligent agent technology has been touted as becoming the paradigm of choice for the development of complex distributed systems and as the natural progression to object oriented programming. Learning has an important role to play in both cooperative and autonomous systems. Agents with predefined behaviours based on a priori knowledge of the system that is modified using feedback from experience will continue to mature. Rather than having purely agent-based applications, we then have cooperative applications involving teams of agents and humans. We expect that intelligent agents will retain their architectural foundations but the availability of more appropriate reasoning models and better design methodologies will see them being increasingly used in mainstream software development. Furthermore, better support for human-agent teams will see the development of a new class of intelligent decision support applications.

1.2 Intelligent Agents and Web

Simply speaking, an Artificial Intelligence application is a system that possesses knowledge about an application domain that takes in data from its environment, and reasons about that data to derive information. The system may be combined with various algorithms for reasoning, learning, planning, speech recognition, vision, and language understanding. For example, knowledge can be represented procedurally or as a set of logical conditions. The rules formed the initial inference conditions in expert systems used to determine a conclusion. If all of conditions are true, then the conclusion holds. When reasoning, an expert system, forward chaining is used. A tree or path is built by which reverse chaining can be used to derive estimates or values from which facts could be extrapolated to validate the conclusions. The conclusion from one rule can be compounded to form part of another condition to build more complex decision support.

A blackboard system may be thought of as a componentized system, where each box could function as a database, series of "pigeon holes" or behave with an unknown black box behaviour that represents a specific aspect of a system or sub-systems engaging a problem. This needs to occur in an environment where experts and modular software subsystems, called knowledge repositories, capable of representing different points of view, strategies, and knowledge formats, required to solve a problem. These problem-solving paradigms may include:

- Rule-Based Systems,
- Case-Based Systems,
- Bayesian Networks,
- Genetic algorithms and programs,
- Neural Networks,
- Fuzzy Logic Systems,
- Legacy (traditional or formal procedural) software systems, and
- Hybrid systems, consisting of Teams, Multi-Agent System (MAS) and complete agent oriented systems.

As can be seen, this metaphor is used as a 'meta-architecture', by which a group of experts gathers around a blackboard to collaboratively solve a complex problem. They form a good way of communicating heterogeneously, however all sources do not need to know about each other. The black box model has been chosen to wrap our project to ensure that it has structure and is able to comply with the rigor of model agency.

We expect Intelligent Agent (IA) will retain their architectural foundations, but that the availability of more appropriate reasoning models and better design methodologies will assist them being used increasingly in mainstream software development. Furthermore, better support for human-agent teams will see the development of a new class of intelligent decision support applications. Autonomous connectivity using one of the existing communication modes will be required to assist with the higher level interoperability.

1.3 Chapters Included in This Book

The book includes thirteen chapters. Chapter one introduces intelligent agents and their impact on the evolution of web. Chapter two proposes a multi-agent based system for the semi-automated study of the usability of the websites. The architecture of the system uses rule-based agents in the evolutionary environment. Chapter three is on the norm-governed behaviour in virtual enterprises. The analysis of a number of artificial societies and normative systems is presented.

Chapter four presents e-JABAT: a novel middle ware for the design and implementation of A-Team for solving optimization problems. Chapter five presents a resource discovery method based on multiple mobile agents in P2P systems. The method is validated by comparing the quantity of message data transmitted in the P2P system. Chapter six is on the use of agents in web-based education systems. The authors have divided the architecture of the system into the client and server to facilitate adaptability. Chapter seven is on browsing agent-based assistant for changing pages. The assistant adapts views of revisited pages, searches history and produces summary of page history. Chapter eight is on resource management in agent-based virtual organization. The authors have presented their agent-based system used to fulfill the specific needs of the users.

Chapter nine presents the architecture for developing mixed initiative Beliefs, Desires, Intentions (BDI) Multi-Agent System (MAS) for supporting reasoning, problem solving and adaptation. Chapter ten is on ontology for agents and the web. The authors have presented guidelines related to the use of ontology for agents and the web. Chapter eleven is also about ontology for agents and their applications in the web-based adaptive and intelligent education systems. Chapter twelve presents an evolutionary approach for intelligent negotiation agents in e-market places. The final chapter presents the security friendly architectures of the intelligent agents for the web-based applications.

1.4 Summary

Agent technology is still in need of the killer application and more mature tools to promote its acceptance in commercial circles. Intelligent agent technology is an extension of agent research and currently appears to be maturing [11]. The development of intelligent agent applications using current generation agents is not yet routine, however commercial strength agent applications are increasingly being developed in domains as diverse as meteorology, manufacturing, war gaming, capability assessment and UAV mission management. The distinguishing feature of this paradigm is that an agent can have autonomy over its execution. Furthermore commercially supported development environments are becoming available and design methodologies, are being supported by reference architectures and standards. The uptake of the technology is not as rapid or as pervasive as its advocates would have expected, predominantly due to a number of technological reasons, such as; the architecture (Harvard and Von Neuman microprocessors), parallelism, speed and capacity. Although there are current impediments, the future is positive and it is believed that Multi-Agent System (MAS) technology will become another disruption in the technology continuum.

1.5 Resources

Following is a sample of resources on intelligent agents and Web. It includes a non-exclusive list of significant resources from journals, special issues of specific journals, conferences, conference proceedings, conference notes, book series and books. The reader will also find the Knowledge-Based Intelligent Information and Engineering Systems (KES) web site[1] for additional material and resources.

1.5.1 Journals

- IEEE Internet Computing, IEEE Press, USA
  www.computer.org/internet/
- EEE Intelligent Systems, IEEE Press, USA
  www.computer.org/intelligent/

[1] http://www.unisa.edu.au/kes/International_conference/default.asp.

- Web Intelligence and Agent systems, IOS Press, The Netherlands.
- International Journal of Knowledge-Based intelligent Engineering systems, IOS Press, The Netherlands.
  http://www.kesinternational.org/journal/
- International Journal of Hybrid Intelligent Systems, IOS Press, The Netherlands.
  http://www.iospress.nl/html/14485869.html
- Intelligent Decision Technologies: An International Journal, IOS Press, The Netherlands.
  http://www.iospress.nl/html/18724981.html

1.5.2 Special Issue of Journals

- Wade, V.P. and Ashman, H., Evolving the Infrastructure for Technology Enhanced Distance Learning, IEEE Internet Computing, Volume 11, Number 3, 2007.
- Tu, S. and Abdelguerfi, M., Web Services for GIS, IEEE Internet Com-puting, Volume 10, Number 5, 2006.
- Cahill, V. and Clarke, S., Roaming: Technology for a Connected Society, IEEE Internet Computing, Volume 11, Number 2, 2007.
- Abraham, A. and Jain, L.C., Computational Intelligence on the Internet, Journal of Network and Computer Applications, Elsevier Publishers, Volume 28, Number 2, 2005.
- Abraham, A. and Jain, L.C., Knowledge Engineering in an Intelligent Environment, Journal of Intelligent and Fuzzy Systems, The IOS Press, The Netherlands, Volume 14, Number 3, 2003.

1.5.3 Conferences

- IEEE/WIC/ACM International Conferences on Web Intelligence
- AAAI Conference on Artificial Intelligence
  www.aaai.org/aaai08.php
- KES International Conference Series
  www.kesinternational.org/
- European Conferences on Artificial Intelligence (ECAI)
- Australian World Wide Web Conferences
  http://ausweb.scu.edu.au

1.5.4 Conference Proceedings

- Apolloni, B., Howlett, R.J. and Jain, L.C. (Editors), Knowledge-Based Intelligent Information and Engineering Systems, Lecture Notes in Artificial Intelligence, Volume 1, LNAI 4692, KES 2007, Springer-Verlag, Germany, 2007.

- Apolloni, B.,Howlett, R.J.and Jain, L.C. (Editors), Knowledge-Based Intelligent Information and Engineering Systems, Lecture Notes in Artificial Intelligence, Volume 2, LNAI 4693, , KES 2007, Springer-Verlag, Germany, 2007.
- Apolloni, B.,Howlett, R.J.and Jain, L.C. (Editors), Knowledge-Based Intelligent Information and Engineering Systems, Lecture Notes in Artificial Intelligence, Volume 3, LNAI 4694, KES 2007, Springer-Verlag, Germany, 2007.
- Nguyen, N.T., Grzech, A., Howlett, R.J. and Jain, L.C., Agents and Multi-Agents Systems: Technologies and Applications, Lecture Notes in artificial Intelligence, LNAI 4696, Springer-Verlag, Germany, 2007.
- Howlett, R.P., Gabrys, B. and Jain, L.C. (Editors), Knowledge-Based Intelligent Information and Engineering Systems, Lecture Notes in Artificial Intelligence, KES 2006, Springer-Verlag, Germany, Vol. 4251, 2006.
- Howlett, R.P., Gabrys, B. and Jain, L.C. (Editors), Knowledge-Based Intelligent Information and Engineering Systems, Lecture Notes in Artificial Intelligence, KES 2006, Springer-Verlag, Germany, Vol. 4252, 2006.
- Howlett, R.P., Gabrys, B. and Jain, L.C. (Editors), Knowledge-Based Intelligent Information and Engineering Systems, Lecture Notes in Artificial Intelligence, KES 2006, Springer-Verlag, Germany, Vol. 4253, 2006.
- Liao, B.-H., Pan, J.-S., Jain, L.C., Liao, M., Noda, H. and Ho, A.T.S., Intelligent Information Hiding and Multimedia Signal Processing, IEEE Computer Society Press, USA, 2007. ISBN: 0-7695-2994-1.
- Khosla, R., Howlett, R.P., and Jain, L.C. (Editors), Knowledge-Based Intelligent Information and Engineering Systems, Lecture Notes in Artificial Intelligence, KES 2005, Springer-Verlag, Germany, Vol. 3682, 2005.
- Skowron, A., Barthes, P., Jain, L.C., Sun, R.,Mahoudeaux, P., Liu, J. and Zhong, N.(Editors), Proceedings of the 2005 IEEE/WIC/ACM International Conference on Intelligent Agent Technology, Compiegne, France, IEEE Computer Society Press, USA, 2005.
- Khosla, R., Howlett, R.P., and Jain, L.C. (Editors), Knowledge-Based Intelligent Information and Engineering Systems, Lecture Notes in Artificial Intelligence, KES 2005, Springer-Verlag, Germany, Vol. 3683, 2005.
- Khosla, R., Howlett, R.P., and Jain, L.C. (Editors), Knowledge-Based Intelligent Information and Engineering Systems, Lecture Notes in Artificial Intelligence, KES 2005, Springer-Verlag, Germany, Vol. 3684, 2005.
- Khosla, R., Howlett, R.P., and Jain, L.C. (Editors), Knowledge-Based Intelligent Information and Engineering Systems, Lecture Notes in Artificial Intelligence, KES 2005, Springer-Verlag, Germany, Vol. 3685, 2005.
- Negoita, M., Howlett, R.P., and Jain, L.C. (Editors), Knowledge-Based Intelligent Engineering Systems, KES 2004, Lecture Notes in Artificial Intelligence, Vol. 3213, Springer, 2004
- Negoita, M., Howlett, R.P., and Jain, L.C. (Editors), Knowledge-Based Intelligent Engineering Systems, KES 2004, Lecture Notes in Artificial Intelligence, Vol. 3214, Springer, 2004

- Negoita, M., Howlett, R.P., and Jain, L.C. (Editors), Knowledge-Based Intelligent Engineering Systems, KES 2004, Lecture Notes in Artificial Intelligence, Vol. 3215, Springer, 2004
- Murase, K., Jain, L.C., Sekiyama, K. and Asakura, T. (Editors), Proceedings of the Fourth International Symposium on Human and Artificial Intelligence Systems, University of Fukui, Japan, 2004.
- Palade, V., Howlett, R.P., and Jain, L.C. (Editors), Knowledge-Based Intelligent Engineering Systems, Lecture Notes in Artificial Intelligence, Vol. 2773, Springer, 2003
- Palade, V., Howlett, R.P., and Jain, L.C. (Editors), Knowledge-Based Intelligent Engineering Systems, Lecture Notes in Artificial Intelligence, Vol. 2774, Springer, 2003
- Damiani, E., Howlett, R.P., Jain, L.C. and Ichalkaranje, N. (Editors), Proceedings of the Fifth International Conference on Knowledge-Based Intelligent Engineering Systems, Volume 1, IOS Press, The Netherlands, 2002.
- Damiani, E., Howlett, R.P., Jain, L.C. and Ichalkaranje, N. (Editors), Proceedings of the Fifth International Conference on Knowledge-Based Intelligent Engineering Systems, Volume 2, IOS Press, The Netherlands, 2002.
- Baba, N., Jain, L.C. and Howlett, R.P. (Editors), Proceedings of the Fifth International Conference on Knowledge-Based Intelligent Engineering Systems (KES2001), Volume 1, IOS Press, The Netherlands, 2001.
- Baba, N., Jain, L.C. and Howlett, R.P. (Editors), Proceedings of the Fifth International Conference on Knowledge-Based Intelligent Engineering Systems (KES2001), Volume 2, IOS Press, The Netherlands, 2001.
- Howlett, R.P. and Jain, L.C.(Editors), Proceedings of the Fourth International Conference on Knowledge-Based Intelligent Engineering Systems, IEEE Press, USA, 2000. Volume 1.
- Howlett, R.P. and Jain, L.C.(Editors), Proceedings of the Fourth International Conference on Knowledge-Based Intelligent Engineering Systems, IEEE Press, USA, 2000. Volume 2.
- Jain, L.C.(Editor), Proceedings of the Third International Conference on Knowledge-Based Intelligent Engineering Systems, IEEE Press, USA, 1999.
- Jain, L.C. and Jain, R.K. (Editors), Proceedings of the Second International Conference on Knowledge-Based Intelligent Engineering Systems, Volume 1, IEEE Press, USA, 1998.
- Jain, L.C. and Jain, R.K. (Editors), Proceedings of the Second International Conference on Knowledge-Based Intelligent Engineering Systems, Volume 2, IEEE Press, USA, 1998.
- Jain, L.C. and Jain, R.K. (Editors), Proceedings of the Second International Conference on Knowledge-Based Intelligent Engineering Systems, Volume 3, IEEE Press, USA, 1998.
- Jain, L.C. (Editor), Proceedings of the First International Conference on Knowledge-Based Intelligent Engineering Systems, Volume 1, IEEE Press, USA, 1997.

- Jain, L.C. (Editor), Proceedings of the First International Conference on Knowledge-Based Intelligent Engineering Systems, Volume 2, IEEE Press, USA, 1997.
- Narasimhan, V.L., and Jain, L.C. (Editors), The Proceedings of the Australian and New Zealand Conference on Intelligent Information Systems, IEEE Press, USA, 1996.

1.5.5 Book Series

- Advanced Intelligence and Knowledge Processing, Springer-Verlag, Germany `www.springer.com/series/4738`
- Computational Intelligence and its Applications Series, Idea group Publishing, USA
  `http://www.igi-pub.com/bookseries/details.asp?id=5`
- The CRC Press International Series on Computational Intelligence, The CRC Press, USA
- Series on Innovative Intelligence
  `http://www.worldscientific.com.sg/books/series/sii_series.shtml`
- Advanced Information Processing, Springer-Verlag, Germany.
- Knowledge-Based Intelligent Engineering Systems Series, IOS Press, The Netherlands.
  `http://www.kesinternational.org/bookseries.php`
- International series on Natural and artificial Intelligence, AKI.
  `http://www.innoknowledge.com`

1.5.6 Books

- Henninger, M., The Hidden Web, Second Edition, University of New South Wales Press Ltd, Australia, 2008.
- Jarvis, J., Ronnquist, R, Jarvis, D. and Jain, L.C., Holonic Execution: A BDI Approach, Springer-Verlag, 2008.
- Tsihrintzis, G. and Jain, L.C., Multimedia Services in Intelligent Environments, Springer-Verlag, 2008.
- Jain, L.C., Srinivasan, D. and Palade, V. (Editors), Advances in Evolutionary Computing for system Design, Springer-Verlag, 2007.
- Khosla, R., Ichalkaranje, N. and Jain, L.C.(Editors), Design of Intelligent Multi-Agent Systems, Springer-Verlag, Germany, 2005.
- Nikravesh, M., et al. (Editors), Enhancing the power of Internet, Springer-Verlag, Germany, 2004.
- Howlett, R., Ichalkaranje, N., Jain, L.C. and Tonfoni, G. (Editors), Internet-Based Intelligent Information Processing, World Scientific Publishing Company Singapore, 2002.
- Jain, L.C., et al. (Editors), Intelligent Agents and Their Applications, Springer-Verlag, Germany, 2002.
- Deen, S.M.(Editor), Agent-Based Manufacturing, Springer-Verlag, Germany, 2003.

- Meisels, A., Distributed Search by Constrained Agents, Springer-Verlag, London, 2008.
- Ko, C.C., Creating Web-based Laboratories, Springer-Verlag, London, 2004.
- Stuckenschmidt, H. and Harmelen, F.V., Information Sharing on the Semantic Web, Springer-Verlag, London, 2005.
- Namatame, A., et al. (Editors), Agent-Based Approaches in Economic and Social Complex Systems, IOS Press, The Netherlands.

References

1. Franklin, S., Graesser, A.: Is it an agent, or just a program?: A taxonomy for autonomous agents. In: Proceedings of the Third International Workshop on Agent Theories, Architectures and Languages. Budapest, Hungary, pp. 193–206 (1996)
2. Nwana, H.S.: Software agents: An overview. In: McBurney, P. (ed.) The Knowledge Engineering Review, Cambridge Journals, Simon Parsons, vol. 11(3), pp. 205–244. City University of New York, USA (1996)
3. Russel, S., Norvig, P.: Artificial Intelligence: A Modern Approach, 2nd edn. Prentice-Hall, Inc., Englewood Cliffs (2003)
4. Rudowsky, I.: Intelligent agents. Communications of the Association for Information Systems, 275–190 (2004)
5. Shoham, Y.: An overview of agent-oriented programming. In: Bradshaw, J.M. (ed.) Software Agents, vol. 4. AAAI Press, Menlo Park (1997)
6. Wooldridge, M., Jennings, N.R.: The cooperative problem-solving process. Journal of Logic and Computation 9(4), 563–592 (1999)
7. Grevier, D.: AI – The Tumultuous History of the Search for Artificial Intelligence. Basic Books, New York (1993)
8. Callan, R.: Artificial Intelligence. Palgrave MacMillan, Hampshire (2003)
9. Minsky, M.: Society of Mind. Simon and Schuster, Pymble (1985)
10. Thagard, P.R.: Computational Philiosphy of Science. MIT Press, Cambridge (1993)
11. Tweedale, J., Ichalkaranje, N., Sioutis, C., Jarvis, B., Consoli, A., Phillips-Wren, G.: Innovations in multi-agent systems. Journal of Network Computing Applications 30(3), 1089–1115 (2007)

2

A Multi-agent System Based on Evolutionary Learning for the Usability Analysis of Websites

Eduardo Mosqueira-Rey, David Alonso-Ríos, Ana Vázquez-García, Belén Baldonedo del Río, and Vicente Moret-Bonillo

Department of Computer Science, University of A Coruña, Campus de Elviña, 15071, A Coruña, Spain
{eduardo, dalonso, mavazquez, belen, civmoret}@udc.es

Abstract. In this chapter we propose a novel multi-agent system for the semi-automated study of the usability of websites, an increasingly critical issue given the ubiquity of the Web and its technological evolution. The proposed system constructs a key phrase-based model of the users trying to reach one URL from another, simulates the browsing process, and analyses the web pages in the path. The resulting usability analysis is focused on issues such as navigation paths, links, page content, HTML coding, and accessibility. Our system automatically draws usability conclusions and suggestions and also presents significant data in support of the human usability expert. The architecture of the system consists of rule-based reactive agents subject to evolutionary processes. The application of evolution allows the agents to explore possible solutions in a more realistic way than either exhaustive or arbitrary examinations.

Keywords: Usability, websites, intelligent agents, evolutionary learning, user modelling, HTML analysis, genetic algorithms.

2.1 Introduction

This chapter presents a novel evolutionary multi-agent system for the semi-automated study of the usability of websites. The system constructs a model of the users trying to reach one URL from another, simulates the browsing process, and analyses the web pages that make up possible paths between source and destination. As a result, the system is capable of automatically making suggestions and critiques with respect to usability aspects. Finally, the system selects and suitably presents relevant information in support of the human usability expert whose task it is to analyse usability.

The chapter is structured as follows: section 2.2 defines the concept of usability, places it in the context of the World Wide Web, and summarises the existing techniques for the study of usability; section 2.3 introduces the concept of evolutionary intelligent agents, which constitute the basis of our system; section 2.4 presents GAEL, the programming framework we created to facilitate the development of evolutionary multi-agent systems; section 2.5 describes in detail our evolutionary multi-agent system for the study of the usability of websites; section 2.6 shows an example of the system at work; section 2.7 provides a discussion and a summary of our conclusions; finally, the chapter ends with some proposals for future work.

N.T. Nguyen, L.C. Jain (Eds.): Intel. Agents in the Evol. of Web & Appli., SCI 167, pp. 11–34.
springerlink.com

2.2 The Study of Web Usability

In this section we firstly define the concepts of usability and User-Centred Design, considering them in light of the technological and social implications of the evolution of computers and the World Wide Web. We then examine the usability results found in the real world and the actual role played by usability studies in web development. We finish the section with a summary of usability study methods and their corresponding automation techniques.

2.2.1 The Evolution of the Web and Its Users

The spectacular evolutions of the World Wide Web in the past 15 years and its corresponding popularisation [1] have brought with them a remarkable diversification of the types of web users. At present, websites and web applications are used on a daily basis by persons from different age groups, from diverse cultures, and with different computer skills. Moreover, technology is becoming increasingly available to people with different types and varying levels of disabilities. Thus, the potential users of a website can be so heterogeneous that it can be impossible to know their characteristics in advance.

At the same time, technological progress has brought about a significant change of mentality in users, compelling us to become more demanding and to get what we want with the least effort. As a consequence, we are quick to become impatient with user-unfriendly applications.

Our society has progressively become aware of this new situation, which has finally been legislatively recognised with examples like the 90/270/EEC Council Directive of the European Union [2] and the Section 508 of the government of the United States of America [3].

Naturally, website companies have also had to pay attention to all these issues. These companies operate in a very competitive environment and realise that they cannot afford to ignore the needs of their users. Doing so would have a negative impact on the image and the credibility of both the product and the company. In fact, software vendors have lately been highlighting the user-friendliness of their products in their advertisements, which is a significant change of policy. Formerly, when applications were simpler and more expensive, the number of functionalities was by far the most decisive factor when choosing a product, and it was considered perfectly acceptable to ask users to adjust their needs to technological limitations [4].

2.2.2 User-Centred Design and Usability

User-Centred Design (UCD) is a new design philosophy that offers a user-aware alternative to conventional "system-centred" design paradigms. In these traditional paradigms, design decisions were guided by the functionalities and the internal architecture of the system, and technological attributes such as performance were given maximum priority. In UCD, on the other hand, the needs and the limitations of the users are the factors that guide design decisions.

One of the key concepts in UCD is usability, which is defined by the International Organization for Standardization as the "extent to which a product can be used by specified users to achieve specified goals with effectiveness, efficiency and satisfaction in a specified context of use" [5]. Effectiveness is defined as the "accuracy and completeness with which users achieve specified goals", efficiency as the "resources expended in relation to the accuracy and completeness with which users achieve specified goals", and satisfaction as the "freedom from discomfort, and positive attitudes towards the use of the product". As indicated by its definition, usability is not an inherent property of a system. Rather, it is relative to its context of use, which is composed of users, tasks, equipment and environments.

2.2.3 Usability in the Real World

In spite of the circumstances described in section 2.2.1, the consensus among usability experts is that the majority of websites have usability or accessibility problems [6] [7] [8].

This situation has many negative consequences. On an individual level, users are not able to find information or carry out tasks in an efficient or satisfying manner. As a result, both their productivity and their interest in exploring the systems decrease. On a group work level, the work flow can become unjustifiably interrupted, and it is also necessary to dedicate more resources to user training and support. In short, performance may be so negatively affected that it can be preferable not to use the system at all.

The poor usability of web software is not a consequence of inherent, unavoidable technological flaws. Rather, it is frequently a product of bad management policies or simply the lack of necessary resources. Bad usability policies are often caused by a poor understanding of the real importance and complexity of the problem, but they can also be motivated by some kind of misguided pragmatism that prioritises more "urgent" and "concrete" matters, such as economic costs or the reuse of resources. In the real world, the actual nature of the situation remains largely unknown, so, even today, usability problems are widely considered inevitable—or, even worse, the blame is placed on the users.

Unfortunately, the benefits of paying attention to usability tend to be cumulative, difficult to measure, and only apparent after the system is released, which makes it very hard to justify the necessary investment. The task of solving usability problems is therefore considered a low-priority issue and is normally carried out through ad hoc techniques. These kinds of techniques are very popular because they are simple, quick and cheap, but they tend to provide only partial and limited solutions. In order to properly solve usability problems, it is necessary to apply more rigorous methods of study.

2.2.4 Usability Study Methods

A proper study of the usability of a system is a complex task involving both objective and subjective elements. The former can be evaluated using metrics and rules, whereas the latter can be assessed through the opinions of users and the judgements of usability experts. [9]

The fact that usability studies can focus on different aspects and approaches leads to different methods, which can be classified into testing, inspection, inquiry, analytical modelling, and simulation [10]. Testing is based on the empirical observation of real users interacting with the system. Inspection methods are based on heuristic studies of the system concerning aspects such as the conformity with guidelines and standards. Inquiry consists in asking users to answer questions, surveys, and so forth. Analytical modelling, as its name suggests, uses formal models to analyse the system with the aim of identifying usability problems and predicting usability aspects. Finally, simulation tries to emulate the interactions between real users and the system.

The implementation of usability study methods leads in turn to different kinds of activities, which can be divided into capture, analysis, and critique [10]. Capture consists in the collection of usability data (e.g., execution times, subjective ratings). Analysis activities comprise both the interpretation of usability data and the examination of specific aspects of the system. Finally, the goal of critique activities is to suggest ways of improving the usability of the system. Critique is, of course, the most difficult activity to carry out, and therefore the study of usability tends to be focused mainly on capture and analysis.

Given the variety and complexity of usability study methods, it is advisable to perform a reasonably high number of studies and to employ different methods. A diversity of methods not only allows us to analyse a great variety of aspects but also to take advantage of the synergies between complementary methods.

Depending on their type, the distinct usability study methods can be applied in different phases of the life cycle of a project. All stages can be covered, from requirements analysis to maintenance. For example, inspection methods can be used during the design phase to identify design flaws, whereas testing can be performed when functional prototypes exist. One of the main ways of integrating the study of usability in the life cycle is to use iterative design, applying the methods and subsequently evaluating the results in later cycles.

The ideal way of obtaining realistic usability assessments would be to analyse, in the most objective way possible, the functioning of a final version of the system when used by a large enough sample of representative users. Naturally, this situation is rarely possible, so in practice it is usual to work with approximations of this ideal scenario. One way of facilitating this task is to use computer techniques to automate—partly or fully—these activities. This practice has lately become a thriving field of research.

2.2.5 The Automation of the Study of Usability

The automation of the study of usability offers many benefits compared to manual methods. Firstly, it helps to reduce costs and human effort. It also allows to cover a wider range of possibilities, making it easy to perform some tasks which are difficult or impossible to carry out manually (for example, pattern recognition or the simultaneous comparison of different alternatives). Another advantage is that it permits a reduction in the amount of knowledge and experience required of the human evaluators.

Usability study methods can be automated through a vast array of techniques. In testing, there are tools to implement activities such as communication, test recording, data collection, and log analysis. Inspection tasks can be automated by using systems capable of performing analyses and critiques. In inquiry, there exist tools to simplify the creation and use of questionnaires. Analytical modelling uses computational techniques to construct formal models of the user or the system. Finally, simulation methods emulate the interaction between the user and the system through techniques such as user models which operate the system [10].

The automation techniques for the study of usability have progressed remarkably, but they still have not become as widespread as they should. The techniques developed have often been restricted to the academic community, whose research has remained largely theoretical in nature. Software development organisations have also contributed to this field, but they have generally kept their research private, creating their own in-house tools or even reinventing the whole usability engineering life cycle in the process [11].

So far, the most sophisticated automation techniques have been based on artificial intelligence concepts. For example, expert systems are particularly appropriate for modelling the knowledge and the decision methods of human experts, and intelligent agents and genetic algorithms can be used to simulate the behaviour and the learning processes of real users.

The remainder of this chapter is dedicated to describing our particular approach to the study of web usability. We begin by introducing the necessary concepts and background, and then we proceed to explain in detail the architecture of our system.

2.3 Evolutionary Intelligent Agents

This section describes the specific artificial intelligent technique that constitutes the basis of our system, namely, evolutionary intelligent agents. Firstly, we define the concept of intelligent agent and its general characteristics. Secondly, we describe how evolutionary computation can be used to improve the learning mechanism of intelligent agents.

2.3.1 Agents and Intelligent Agents

An agent, in a general sense, can be described as "a computer system that is situated in some environment, and that is capable of autonomous action in this environment in order to meet its design objectives" [12]. An intelligent agent is a specific type of agent that is able to exhibit characteristics such as being capable of perceiving and reacting to the environment, showing proactive behaviour, and being able to interact with other agents or with humans.

The functioning of an intelligent agent is as follows: faced with a specific situation, an intelligent agent must choose a particular action from its repertory of possible actions, bearing in mind that some of these actions will only apply when particular preconditions are met. A typical way of implementing this feature is to use a rule base, that is, a knowledge base that stores antecedent-consequent rules. It is important to note that, in relatively complex domains, environments are typically non-deterministic, which means that agents do not have complete control over the environment and therefore do not always succeed at their goals.

Traditionally, there have been three intelligent agent architectures: deliberative, reactive, and hybrid. Deliberative agents possess a symbolic representation of the world in which they operate, and their decision-making behaviour is always explicitly programmed. In contrast, reactive agents are not programmed beforehand to behave in a certain way; they act by reflex instead, performing the action they believe most suitable. Finally, hybrid agents integrate both approaches.

Intelligent agents offer several advantages over traditional computational techniques. They are more flexible than standard computer programmes, which tend to be unable to deal with changes in the assumptions that underlie their actions. Intelligent agents bear some similarities to conventional objects—and can, in fact, be developed using classical object-oriented languages—but are more autonomous in that they have control over their own behaviour. The main difference between intelligent agents and expert systems is that the latter are generally more passive—they are not capable of reactive, proactive, or social behaviour, and they do not interact with an environment, properly speaking [12].

2.3.2 Evolutionary Learning in Intelligent Agents

One of the main characteristics of intelligent agents—and an essential element of this research—is their ability to learn. Learning enables agents to be more autonomous and adaptable. A typical learning model is based on the use of reinforcement, which is periodically generated to indicate whether or not the actions implemented by an agent have helped it achieve its goals.

However, learning based solely on reinforcement has two significant limitations: it can be slow and it is not always appropriate for circumstances which are subject to change. It is important, then, to find a way to enhance the learning mechanism of the agents. Evolutionary computation allows us to share genes (in our case, rules) among

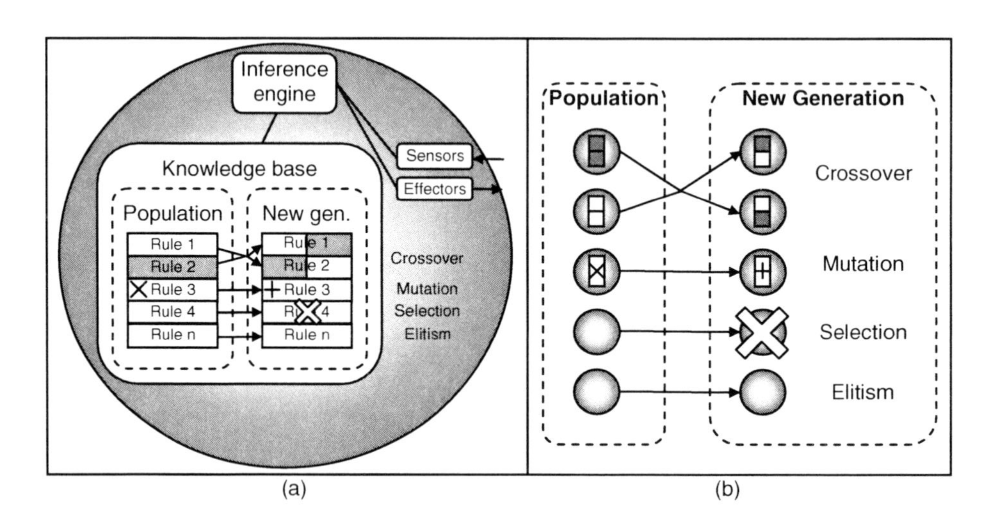

Fig. 2.1. Two approaches for introducing evolutionary computation into an agent architecture. (a) The population is composed of the rules of a single agent. (b) The population is composed of a set of agents.

the population and to pass on the best ones to subsequent generations, thus providing a means for accelerating the learning process and exploring the space of possible solutions.

There exist a number of approaches for introducing evolutionary computation into an agent architecture. One of them is to assume that the population to be evolved is composed of the production rules of a single agent (such as, for example, dynamic rule prioritisation [13]). This option is appropriate when it is not possible to have several agents acting simultaneously in the same environment—as their actions would be incompatible—or when it is not possible to simulate several parallel environments.

Another approach is to have a population of agents which are evolved at specific intervals. Some examples are FuzzyEvoAgents [14], the Amalthaea system [15], and the InfoSpider system [16]. This approach is suitable when agents act simultaneously in the environment or when the environment can be simulated computationally. Fig. 2.1 shows a schematic representation of both architectures.

2.4 The GAEL Framework

In our previous work [18], we created a generic programming framework with the aim of facilitating the development of evolutionary multi-agent systems. Our goal was to create multi-agent systems suitable for problems in which knowledge of the environment is minimal and where conditions can be constantly changing. Another characteristic of our framework is that it can be applied to situations where multiple agents are acting in the same environment in parallel, trying to reach the same goal. Therefore, with the aim of achieving greater flexibility and efficiency, we have chosen the second approach described in section 2.3.2.

The framework provides all the basic infrastructure required to create evolutionary multi-agent systems that can be then applied to specific problems. It is thus only necessary to write code for the elements that characterise the problem in question. We named this framework GAEL (Generic Agents with Evolutionary Learning).

2.4.1 Agent Architecture of the Framework

The multi-agent systems of our framework consist of reactive agents, each of which possesses a situation-action rule base. A rule has an associated weight which represents the quality of that rule when trying to reach the goal of the agent. The agents know what actions they can perform in the environment, but have no way of determining their correctness in advance. Agents are capable of acquiring knowledge by means of a learning process. This learning process is implemented at two levels: an individual level and a population level.

At the first level, the agents learn individually by reinforcement. The process is as follows: the agent acts by observing the environment and reacting to the stimuli that it receives. When a stimulus is received, the agent first of all determines if it is going to explore new rules (useful for escaping dead ends) or if it is going to exploit the pre-existing rules in its rule base (in order to test their quality). In the latter case the system constructs a conflict set composed of active rules (i.e., those which refer to the actual state of the environment). Once a conflict set has been created, the agent has two possible choices: either to choose the best rule in the set (that with the greatest weight) or to statistically choose on the basis of the weights (although the best rule has the greatest probability of being chosen, this is not guaranteed). In the event that the agent decides to explore or the conflict set is empty, a new rule needs to be created for the current state of the environment. The new rule will be created randomly, but bearing in mind, firstly, that the action defined by the rule should be valid for the current state of the environment, and secondly, that the created rule is not already included in the knowledge base of the agent. After executing a series of rules, at pre-determined intervals the system receives a reinforcement that indicates whether these rules have enabled it to comply with its aims. If so, the weight of the rule is increased, and if not, the weight is decreased.

Finally, a genetic algorithm (learning at the agent population level) is executed regularly, applying techniques like natural selection, crossover, mutation, and so forth. During the application of those techniques, it is considered that the best agents are those having the highest fitness. The fitness of an agent is based on the weights of its rules.

According to the Goonatilake and Khebbal classification of hybrid systems [17], a hypothetical system of agents undergoing evolutionary learning corresponds to hybridisation by function replacement. In this kind of hybridisation, a main function of a specific intelligent technique is replaced by another technique. In our case, the learning function of the agents is not quite replaced but is, rather, improved by the use of a genetic algorithm.

2.4.2 Design of the Framework

The framework was designed following an object-oriented approach, which offers several advantages such as extensibility, reusability, ease of maintenance, and scalability. A simplified class diagram of the framework is shown in Fig. 2.2.

The System class manages the configuration and functioning of the system. The learning process is controlled by a set of parameters, described as follows:

- Number of agents: the quantity of intelligent agents in the population that is to evolve. Although a large number means better parallelism in the search for a solution, it also implies a greater expenditure of resources.
- Number of rules: the maximum number of rules for each agent. Limiting the number of rules prevents an agent from dedicating time and resources to rules that are never likely to be used.
- Reproduction rate: proportion of agents to be selected for reproduction in the genetic algorithm.
- Type of reproduction: this indicates how to select the agents for reproduction or how to replace agents already in existence with new agents.
- Mutation rate: the probability that an agent will undergo a mutation. If this probability is high, randomness is added to the learning process.
- Exploration rate: the probability of searching for a new rule rather than using an existing one.
- Rule selection strategy: the decision as to which rule to select during exploitation.

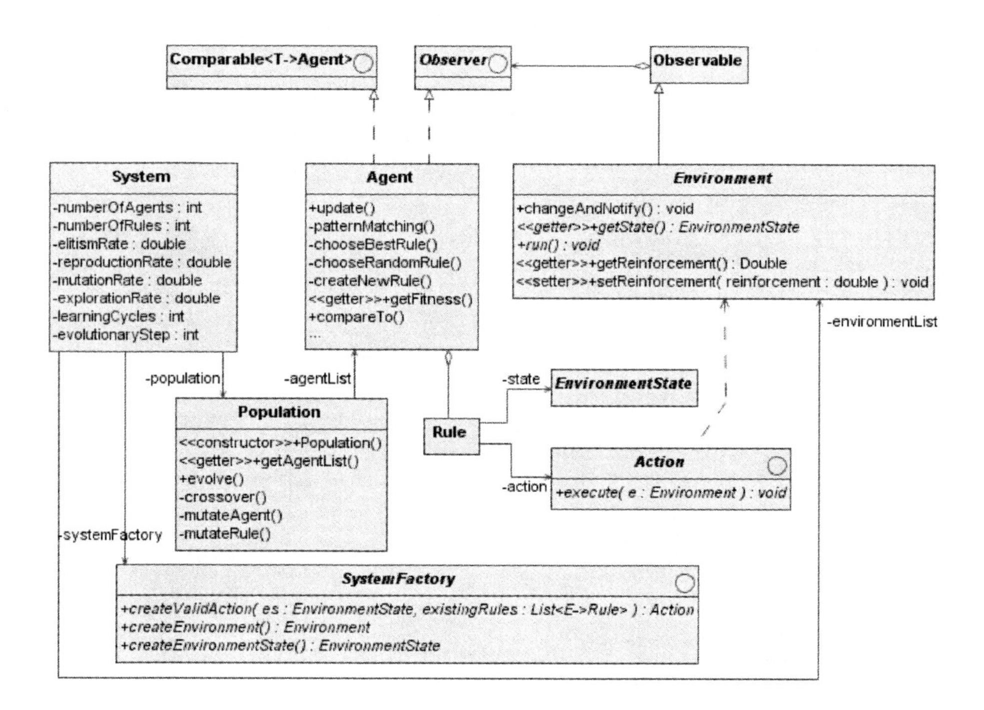

Fig. 2.2. Simplified class diagram of the framework (details have been omitted for clarity)

- Learning cycles: number of learning cycles implemented, during which learning by reinforcement occurs.
- Evolutionary step: the frequency with which the evolutionary algorithm acts in terms of number of learning cycles.

The Population class (which models the population of intelligent agents) includes all the methods required to execute the evolutionary algorithm. The individual agents are instances of the Agent class. The agents possess situation-action rules (Rule class). The antecedent of these rules is a specific state of the environment (EnvironmentState class), and the consequent is an action (Action class) performed on the environment (Environment class).

The framework can be applied to different types of problems, as global elements are defined in a generic way. For application to a specific problem, the only elements that need to be developed are problem-specific elements, as follows:

- Environment: the element that generates the events to which the agents react.
- Environment state: the identifiable state of the environment that the agent uses to implement pattern matching with rule antecedents.
- Action: an operation performed by the agent on the environment, which is the consequent of the rules of the agent.

It should be kept in mind that correct functioning will require a study of the characteristics of the problem in order to adjust the parameters which control the learning mechanism.

The design of the framework is also based on design patterns, which are effective and reusable solutions to recurring programming problems [19].

The interaction between agent and environment has been modelled using the Observer pattern. This pattern defines a one-to-many dependency between objects in which, when the state of the observed object changes, all its dependent objects are notified and updated.

The actions that the agents perform on the environment have been modelled using the Command pattern. This pattern declares an interface for the execution of actions, encapsulating them as objects, and allowing functionalities like the composition of actions and the addition of new actions without modifying the existing classes. Every action in the pattern is performed on a receptor (in our case, the environment) which knows how to perform said action.

Finally, the creation of objects is modelled using the Abstract Factory pattern, which provides an interface (SystemFactory class) that allows relationships between families of objects to be created without specifying which particular classes are involved. This pattern is fundamental to our aim of abstracting the framework classes from the specific classes that solve the problem in question, which cannot be known in advance by the framework.

2.5 An Evolutionary Multi-agent System for Studying the Usability of Websites

This section describes in detail the architecture of our multi-agent system for the study of the usability of websites. In order to detect usability problems, the system models the goals of the users from a collection of key phrases, simulates the browsing process, and analyses the HTML code of web pages. The multi-agent system is based on the evolutionary agent architecture and was developed using the GAEL framework (see section 2.4).

2.5.1 Related Work

Our system is founded on a combination of usability study methods that can be classified into inspection, analytical modelling, and simulation (see section 2.2.4). What follows is a summary of tools found in the literature which partially automate those methods and to some extent cover similar ground to our work.

Several inspection tools exist which analyse HTML code with the purpose of identifying typical usability problems. These applications focus mainly on validating code syntax and checking the compliance with accessibility guidelines, especially the World Wide Web Consortium's WCAG [20] and the above-mentioned Section 508 [3]. Depending on the tool, the user has the possibility of either analysing an individual web page or performing an exhaustive inspection of an entire website. Popular tools include WAVE [21], WebXACT [22], FAE [23], TAW [24], CSE HTML Validator [25], the W3C Markup Validation Service [26], and the W3C CSS Validation Service [27].

Chi et al. [28] developed a system for simulating and measuring how the text content of a website would help users find information. Starting from a particular web page, the system has as its goal to enter pages containing specific keywords. Their approach is based on the hypothesis that users follow a sequence of pages that have the aforementioned keywords in common. This assumption is not applied in a completely deterministic way but probabilistically, in order to also consider less similar routes. Aspects like density of information or reading complexity are not taken into account. The system automates capture—storing navigation paths, and calculating the proportion of users that would theoretically reach the goal. The results must be manually interpreted by a human evaluator [10].

The AMME tool [29] constructs a mental model of the user from actual logged data. This model consists in a Petri network that establishes relations between system states and user interactions. AMME obtains usability information on the complexity of the system, the profiles of the users, and the occurrence of repetitive actions.

Programmable User Models (PUM) [30] [31] are an analytical modelling technique which examines what knowledge the user needs in order to interact with a system. The users are modelled by means of a program which is given instructions on the operations for each task. These operations are analysed paying special attention to the psychological limitations of the users. If the limits are exceeded, it is probably because of usability problems.

Information Processor Modelling [10] is another technique based on psychological principles. Cognitive architectures are used to simulate system use, focusing on cognitive, motor or perceptional aspects, and obtaining usability predictions about the execution of tasks, the memory load, the learning process, or the behaviour of the users.

With regards to the use of genetic algorithms for usability studies, Kasik and George [32] developed an automated technique to genetically manipulate data on the use of a system, thus generating new data. This intelligent, non-arbitrary manipulation permits simulating the behaviour of inexpert users who learn on a trial-and-error basis.

It can be observed that most of these usability study techniques automate capture and analysis (see section 2.2.4). However, our aim was to develop a system which also automates critique, that is, indications with regards to solving usability problems detected in the analysis process.

Finally, our approach of using key phrases to model the browsing process could be considered the reverse of extracting the keywords or key phrases that best represent a website. In fact, both approaches are complementary: The extracted key phrases could be used as an input to our system in order to test their usefulness. The existing techniques of keyword extraction have been mainly based on frequency of occurrence—for example, analysing user queries [33] [34], or calculating the frequency of occurrence of a word in a specific page compared to the rest of the Web [35]. Other, more sophisticated keyword extraction techniques have been based on applying clustering algorithms to word similarity [36] and to user behaviour [37], or on identifying candidate key phrases through lexical methods and using machine learning to select the most relevant ones [38] [39].

2.5.2 System Architecture

Our system performs two distinct tasks: analysing the usability of web pages and simulating the browsing process of users. The usability analysis is carried out through inspections of the HTML code in which relevant data is extracted and examined. The simulation of the browsing process is based on modelling users who try to reach one URL from another and are guided by a set of goals (e.g., desired information or intended actions) that are represented by key phrases.

Given our requirements, intelligent agents are particularly appropriate to our problem. The agents can be used to model both the usability experts who study the usability of a website and the human visitors. The general environment in which they all operate is, of course, the website itself.

The resulting multi-agent system is thus based on two types of specialist agents: an HTML analyser agent and a population of user agents. Detailed descriptions of both types of agents appear in Section 2.5.3 and 2.5.4, respectively.

The system functions on the basis of a division of tasks among these two types of agents: whereas the user agents test navigability, the HTML analyser agent analyses code. The HTML analyser agent can operate on its own—examining a given set of pages—or in cooperation with user agents. Their cooperation allows us to focus the usability study on specific sequences of pages which constitute significant navigation paths. This user-goal-oriented approach contrasts with that of the inspection tools mentioned in section 2.5.1, which are only capable of blindly performing exhaustive, contextless analyses.

As Fig. 2.3 shows, when a user agent arrives to a new page, it requests information from the HTML analyser agent on the available links (link text, text surrounding the link, target URL). The HTML analyser agent acts on demand, examining the pages as they are requested by the user agents, and storing the information for future requests.

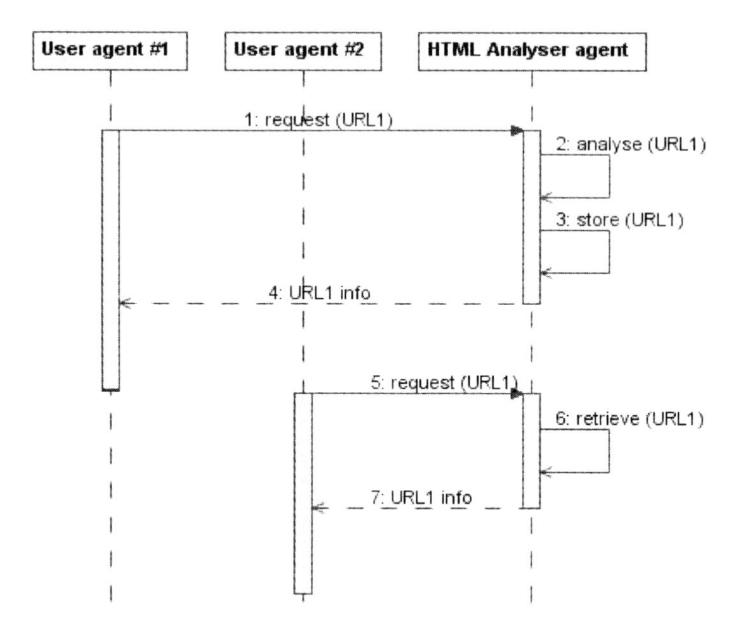

Fig. 2.3. User agents request information from the HTML analyser agent on the available links

The user agent then checks this information against its rules. Fig. 2.4 shows another example: when a user agent clicks on a broken link, this information should be notified to other agents to avoid time wasting. One way to do this is to use the HTML analyser agent as an intermediary.

Our system contributes to usability studies in two ways. Firstly, for aspects of usability that can be analysed automatically, the system itself draws conclusions and implements a critique based on principles documented in the usability literature. Secondly, for aspects of usability that depend on subjective evaluation, the system assists the expert—or, in fact, anybody with knowledge of the principles of usability—with the costly task of analysis by means of an intelligent selection of the most relevant information presented in a way that is easily digested.

This multi-agent system was developed using the GAEL framework (see section 2.4), defining the following problem-specific elements:

- Environment: the work area of the web browser.
- Environment state: any URL that could be visited by the agents.
- Action: clicking on a link.

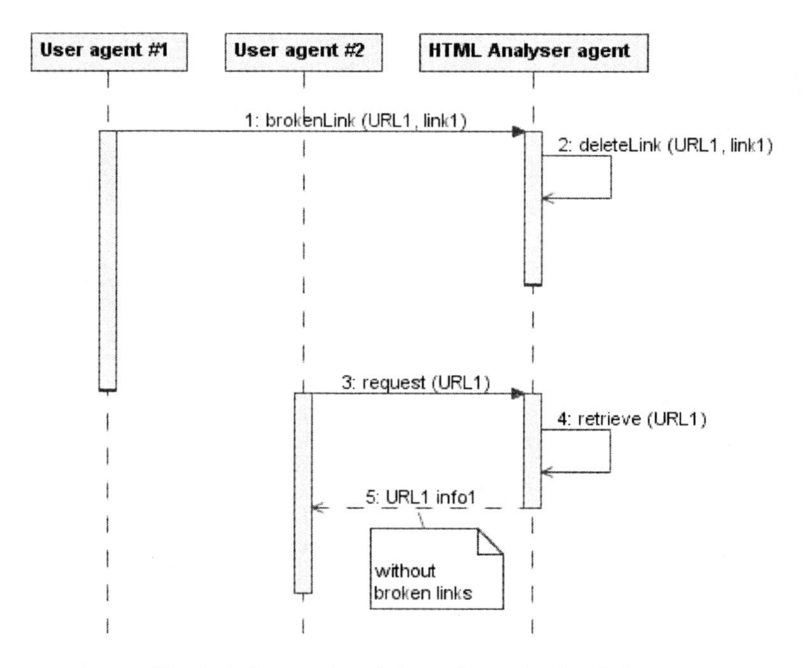

Fig. 2.4. Agents share information on broken links

2.5.3 The HTML Analyser Agent

The HTML analyser agent examines the HTML code of a single page, a group of pages, or an entire website, with the aim of detecting usability problems. Even though HTML code is relatively transparent and consists of a reduced set of elements, inferring usability issues from it can be a complex task involving all kinds of ambiguities. In some cases, a computer programme cannot resolve these ambiguities, and the judgement of a human expert is required. The HTML analyser agent, thus, uses heuristics to decide whether a usability problem actually exists, and reports every problem found indicating its estimated criticality.

In order to examine a web page, the HTML analyser agent creates a structured model of its HTML elements. As an example, Fig. 2.5 shows a part of this model, namely, the text elements.

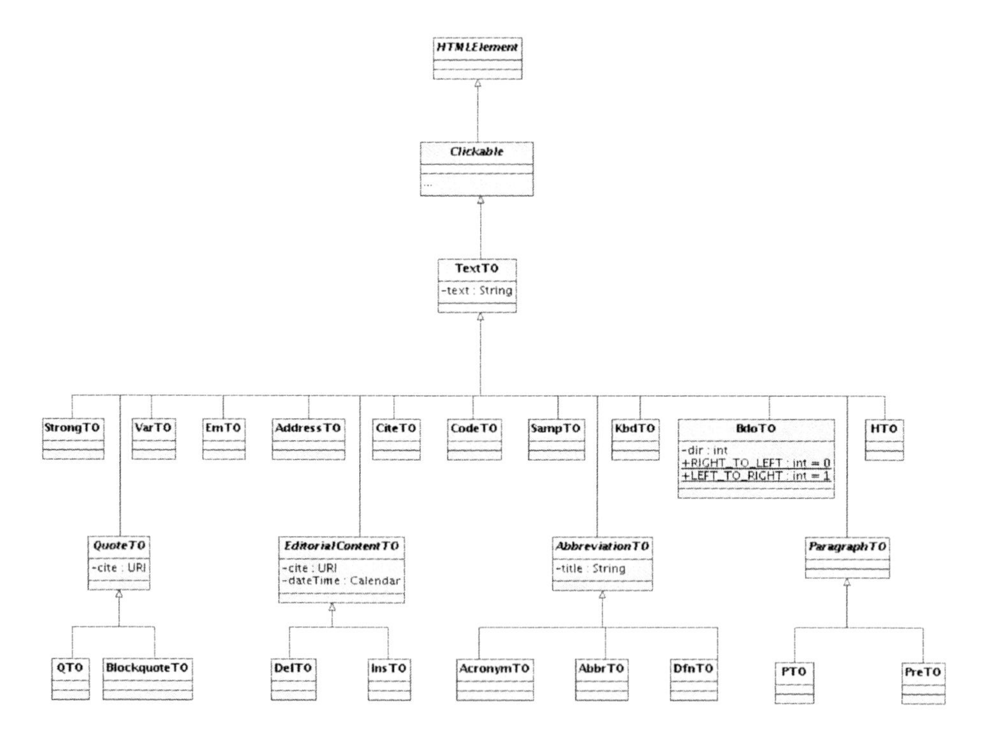

Fig. 2.5. Class diagram representing the structured model of text elements used by the HTML analyser agent

Below is a non-exhaustive list of HTML usability aspects analysed by the agent. Some of these aspects are always symptoms of usability problems, whereas others are significant issues (some positive, most negative) that ought to be notified to the usability expert. The aspects are classified into the following groups:

- Web page usability issues:
  - Estimated page size.
  - Programming problems such as browser-specific tags, deprecated tags, the lack of text encoding, and the presence of special characters that should have been replaced by HTML entities.
  - Flexibility problems, such as fixed font sizes and fixed-width elements.

  - The use of elements which can be problematic in certain contexts, such as: frames, cookies, CSS formatting, Javascript code, events, and animations.
  - The presence of search engines. These elements are virtually mandatory for big websites. If search engines are used, they should preferably be located at the home page.
- Image-specific usability issues:
  - Size (i.e., width and height).
  - Whether image size is specified in the HTML code.
  - Weight (i.e., amount of kilobytes), which influences download time.
  - Accessibility features such as alternative text (which should exist, but should not be too long) and the "long description" attribute (which should never be confused with alternative text).
  - The use of image maps.
- Form-specific usability issues:
  - Number of elements.
  - The existence of elements that are subject to some type of validation.
  - Accessibility features such as easily clickable controls.
- Table-specific usability issues:
  - Size.
  - Whether tables are used for presenting data or for establishing the page layout (the former is their intended use, whereas the latter is considered problematic).
  - The existence of recommended elements such as headers, captions, and summaries.
- Link-specific usability issues:
  - Non-standard representations that can cause the user to ignore the links (e.g., non-underlined links).
  - Broken links.
  - Badly constructed links.
  - Inappropriate link texts. For example, non-descriptive phrases and jargon words. These links do not represent actual content or actions, and may even be confusing in themselves. Another example of inappropriate link text is a page with links that have identical text but lead to different URLs.
  - The use of anchors (i.e., links to a specific area of a page), distinguishing between same-page and different-page anchors.
  - The existence of links to non-HTML files. The most common types are document files (especially PDF files, which are often fine for printing but not for reading), multimedia files, and archive files. While all these kinds of files are not necessarily bad in themselves, they can have a negative impact on usability in many ways. The main problems arise when these kinds of files are used as a substitute for text content that should have been available in the more flexible and manageable HTML format. Furthermore, lack of consistency in presentation and the switching between different formats make the navigation more awkward. Many additional problems may occur: some formats are proprietary (e.g., Microsoft Word,

MP3, and RAR); the necessary software may not be installed (and could even require the purchase of a commercial application); text browsers are unable to display images; and so on.

2.5.4 The User Agents

User agents model the browsing process of the human users. Each agent has as its goal to arrive to a destination URL from an initial URL, and possesses a set of key phrases of potential use in achieving this goal. The motivation for this word-based approach is that the Web is primarily a linguistic medium that involves browsing through text pages and examining text content.

Our multi-agent system contains a population of non-identical user agents that model different types of human users and obtain different results. User agents are implemented using the evolutionary agent architecture described in section 2.4.1—that is, rule-based reactive agents. The rationale for choosing the reactive architecture was that our agents are not designed beforehand to behave in a certain way. It should be borne in mind that the users who visit a website often have little previous knowledge of how it works. The navigation of a website, then, should be as intuitive as possible: users should not have to be taught how to use it. Another reason is that the structure of a website is generally freer than that of traditional interfaces, and can also be constantly changing. All these properties require the dynamic and flexible behaviour typical of reactive agents.

Thus, instead of following a predetermined set of instructions, the agents act reactively, matching their key phrases against page content in order to decide which links to click. Furthermore, we also impose constraints on characteristics such as user fatigue and text visibility (see below) with the aim of modelling agent behaviour more realistically.

The motivation behind the fallible, non-deterministic behaviour of the agents is to decide if the text content and the links of the website help users in performing tasks and finding the information they seek. That is, if an agent fails, it is probably because of usability problems in the website. In a product with a good level of usability, computerised task implementation will be as close as possible to the mental model of the user—in our case, the structure of the web site, the text content, and the link labels should be as intuitive as possible.

The agents are defined by a series of parameters that model different aims and user profiles, namely:

- Initial URL.
- Destination URL (which does not necessarily have to belong to the same domain as the initial URL).
- List of initial key phrases: these phrases model the aims of the user, and may describe concepts, actions, and so forth. They can be obtained from several sources: from a description of the functionalities or the requirements of the product, from surveys of users, from keyword extraction tools (see section 2.5.1), etc. It is also possible to create agents with no initial key phrases, in which case their aims would not be those of human users; they would, rather, be dedicated purely to exploration.

- Awareness of surrounding text: usability studies conclude that 79% of users only glance at web pages, and very few users read pages word for word [40]. When selecting links, it is typical to focus on the highlighted words and to ignore the text surrounding the link.
- Quantity of links viewed: similarly to the previous comment, users tend to focus solely on the most visible sets of links.
- Quantity of links visited before giving up: when a user seeks certain information, the physical or mental effort invested in the search will depend on the value assigned to achieving an aim.

User agents are subject to an evolutionary process in which the best rules are passed on to the next generation. Because the GAEL evolutionary process is fitness-based, we need to implement a reinforcement system that will reward or penalise agent actions. It should be kept in mind that the reinforcement is necessarily heuristic—that is, it does not always reflect the best way of achieving a goal. The different kinds of reinforcement are as follows:

- Positive reinforcement, on being able to use the initial key phrases.
- Positive reinforcement, for similarities between the current URL and the destination URL.
- Negative reinforcement, on being obliged to return to the previous page.

To sum it up, different agents defined by different parameters try to reach the destination URL. Those that perform reasonably well are given positive reinforcement and have a higher probability of propagating their rules to the subsequent generations. In addition to ensuring that good rules are retained, the evolutionary algorithm also enhances the rule bases of the agents through techniques such as crossover and mutation. As a result, the learning process of the agents is faster and richer. The system ends up generating a population of intelligent agents that represent distinct variations of how human users would browse the website under the specified conditions.

The purpose of user agents is to facilitate the detection of typical symptoms of usability problems in web browsing, such as:

- The impossibility of reaching the destination by using the initial key phrases.
- Confusing navigation.
- Inter-page usability aspects, such as:
  - Composition of page titles.
  - Consistency in appearance.

Other usability problems can be identified by taking into account metrics such as:

- Number of pages visited.
- Number of different paths between source and destination.
- Length of said paths.
- Number of key phrases found in the links.
- Number of key phrases found in the text surrounding the links.

2.6 The System at Work

In order to test the prototype of our system we chose the Project Gutenberg website [41]. This is one of the oldest and best known Internet sites, and claims to have the first and largest single collection of free electronic books. In a website of this kind, which aims to store a large amount of material and receive many casual visitors, navigation should be simple and intuitive. This was the main reason why this site was considered to be ideal for the first application of our system.

Our agents model users who visit the Project Gutenberg website on having learned that they can download a CD-ROM containing a collection of science fiction books. The list of key phrases selected was "download", "science fiction", "sci-fi", "sf", "cd-rom", "cd", "compact disc".

The destination URL has two features of interest: firstly, it is a .torrent file, and not a web page; secondly, it is not located in the Project Gutenberg domain but on a remote server.

Below is a summary of the main usability issues detected by our multi-agent system. We have limited our description to usability problems related to the browsing process, both for the sake of conciseness and because this is where the main interest of our approach lies. Thus, less complex usability issues such as deprecated tags, same-page anchors, or the use of Javascript code have been omitted from our description.

In our first test, we decided to parameterise the system in such a way that the agents would go directly to the links and ignore the surrounding text. The outcome, even after a reasonable number of cycles, was that no agent achieved the goal. We then parameterised the system so that the agents would read the surrounding text. On this occasion the agents reached the destination URL on the basis of the key phrases provided.

The minimum sequence of links in order to achieve the goal was:

- "Project Gutenberg BitTorrent Tracker" — "DL" (only the first link)

Another sequence of valid links—more intuitive but longer—was:

- "Project Gutenberg BitTorrent Tracker" — "Project Gutenberg Science Fiction CD - Mar. 2007.zip.torrent" — "DOWNLOAD TORRENT"

The system automatically drew the following usability conclusions:

- The destination URL can be reached by following a sequence of only two links (this is considered to be perfectly acceptable [42]).
- There are different paths to the destination (flexibility is important for usability).

Additionally, the system also suggested rewriting the labels for the following reasons:

- The optimal link path does not contain any of the key phrases, although some of these do appear in the surrounding text (see Fig. 2.6b and Fig. 2.7b).
- The fact that an intermediate page contains several links with identical text ("DL", "Link") but leading to different URLs might make navigation confusing (see Fig. 2.7c and Fig. 2.7d).
- Some links for visited pages contain words that are usually indicative of inappropriate nomenclature—that is, they are computer jargon words ("http") or words that say nothing about the task ("start here", "here", "link"), as Fig. 2.6a, Fig. 2.7a and Fig. 2.7d show.

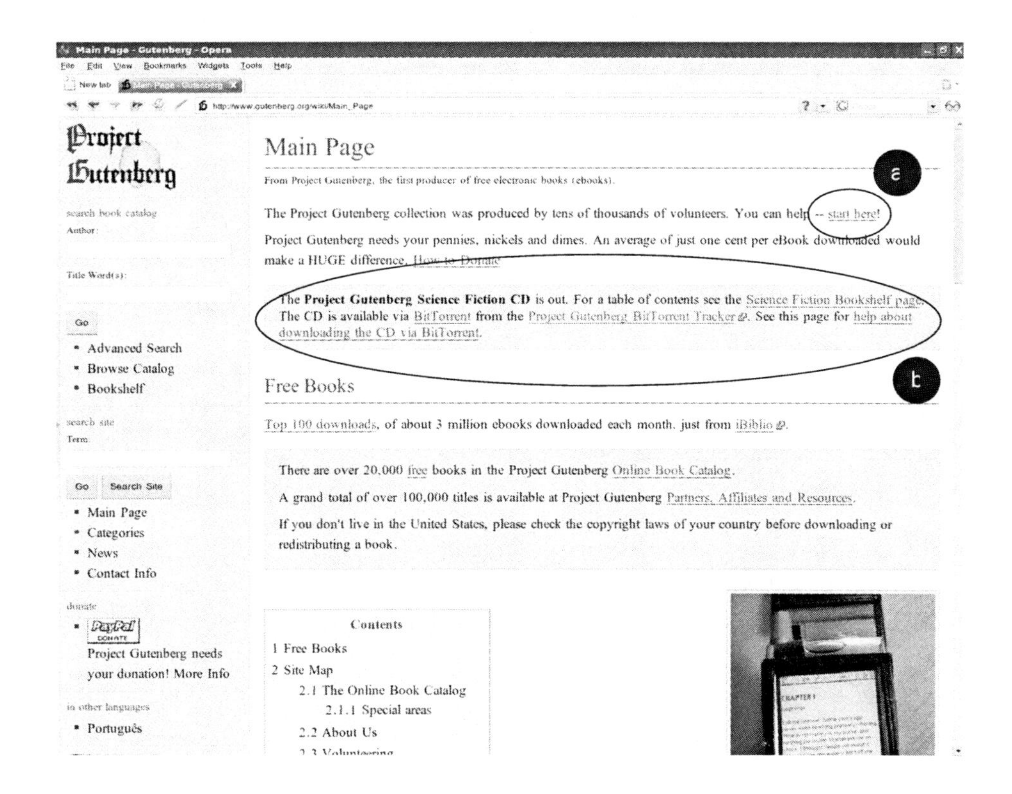

Fig. 2.6. Project Gutenberg main page. (a) Example of inappropriate link text. (b) Text containing links to download the science fiction CD.

Observing the results of the executions modelled by the system, usability experts made the following observations on the website content semantics:

- The way in which the website is written is not appropriate given its readership. The links in the website focus on computerised implementation rather than on the aims of users. For example, in order to achieve their goal, users need to know what a BitTorrent Tracker is (see Fig. 2.6b) and how it works. Furthermore, the Tracker page makes a questionable use of abbreviations: in order to be able to follow the optimal path, the user needs to infer that "DL" in this context is the abbreviation for download (see Fig. 2.7c).
- Project Gutenberg is available in Portuguese as well as in English, and when testing the site for this language, our experts observed that at one point the link path changed language without warning. Given that this may confuse users—they may think they did something wrong, or they may search in vain for a language change button—, the language shift should be indicated by adding, for example, "(in English)" to the end of the label for the link that leads to a page in the other language.

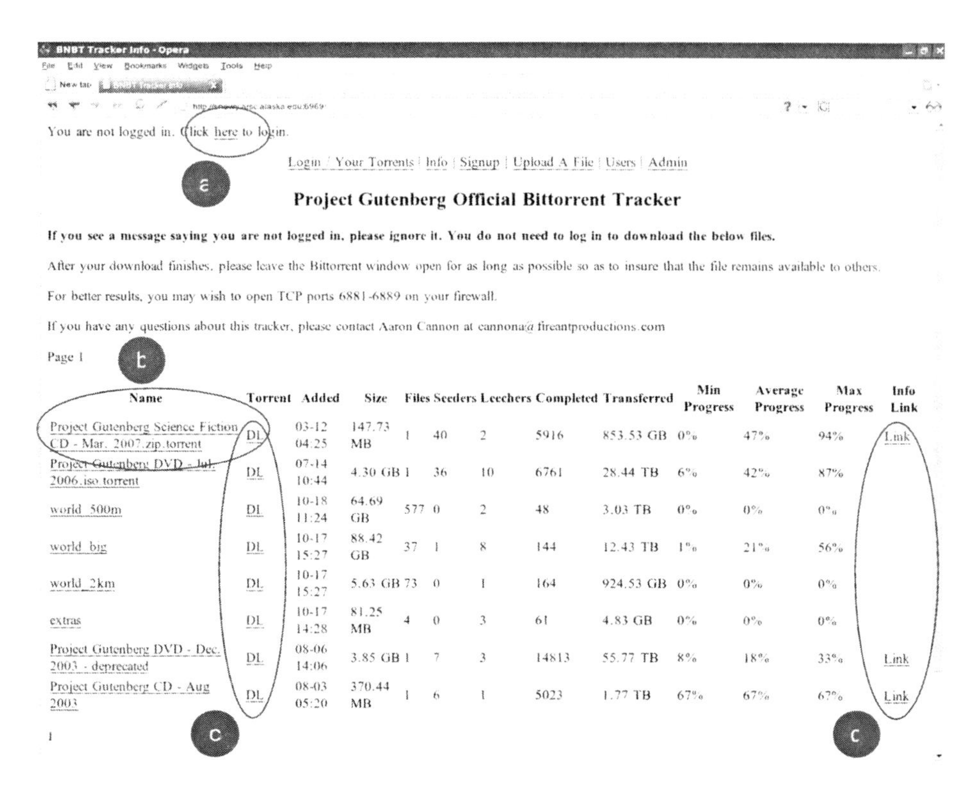

Fig. 2.7. Project Gutenberg BitTorrent Tracker. (a) Example of inappropriate link text. (b) Text containing links to download the CD. (c, d) Links with identical text leading to different URLs.

2.7 Discussion and Conclusions

During the last 15 years, the World Wide Web has evolved rapidly to become the enormous, ubiquitous resource that it is today. Web sites and applications are reaching a wider audience than ever before, and are being used daily by people from different age groups, from diverse cultures, with different computer skills, and a wide range of potential disabilities. Web developers and application vendors operate in a very competitive environment, and cannot afford to ignore the diverse needs of their audience. Moreover, technological progress has conditioned us all to demand instant gratification, and poor usability becomes annoying very quickly.

In spite of all of the above-mentioned reasons, numerous studies show that users still find most web sites—and computer applications in general—overly complicated to use and understand, and this causes a great amount of frustration and losses in productivity. To this day, these problems are often still considered inescapable—or, worse, the user's fault—but, actually, they are often simply a result of poor design decisions.

The importance of performing usability studies is becoming increasingly apparent, but, at present, they are mostly carried out in an ad hoc manner. Their automation is therefore a desirable goal, and consequently has become a thriving area of research.

In the field of websites and web applications, one of the most popular automated techniques of usability study is HTML analysis. The main reasons are that HTML consists of a relatively limited set of elements, and the code structure is linear and transparent. In this context, it becomes perfectly feasible to perform usability analyses that are both relatively sophisticated and generalisable to any website. This contrasts with traditional programming languages, where automated analyses are more complicated to perform and the results are, of necessity, more limited.

In this work, we combine HTML analysis with intelligent agents (chosen for being capable of modelling human behaviour and processing large amounts of information) to create a multi-agent system for the semi-automated study of the usability of websites. The system constructs a key phrase-based model of the users trying to reach one URL from another, simulates the browsing process, and analyses the web pages that make up possible paths between source and destination. The system automatically draws conclusions and makes suggestions with the aim of improving the usability of the website. It focuses on aspects such as:

- The path between two URLs with the minimum number of links.
- The existence of alternative paths.
- Easily locating paths on the basis of a set of phrases that represent a mental model of the aims of the user.
- The length of navigation paths.
- Problems with the HTML coding for the web pages.
- Problems with the content of web pages.
- Problems with links between web pages.

Likewise, the system selects and presents additional usability data with the aim of facilitating the human expert the task of evaluating the usability of a website.

The system consists of two types of intelligent agents: user agents and an HTML analyser agent. User agents have as their goal to model user behaviour attempting to arrive to a given destination URL, and the HTML analyser agent examines web pages to extract useful data for the study of usability.

Our system is based on GAEL (Generic Agents with Evolutionary Learning), a framework designed for the development of evolutionary multi-agent systems composed of rule-based reactive agents. Evolutionary processes allow us to model the exploration of the space of possible solutions more realistically than when using traditional methods. Reactive behaviour is particularly appropriate for our problem due to the fact that users who visit a website often have little previous knowledge of how the website works. Thus, it is important for the navigation of a website to be an intuitive process. Furthermore, websites generally have a freer structure than traditional interfaces [10], and this requires a particularly dynamic and adaptable approach.

Key phrases are useful to model the behaviour of the users because the Web is primarily a text-based medium. Furthermore, given the increasing popularity of search engines, the Web has lately become very search-oriented. It is therefore quite important to optimise text content for maximum searchability. "Speak the user's language" is one of the oldest usability guidelines [43], and it is becoming more relevant every day. It is also important to think in terms of key phrases, not key words, because isolated words are rarely capable of modelling the goals of the user.

The combination of intelligent agents and the key phrase-based modelling of user goals constitutes a more flexible approach than that offered by conventional inspection methods. More specifically, the exhaustive, "blind" inspections performed by standard HTML analysis tools fail to take into account the aims of the users, whereas, on the other hand, task-oriented techniques such as GOMS modelling are excessively rigid and limited [10].

Although implementation using intelligent agents is a logical solution to automating tasks that are normally carried out by humans, human experts often continue to be necessary. First of all, because usability is relative to the context of use; secondly, because the study of usability is full of ambiguities that cannot be resolved through heuristic means. Despite their many advantages, intelligent agents remain far from being able to completely replace humans when it comes to information comprehension and disambiguation [44].

The scope of our inspections is restricted to HTML code, which to some extent limits the breadth of our usability analyses. The use of Javascript code is taken into account in our heuristics, but the Javascript code itself is not inspected for usability problems. Likewise, we do not analyse how CSS affects the presentation of the pages.

Because our system is only capable of examining client-side code, it is therefore unable to analyse the possible impact that server-side code (e.g., JSP, ASP.NET) might have on usability. Nevertheless, the bulk of the usability problems of a website can be attributed to client-side code. In any case, it is impossible to obtain the server-side code of a website unless it is provided by its owners. Custom-built solutions ultimately amount to a completely different approach, and thus fall outside the scope of this work.

2.8 Future Work

Finally, there are some open issues which we aim to address in our future work.

Firstly, we intend to investigate new techniques of HTML analysis, especially ones based on heuristics and other artificial intelligence techniques. We also aim to incorporate into our system the capabilities for performing Javascript and CSS inspections.

Secondly, we intend to make the modelling of key phrases more flexible, both lexically and syntactically. We also need to classify key phrases into groups and to create a phrase hierarchy. Since our system is capable of identifying the existence of search engines in web pages, these classifications will also allow the agents to enter appropriate keywords into the search engine, thus making the browsing process more powerful and realistic.

Acknowledgments. The research has been partly funded by the Ministerio de Educación y Ciencia (Project TIN2005-04653) and the Xunta de Galicia (Project PGIDT06PXIB105205PR).

References

1. Internet World Stats, http://www.internetworldstats.com/
2. The Council of the European Communities/European Economic Community. Council Directive 90/270/EEC of 29 May 1990 on the minimum safety and health requirements for work with display screen equipment (The fifth individual Directive within the meaning of Article 16(1) of Directive 89/391/EEC). Official Journal of the European Communities L156, 14–18 (1990)
3. Section 508, http://www.section508.gov/

4. Nielsen, J.: Usability Engineering. Academic Press, San Diego (1993)
5. International Organization for Standardization: ISO 9241-11, Ergonomic Requirements for Office Work with Visual Display Terminals (VDTs), Part 11: Guidance on Usability. International Organization for Standardization, Geneva, Switzerland (1998)
6. Nielsen, J.: The Web Usage Paradox: Why Do People Use Something This Bad? (1998), http://www.useit.com/alertbox/980809.html
7. Ivory, M.Y.: Preliminary Findings on Quantitative Measures for Distinguishing Highly Rated Information-Centric Web Pages. In: 6th Conference on Human Factors and the Web (HFWeb) (2000)
8. Montero, F., González, P., Lozano, M., Vanderdonckt, J.: Quality Models for Automated Evaluation of Web Sites Usability and Accessibility. In: International COST294 Workshop on User Interface Quality Models, Interact 2005, pp. 37–43 (2005)
9. Adelman, L., Riedel, S.: Handbook for Evaluating Knowledge-Based Systems: Conceptual Framework and Compendium of Methods. Kluwer Academic Publishers, Norwell (1997)
10. Ivory, M.Y., Hearst, M.A.: The State of the Art in Automating Usability Evaluation of User Interfaces. ACM Computings Surveys 33(4), 470–516 (2001)
11. Seffah, A., Metzker, E.: The Obstacles and Myths of Usability and Software Engineering. Communications of the ACM 47(12), 71–76 (2004)
12. Wooldridge, M.: An Introduction to MultiAgent Systems. John Wiley & Sons Ltd., Chichester (2002)
13. Nonas, E., Poulovassilis, A.: Optimisation of Active Rule Agents Using a Genetic Algorithm Approach. In: Quirchmayr, G., Bench-Capon, T.J.M., Schweighofer, E. (eds.) DEXA 1998. LNCS, vol. 1460, pp. 332–341. Springer, Heidelberg (1998)
14. Di Nola, A., Gisolfi, A., Loia, V., Sessa, S.: Emerging Behaviors in Fuzzy Evolutionary Agents. In: 7th European Congress on Intelligent Techniques and Soft Computing, EUFIT 1999 (1999); Published on CD-ROM
15. Moukas, A.: Amalthaea: Information Discovery and Filtering Using Multiagent Evolving Ecosystem. In: Applied Artificial Intelligence, vol. 11(5), pp. 437–457. Taylor and Francis Ltd., Philadelphia (1997)
16. Menczer, F., Monge, A.E.: Scalable Web Search by Adaptive Online Agents: an InfoSpiders Case Study. In: Klusch, M. (ed.) Intelligent Information Agents: Agent-Based Information Discovery and Management on the Internet, pp. 323–347. Springer, Berlin (1999)
17. Goonatilake, S., Khebbal, S. (eds.): Intelligent Hybrid Systems. John Wiley & Sons, England (1995)
18. Mosqueira-Rey, E., Alonso-Ríos, D., Vázquez-García, A., Baldonedo del Río, B., Alonso-Betanzos, A., Moret-Bonillo, V.: An Evolutionary Approach to Including Learning Mechanisms in Multi-Agent Systems. In: Dynamics of Learning Behavior and Neuromodulation Workshop, 9th European Conference on Artificial Life (ECAL) (2007)
19. Gamma, E., Helm, R., Johnson, R., Vlissides, J.: Design Patterns: Elements of Reusable Object-Oriented Software. Addison-Wesley, Reading (1995)
20. W3C: Web Content Accessibility Guidelines 2.0 (WCAG 2.0). In: Caldwell, B., Cooper, M., Guarino Reid, L., Vanderheiden, G. (eds.), http://www.w3.org/TR/WCAG20/
21. WAVE: Web Accessibility Evaluation Tool, http://wave.webaim.org/
22. Watchfire WebXACT, http://webxact.watchfire.com/
23. FAE: Functional Accessibility Evaluator, http://devserv.rehab.uiuc.edu/fae/
24. TAW: Web Accessibility Test, http://www.tawdis.net/taw3/cms/en
25. CSE HTML Validator, http://www.htmlvalidator.com/
26. The W3C Markup Validation Service, http://validator.w3.org/

27. The W3C CSS Validation Service, http://jigsaw.w3.org/css-validator/
28. Chi, E.H., Pirolli, P., Pitkow, J.: The Scent of a Site: A System for Analyzing and Predicting Information Scent, Usage, and Usability of a Website. In: Conference on Human Factors in Computing Systems, pp. 161–168. ACM Press, New York (2000)
29. Rauterberg, M., Aeppili, R.: Learning in Man-Machine Systems: The Measurement of Behavioural and Cognitive Complexity. In: IEEE Conference on Systems, Man and Cybernetics, pp. 4685–4690. IEEE, New York (1995)
30. Young, R.M., Green, T.R.G., Simon, T.: Programmable User Models for Predictive Evaluation of Interface Designs. In: Conference on Human Factors in Computing Systems, pp. 15–19. ACM Press, New York (1989)
31. Butterworth, R., Blandford, A.: Programmable User Models: The Story So Far. Middlesex University, London (1997)
32. Kasik, D.J., George, H.G.: Toward Automatic Generation of Novice User Test Scripts. In: Tauber, M.J., Bellotti, V., Jeffries, R., Mackinlay, J.D., Nielsen, J. (eds.) Proceedings of the ACM CHI 1996 Human Factors in Computing Systems Conference, pp. 244–251. ACM Press, New York (1996)
33. Baeza-Yates, R.: Web Usage Mining in Search Engines. In: Scime, A. (ed.) Web Mining: Applications and Techniques, pp. 307–321. Idea Group Publishing, Hershey (2004)
34. GoodKeywords, http://www.goodkeywords.com/
35. Buyukkokten, O., Garcia-Molina, H., Paepcke, A.: Seeing the Whole in Parts: Text Summarization for Web Browsing on Handheld Devices. In: 10th International Conference on World Wide Web, pp. 652–662. ACM Press, New York (2001)
36. Runkler, T.A., Bezdek, J.C.: Automatic Keyword Extraction with Relational Clustering and Levenshtein Distances. In: Ninth IEEE International Conference on Fuzzy Systems, pp. 636–640. IEEE, San Antonio (2000)
37. Velásquez, J.D., Fernández, J.I.: Towards the Identification of ImportantWords from the Web User Point of View. In: International Workshop on Intelligent Web Based Tools (IWBT-2007), pp. 17–26. CEUR-WS (2007), http://CEUR-WS.org/Vol-302/
38. Kea: Keyphrase Extraction Algorithm, http://www.nzdl.org/Kea/
39. Witten, I.H., Paynter, G.W., Frank, E., Gutwin, C., Nevill-Manning, C.G.: Kea: Practical Automatic Keyphrase Extraction. In: Theng, Y.L., Foo, S. (eds.) Design and Usability of Digital Libraries: Case Studies in the Asia Pacific, pp. 129–152. Information Science Publishing, London (2005)
40. Nielsen, J., Loranger, H.: Prioritizing Web Usability. New Riders Press, Berkeley (2006)
41. Project Gutenberg, http://www.gutenberg.org/
42. Nielsen, J.: Designing Web Usability. New Riders Publishing, Indianapolis (2000)
43. Nielsen, J.: Use Old Words When Writing for Findability (2006), http://www.useit.com/alertbox/search-keywords.html
44. Lesnick, L., Moore, R.: Creating Cool Intelligent Agents for the Net. IDG Books Worldwide Inc., Foster City (1997)

3

Towards Norm-Governed Behavior in Virtual Enterprises

Paul Davidsson and Andreas Jacobsson

Department of Systems and Software Engineering, School of Engineering,
Blekinge Institute of Technology, SE-372 25 Ronneby, Sweden
{paul.davidsson,andreas.jacobsson}@bth.se

Abstract. An important application of intelligent agents on the World Wide Web is to support Business to Business (B2B) e-commerce, such as, virtual enterprises. A virtual enterprise can be described as a temporary alliance of enterprises that come together to share skills or core competencies and resources in order to better respond to business opportunities, and whose cooperation is supported by computer networks. In order to ensure sound collaboration, it is important that the actors involved, as well as the software representing the actors such as intelligent agents, act according to the explicit and implicit rules, or norms, that the participating enterprises have agreed upon. The purpose of this article is to explore how norm-governed behavior can be achieved in the context of virtual enterprises. We analyse a number of formal models of both artificial societies and normative systems, and compare and try to align them with a formal model of agent-supported virtual enterprises. A general observation is that the models analysed are not concordant with each other and therefore require further alignment. It is concluded that the introduction of different types of norms on different levels can support sound collaboration in virtual enterprises. Moreover, the deployment of norm defender and promoter functionality is argued to ensure norm compliance and impose punishments of norm violations. Finally, a number of additions that may enrich the norm-focused models are suggested.

3.1 Introduction

The concept of *virtual enterprises* in the area of B2B collaboration and interoperability has in a short time become one of the most advanced enterprise paradigms of our time [28]. A virtual enterprise can be described as a temporary alliance of enterprises that come together to share skills or core competencies and resources in order to better respond to business opportunities, and whose cooperation is supported by computer networks [7]. It may include a variety of entities, e.g., people, physical resources, and information systems, which are largely autonomous, geographically distributed, and heterogeneous in terms of their operating environment, culture, social capital, and goals. Because of the dynamic, adaptable, and collaborative nature of virtual enterprises, it provides a promising approach to face the challenges of constantly changing markets.

In order to promote the creation and operation of virtual enterprises, the introduction of *breeding environments* consisting of "virtual enterprise-ready" companies has been

N.T. Nguyen, L.C. Jain (Eds.): Intel. Agents in the Evol. of Web & Appl., SCI 167, pp. 35–55.
springerlink.com

proposed by, e.g., Camarinha-Matos and Afsarmanesh [7]. A breeding environment represents an association of enterprises that have both the potential and the ambition to collaborate with each other through the establishment of long-term cooperation agreements and an interoperable infrastructure.

Plug and Play Business [22] has been suggested as an one way of realizing breeding environments. It is based on an integrated set of ICT-tools that supports the creation and operation of virtual enterprises and it is aimed towards small and medium-sized enterprises (SME) in particular. After having deployed the Plug and Play Business software, companies will be included a networked community where all participants share the goal of increasing business through collaboration. In that way, the purpose of Plug and Play Business is to stimulate the realization of innovations without interfering with the individual goals of the companies of the community. Together with the autonomy, heterogeneity, and possibly conflicting goals of the members of a Plug and Play Business community, this requires ICT-solutions that are able to handle dynamically evolving and distributed business partnerships and processes that cross the borders of various enterprises. An enterprise joins the community by installing and running the Plug and Play Business software and by describing and validating the resources of the enterprise, e.g., production capacity, distribution network, intellectual capital, etc. The community is dynamic in the sense that enterprises may join and leave the community at any time. Each member of the community plays one or more roles, e.g., innovator, supplier/producer (of goods, services, expertise, etc.), distributor, marketer, financier, seller, etc. The choice of role depends on the company's core competencies and business intentions. An important role in the life cycle of businesses is the entrepreneur and one of the main purposes of Plug and Play Business is to automate as much of the entrepreneurial activities as possible, for instance by using intelligent agent technology. A member of the community, typically an innovator, may at any time initiate an attempt to form a collaborative coalition. It may start with just a seed of an innovative idea without any predefined business structure, or it may be a full-fledged business idea with well-defined needs to be met by potential collaborators. The Plug and Play Business software will then support the finding and evaluation of potential partners, as well as the negotiation and contracting process. Moreover, it will support the management of the virtual enterprise, e.g., by enabling interoperability of information systems and by monitoring contract conformance.

It is often claimed that agent technology is useful for the realization of virtual enterprises [19]. This is mainly motivated in that agents can represent real actors, e.g., transporters, producers, etc., take advantage of new business opportunities by organizing themselves in temporary coalitions, adapt to changing circumstances, and achieve common objectives in the presence of individual goals. However, it is crucial that such agents, which can be said to form an artificial society, act according to the explicit and implicit rules, or norms, that the participating enterprises have agreed upon. Artificial societies are typically characterized by agents that interact with each other [13]. Similarly to a human society, agents of the artificial society should be allowed to coexist in a shared environment, and to follow their respective goals in the presence of others. Also, as agents are required to work in increasingly complex environments and interact and coordinate with other agents, there is a need for regulating their behavior in order to avoid disruption and to ensure smooth performance, fairness, stability, and security.

On this theme, norms can be adopted to facilitate the means for basic social interaction in an artificial society [18]. The introduction of norms, broadly interpreted as constraints on the society, is therefore a crucial issue towards the computational representation of artificial societies of agents. The primary reason is that they provide measures to achieve coordination, security and trust, and thereby assist in assuring the desired behavior in a society [26][25]. Also, norms can play an important role on the agent level since they can help to cope with the heterogeneity, autonomy, and diversity of interests among the enterprises involved in a breeding environment of virtual enterprises and virtual enterprise-ready companies.

3.1.1 Approach and Outline

We want to explore how norm-governed behavior can be achieved in the context of virtual enterprises to support B2B collaboration. For this purpose, we analyse and compare a number of formal models of both artificial societies and normative agent-based systems. Of the models compared, one is general and focus on the society aspects, whereas the other models focus on the application of norms. In order to enable cross-comparison, the selected formal models were separately analysed and their descriptions condensed into a comparable format based on a number of critical aspects that were identified. The comparison constitute the foundation for section 3.4 where the means for approaching normative virtual enterprises are investigated.

In the next section, we present a formal model of virtual enterprise breeding environments. Thereafter, we analyse the selected formal models, which are discussed from the starting point of how to achieve norm-governed behavior in virtual enterprises. In the end, we present additional related work, conclusions and pointers for future work.

3.2 A Formal Model of Agent-Based Breeding Environments (ABE) for Virtual Enterprises

An agent-based breeding environment for virtual enterprises, p, can be described as a tuple (see also Figure 3.1 for an illustration):

$$p = \langle A, R, VE, S^t, I, CI \rangle$$

where

- $A = \{a_1, \ldots, a_n\}$ is the set of actors (typically enterprises) in the breeding environment. An actor, a_j, in p can be described as a tuple:

$$a_j = \langle I_j, T_j, C_j, G_j^{actor}, h_j, b_j \rangle$$

where I_j are the relevant information resources of actor a_j, T_j is the set of resources of the actor, C_j is the set of core competencies of the actor, G_j^{actor} is the set of individual goals of a_j, h_j is the person representing the actor/enterprise, and $b_j \in B$ is a set of breeding environment software clients, i.e., an intelligent agent supporting the (agent)

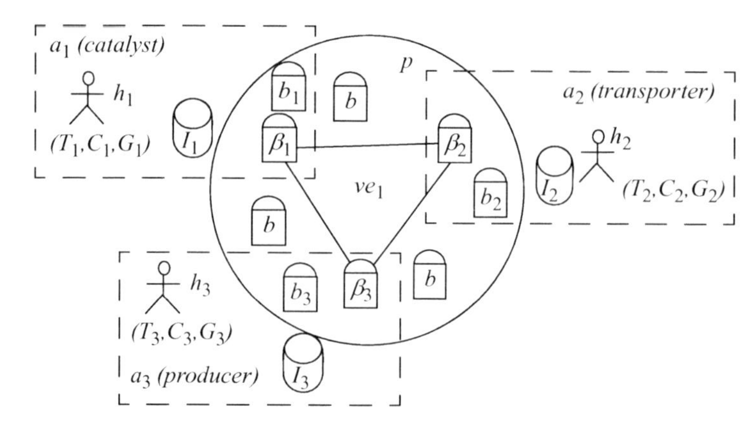

Fig. 3.1. An example of a breeding environment (p) including a number of actors represented by software agents (b), and where three actors (a_1,a_2,a_3), which play the roles of catalyst, producer, and transporter, have formed a virtual enterprise (ve_1) in which they are represented by software agents (β).

communication language, l, acting on behalf of the actor/enterprise. The actors in a virtual enterprise ve_i are a subset of the actors in a breeding environment, that is, $A_i \subseteq A$.

- $R = \{r_1, \ldots, r_m\}$ is the set of roles that the actors in A can play. The roles in a virtual enterprise are a subset of the roles in p, that is, $R_i \subseteq R$.
- $VE = \{ve_1, \ldots, ve_\psi\}$ is the set of virtual enterprises currently active in p. An agent-supported virtual enterprise ve_i can be described as a tuple:

$$ve_i = \langle A_i, R_i, AR_i, CI_i, S_i^t, G_i, \lambda_i \rangle$$

where

- $A_i = \{a_1, \ldots, a_n\}$ is the set of actors (typically enterprises) in ve_i. An actor can be described as a tuple:

$$a_j = \langle I_j, T_j, C_j, G_j^{actor}, \beta_j \rangle$$

- where I_j are the relevant information systems needed in ve_i, T_j is the set of resources of the actor, C_j is the set of core competencies of the actor, G_j^{actor} is the set of individual goals of the actor, and β_j is the agent acting on the behalf of the actor in ve_i.
- $R_i = \{r_1, \ldots, r_m\}$ is the set of roles that the actors can play in ve_i.
- AR_i is a set of triples $\langle a_k, r_j, O_j^k \rangle$ where $a_k \in A_i$ and $r_j \in R_i$, i.e., the actors and their roles in the virtual enterprise, and the set of obligations, O_j^k, that is associated with the actor's role in the virtual enterprise.
- CI_i is the set of communication infrastructures needed for operating the virtual enterprise.
- S_i^t is the set of states of affairs that hold at time, t, in ve_i.

- G_i is the set of goals of the virtual enterprise that is derived from the business opportunities that motivated the initiation of ve_i.
- λ_i is the agent communication language used by the agents β_j. We will assume that λ_i includes a set of relevant interaction protocols, a set of relevant ontologies, and possibly other things necessary to perform useful communication.

- S^t is a set of states of affairs that hold at time t in p. The states of affairs in a virtual enterprise $ve_i \in VE$ are a subset of the states of affairs in p, i.e., $S_i^t \subseteq S^t$.
- l is the agent communication language used by the breeding environment software clients. We will assume that l includes a set of relevant interaction protocols, a set of relevant ontologies, and possibly other things necessary to perform useful communication.
- CI is a set of communication infrastructures needed for operating the breeding environment. The communication infrastructures in a virtual enterprise $ve_i \in VE$ are a subset of the communication infrastructures in p, i.e., $CI_i \subseteq CI$.

A virtual enterprise breeding environment may support three phases of virtual enterprises, which we will now summarize (for a more detailed description, see Jacobsson and Davidsson [22]).

- The definition phase: In the definition phase, a member of the breeding environment, typically an innovator, may at any time initiate an attempt to form a collaborative coalition between the members. In this phase, a "catalyst" Ω_i, where $\Omega_i \in A$, describes the business opportunity in terms of goals G and roles R and a set of requirements Q^j_i for each role r_j of the new virtual enterprise ve_i. Thus, in this phase, the degree of autonomy of $b_{\Omega i}$ is rather low, whereas in other phases it may be higher (cf. adjustable autonomy).
- The formation phase: The formation phase consists of three subtasks and is initiated by Ω_j:
  - The function of finding requires that Ω_i has a list of the roles that must be filled in order to get an operating virtual enterprise. This list is provided by $h_{\Omega i}$, i.e., the person representing Ω_i in the definition phase. Then, for each of the roles, the task for $b_{\Omega i}$ is to find the set of candidate actors K_j, where $K_j \subseteq A$, that are able to play the role r_j where $r_j \in R$.
  - In the evaluation task, Ω_i should rank the actors in K_j according to the set of requirements Q^j_i. Based on this, Ω_i selects the actor $a_k \in K_j$ with the highest rank for negotiating on terms for virtual enterprise operation.
  - The goal of negotiation is to establish an agreement between Ω_i and a_k concerning a_k's set of obligations O_k in ve_i. These obligations should of course be consistent with the set of goals G_i of ve_i and the set of goals G_k of a_k.

When the formation phase is finished and a virtual enterprise is formed, the operation phase begins.

- The operation phase: During this phase, the management of the actual business activities within the virtual enterprise takes place. Collaboration support may be on a quite shallow level, e.g., transactions of information between actors. This may be called the administrational level. On a deeper level, the breeding environment software should support and facilitate complex coordination and synchronization of activities, which is referred to as the operational level. The breeding environment software ideally supports both of these levels of collaboration. They are defined by the type of interaction protocols they support. Administrational collaboration includes only protocols using the "weaker" *performatives*, such as, *ask*, *tell*, *reply*, etc. Operational collaboration supports protocols also using the *performatives* that actually manipulate the receiver's knowledge, such as, *insert*, where the sender requests the receiver to add the content of the message to its knowledge base, and *delete*, where the sender requests the receiver to delete the content of the message from its knowledge base.

We note that this definition of Agent-based Breeding Environments (ABE) lacks dedicated measures for norm enforcement. However, the catalyst, which has the role to initiate a virtual enterprise within the ABE, may perform some norm-related tasks, e.g., by the provision of requirements on candidates and obligations for collaboration in a specific virtual enterprise.

3.3 Formal Models of Agent Societies and Normative Systems

We will now analyse formal models of normative artificial societies from the agent research field of which one model focus on society aspects and two on norms. Even though we make some brief comments to every model, the actual comparison between the models takes place in Section 3.4.

3.3.1 Artificial Societies (AS)

A general theory for different levels of openness in artificial societies including norms, agents and stakeholder perspectives is outlined by Davidsson and Johansson [14]. Based on the work by Artikis and Pitt [1], and Johansson [23], they suggest a formal characterization of agent societies that includes the following entities:

- a set of agents,
- a set of constraints on the society,
- a communication language,
- a set of roles that the agents can play,
- a set of states of affairs that hold at each time in the society,
- a set of owners (of the agents),
- a set of agent designers,
- the environment (computation and/or communication infrastructure),
- an environment owner, and
- an environment designer.

An agent is here defined as "a software entity that typically acts on the behalf of a person or an institution" [14]. Artikis and Pitt [1] describe the set of constraints as "constraints on the agent communication, on the agent behavior that results from the social roles they occupy and on the agent behavior in general". The owner of the agent is the person or institution on whose behalf the agent acts. According to Johansson [23] it has the power to launch the agent, provide it with preferences, as well as make run-time decisions regarding updating of preferences, and when the agent should be terminated. Moreover, he defines the agent designer as the person(s) who has designed (and possibly implemented) the action selection, and execution mechanisms of the agent. Davidsson and Johansson [14] state that the environment owner is the person or organization that has the power to decide which agents may enter the society, which roles they are allowed to occupy, what communication language should be used, and the set of constraints on the society. Similarly, the environment designer is the person(s) who has designed and possibly implemented the conditions (mechanisms for controlling which agents may enter the society, what possible roles they may have, the space of constraints provided, etc.) under which the agents act in the environment.

With respect to norms, an important issue is the level of openness in the artificial society. Davidsson [13] lists four types of artificial society structures, namely open, closed, semi-closed, and semi-open artificial societies. The four categories balance the trade-off between important society properties, such as, openness, flexibility, stability, and trustfulness differently. Open societies support openness and flexibility, but not stability and trustfulness, and the opposite is true for closed societies. In many situations, however, there is a need for societies that promote all those aspects. One such example is in virtual enterprise breeding environments. We will therefore limit our discussion to the two intermediate categories:

- In *semi-closed artificial societies*, external agents are not allowed to enter. However, actors have the possibility to initiate new agents in the society, which will act on behalf of the external actor [14]. In semi-closed societies there is a (central) physical environment, in which the agents (representing their owners) execute on and communicate with other agents in. This requires that the actors' agents can access some level of mutual communication properties, which are included in the breeding environment. Semi-closed societies convey almost the same degree of openness as semi-open societies, but are less flexible. On the other hand, they have a larger potential for implementing important society attributes, such as, security and trust.
- The main difference to semi-closed artificial societies is that, in *semi-open societies, agents* execute locally on the clients individual computer systems [13]. However, another distinction is that the environment owner is no longer in control of the agents even though the environment owner still has the power to, for instance, dictate the rules of engagement within the society. In semi-open societies, there is an institution that functions as a *gate-keeper*, which every agent needs to connect with before participating in the society. When analysing semi-open societies it is useful to make a distinction between two types of such societies, those with a centralized and those with decentralized communication architecture. In the case of *centralized communication*, the agents run on the members' own individual computer systems, but all communication between

the agents is routed via a central server (placed, e.g., in the communication infrastructure). In decentralized communication, the agents run on the members' individual computer systems and all communication is conducted directly between these end nodes. A drawback with this kind of communication architecture is that joint or communal services (e.g., norm-enhancing mechanisms, etc.) can not be provided by the artificial society. Instead, such services must be ensured locally by the individual members and controlled by a gate-keeper facility.

The definition of an artificial society is general, and as such; it could be treated as a reference model to other types of agent-based coalitions. However, it excludes a specification of what constitutes a norm, as well as mechanisms to enforce them, which may be imperative to achieve the desired system behavior – especially in the context of heterogeneity and autonomy of agents.

3.3.2 Normative MAS (NMAS)

López y López et al. [25] present a normative framework for agent-based systems, which, besides providing the means to computationally represent many normative concepts, can be used to give an understanding of norms and normative agent behavior in artificial societies. Their framework is built upon the idea of autonomy of agents, i.e., it is intended to be used by agents that reason about why norms must be adopted, and why an adopted norm must be complied with. In their formal model, a normative multi-agent system consists of the following entities:

- a set of normative agents (*members*), where a normative agent consists of:
  - a set of goals,
  - a set of capabilities (actions that the agent can perform),
  - a set of motivations (preferences),
  - a set of beliefs,
  - ability to rank the goals according to preferences, and
  - a set of adopted *norms*, some of which the agent has decided to comply with (*intended*) and some of which it has decided to reject (*rejected*).
- a set of general norms that govern the behavior of these agents (*generalnorms*),
- a set of norms issued to allow the creation and abolition of norms (*legislationnorms*),
- a set of norms dedicated to enforcing other norms (*enforcenorms*),
- a set of norms directed to encouraging compliance with norms through rewards (*rewardnorms*), and
- the current state of the *environment* represented by the variable environment.

In addition, they identify a number of authorities:

- a set of *legislators* (agents that are entitled to create, modify or abolish norms),
- a set of *defenders* (agents that are directly responsible for the application of punishments when norms are violated), and
- a set of *promoters* (agents whose responsibilities include rewarding compliant addressees).

As the framework has been built upon the idea of autonomy of agents, they reason about what norms to adopt, why, and in what way. *Norms* are formally defined to be composed of the following entities:

- a set of *normative goals*, which capture the purpose of the norm,
- a set of *addressees*, which are the agents directly responsible for the satisfaction of the normative goals,
- a set of *beneficiaries*, which are the agents that benefit from the satisfaction of the normative goals,
- the *context*, which specifies the situations (environmental states) in which addressee agents must fulfil the norm,
- the *exceptions*, which represent the situations in which addressees cannot be punished when they have not complied with the norm, and
- *rewards* (expressed as a set of goals) to be given when normative goals become satisfied, or
- *punishments* to be applied when they are not.

Their model specifies the elements of an artificial society, as well as its normative components, which makes it a potential role model for a formal description of normative agent-supported virtual enterprises. However, a drawback is that components, such as, environment, communication language, and agent owner are excluded from their model.

3.3.3 Normative Systems (NS)

In this model of normative multi-agent systems, regulative and constitutive norms are in focus. According to Boella and van der Torre [2], a normative multi-agent system is composed of the following entities (we are here focusing on entities, not on how they are described, e.g., that a set of literals and rules are used to describe beliefs, desires and goals of the agents, and that there is a function, *MD*, which makes this mapping):

- a set of agents, *A*, where an agent could be either human or artificial. *A* is modeled in terms of:
  - a set of beliefs (*B*),
  - a set of desires (*D*),
  - a set of goals (*G*),
  - a set of decision variables (*X*), which represent an agent's actions,
  - a function agent description (*AD*), which maps each agent to the sets of beliefs, desires, intentions and decision variables, and
  - a priority relation ($\geq$), which expresses each agent's characteristics and how it resolves its conflicts, i.e., rank the importance of the agent's desires and goals.
- a normative agent, **n**, which is a member of *A*,
- a set of roles, *R*, that the agents can play,
- a norm description (*V*) function that represents the norms recognized by the agents, and
- a goal distribution (*GD*) function that corresponds to the goals of the agent that it is responsible for.

Moreover, they distinguish between *regulative* norms, described as obligations, prohibitions and permissions, and *constitutive* norms, such that regulate the creation of institutional facts, as well as the modification of the normative system itself. In particular, regulative norms are formalized as goals, and constitutive norms are formalized as beliefs of the normative system. Regulative norms are based on the notion of conditional obligation with an associated sanction. Obligations are defined in terms of goals of the normative agent, prohibitions are obligations concerning negated variables, and permissions are specified as exceptions to obligations. Constitutive norms introduce new classifications of existing facts and entities, called institutional facts, or they describe the legal consequences of actions on the normative system. Roles are used to specify the powers of agents to create institutional facts or to modify the norms of the system. Thereby, constitutive norms specify both the behavior and the evolution of a system in that they introduce or remove norms from the system.

This model focuses on the normative and logical aspects of artificial systems. Societal entities, such as, environment and agent owners, the physical environment and mechanisms for norm enforcement are, however, excluded. Moreover, given the special emphasis on norms in their model, an interesting aspect of it would be to further study the relationship and balance between the application of regulative and constitutive norms.

3.4 Comparison and Model Alignment

We will now analyse, and try to align the different formal models described above to identify what norms and norm-enhancing mechanisms that should be included in ABE. The parameters for comparison are deducted from the formal models presented in Section 3.2. A more concise version of the comparison can be found in the Appendix.

3.4.1 Agents and Norms

In this analysis, the study object in the first set of models is the society, where norms play a small yet important part. The second set of models reviewed take their starting points from normative perspectives, where the other aspects, e.g., agent ownership, agent roles, system state, etc., play secondary roles. In the normative frameworks, different types of norms for different types of contexts are defined. By contrast, the AS model just considers one type of norms ("constraints") and on only one level. Moreover, the ABE model specifies norms ("obligations") only between actors. An insight is that the use of other types of norms, and on more levels than one, can enrich these formal models of artificial societies so that they are able to capture the types of norm-governed behavior that are necessary in many complex applications. For instance, we intend to introduce norms that regulate the interaction between the agents in the ABE model. These can be both in terms of specific obligations or permissions between individual agents, and in terms of general norms for all the agents.

3.4.2 Agent Ownership and Roles

Although a set of agents is defined in all of the models reviewed, there are some differences in the views on what constitute an agent. Three different perspectives can be identified:

- a norm-autonomous artificial or human entity as in the NSA model,
- a norm autonomous software entity as found in the NMAS model, and
- just a software entity.

These different views obviously affect the treatment of agents, and norms in artificial societies. In most current applications, agents are not norm autonomous, which may either be due to the complexity of implementing norm autonomous agents, or due to that, in some applications, it is desired that the agent owner is involved in decisions regarding what norms to follow, etc. In the ABE model, a human representative is assisting the agent in decision situations. Moreover, it may be possible to use the theories that assume norm autonomous entities also in current applications, but this requires that both the agent and its owner are included in the norm autonomous entity.

The agent owners are not regarded in NMAS and NS, which also may be due to that completely norm autonomous agents are assumed in those models. Since the agent owner has the power to, apart from deciding what norms to follow, release and terminate an agent, and to provide it with goals even during runtime, the inclusion of agent owners would seem to improve those formal models.

All of the models recognize that agents may take on different roles in artificial societies. However, in ABE and NMAS, the occurrence of agent roles are explicitly specified. In NMAS that is a normative framework, agent roles are defined based on the application of norms, e.g., the addressee agents. In ABE, agent roles are founded upon the type of interaction that the actor is involved in – either as a participant, or as an initiator of an virtual enterprise. Even though requirements of flexibility and dynamicity are prominent in the vision of virtual enterprises, the implementation of norm governed breeding environment software agents benefit by learning from the normative frameworks. In ABE, agent owner roles are detailed due to that agents are not assumed to be completely autonomous, e.g., a human representative is included in ABE, which appears to be the case with the other models.

3.4.3 Norm Enforcement

All models except ABE have provided definitions of norms. The most general view is specified in AS where norms are described as constraints on the society. The NS model details regulative (obligations, prohibitions, and permissions) and constitutive norms (institutional facts). In NMAS, norms specify patterns of behavior for the agents of the society and the definition of norms embrace several components, e.g., goals, addressees, beneficiaries, rewards, punishments, etc. As ABE need revision with respect to norms, all of the other formal models may serve as sources of inspiration with respect to defining norms.

There are some suggestions on norm enforcement amongst the reviewed frameworks. For instance, NMAS uses defender agents that are responsible for the application of punishments when norms are violated, and promoter agents that monitor norm compliance. The latter is corresponding to the external observer agent used by Kamara et al. [24] in order to detect whether interacting agents operate in compliance with the norms or not. Moreover, they also use an admission protocol, which allows nodes to create, enter, and exit the agent society. Similarly, a gate-keeper may be used in ABE to regulate the entering to, and leaving from, the breeding environment. An important part of the gate-keeper would be to ensure that information necessary for the formation and operation of virtual enterprises is available to the members of p and to verify the identity of the actors. Possibly, the gate-keeper may also be equipped with capabilities of handling different levels of memberships with different sets of norms in order to cope with the varying needs of potential participants and members. The gate-keeper could also inform the potential members about what general rules that hold in the breeding environment, and require the potential members to comply with them.

Another possible improvement of the ABE would be to include also defender and promoter functionality, as in the NMAS model, to monitor the behavior of members. This can help to ensure norm compliance, and impose punishments to those that violate the norms.

3.4.4 Agent Communication Language

We can observe that an agent communication language is included only in AS and ABE. However, it may be implicitly assumed in the other models. Obviously, the lack of a language would severely limit the application of norms, but even in the case where a language is specified, the language may put restrictions on what norms can be expressed and communicated. The recognition of an agent communication language in formal models of virtual enterprises is important also for achieving interoperable and interorganizational collaboration in the context of heterogeneous and autonomous collaborating parties. However, further investigations into these matters are needed before any conclusions can be drawn.

3.4.5 Goals

Goals are included in all of the formal models except the AS model (agent preferences are discussed in the paper, but not explicitly specified in the formal model). However, the goals in ABE are not associated to agents, but to actors and virtual enterprises. Since goals can be related to norms (cf. the rational choice model as discussed by Boella et al. [5]), the introduction of norms on the agent level should be accompanied by the inclusion of goals in these models.

3.4.6 The Environment and the Owner

It can be observed that the physical environment is included in all models apart from the theoretical norm-oriented frameworks (NMAS and NS). The environment owner is only recognized in the AS model. This is interesting since the environment owner can have a large impact on what norms that will hold for the society. Typically, the environment owner has control over gate-keepers, defenders, promoters, legislators, and any other entities dealing with issues of interaction in the society. Therefore, it is likely that the formal models can be enriched by including the environment owner.

All models specify the occurrence of system state. In AS and ABE, system state is declared to be a set of states that hold at each time in the society, whereas NMAS lets the environment be represented by a variable, and NS outlines parameters that describe both state of the world and institutional facts. By including such components in artificial societies, it is possible to keep track of the events taking place, which might come well at hand in the case of a potential dispute among collaborating parties over, e.g., shared assets, roles, and terms of contracts.

3.5 Related Work

Even though we compared several formal models, there is other work that deserves to be mentioned. Below, we account for these research advancements together with short summaries of their main contributions.

3.5.1 Agent Organizations and Norms

Applying and investigating norms in societies of agents is an extensive but scattered field of research. It includes contributions from a broad spectrum of academic subjects, e.g., computer science, economics, philosophy, and social sciences.

Boella et al. [5] provide an overview of the emerging area of normative multi-agent systems and present definitions of central concepts. In particular, they study the influence of social sciences on agent theory (and vice versa) in the context of norms as a defining characteristic of relations between autonomous agents in a multi-agent system.

Grossi et al. [20][21] analyse the foundations of organizational structures, and contribute to the development of general methods for the assessment of multi-agent systems. They use formal tools to describe the organizational structure, as well as the effect of such structures on various activities in multi-agent systems.

Dignum et al. [17] propose an integrated framework for modelling agent organizations, called OMNI (Organizational Model for Normative Institutions), with the purpose of balancing global organizational requirements with the autonomy of individual agents. OMNI is designed with the purpose of being a general framework. It consists of three dimensions, referred to as normative, organizational and ontological levels, and it is intended to fit all types of agent societies. However, the descriptions do not explicitly include the various entities that make a society.

Fasli [18] presents a logical framework comprising commitments, roles, and obligations for agents equipped with beliefs, desires, and intentions. With respect to such social agents, she also discusses how obligations arise as a result of social commitments and the adoption of roles. Conte and Paolucci [11] present a cognitive model of social responsibility in societies of agents. Their main contribution is the provision of a general and elementary definition of responsibility in multi-agent systems.

Dignum [15] devotes her doctoral thesis to detailing a model for organizational interaction, called OperA (Organizations per Agents), which is based on agents and founded in logic. Among many things, she outlines a formal theory for the OperA framework addressing interaction in multi-agent systems, which, for instance, makes it possible to describe, and verify contracts that regulate interaction between agents.

Kamara et al. [24] propose an agent-based architecture for institutional management of self-organizing ad hoc networks in which a variety of normative aspects are included along with an objective reasoning capacity (that allows agents to reason about normative positions), and a suite of protocols for network management. They conclude by emphasizing the need for a logically sound, and computationally grounded theoretical framework for self-organizing ad hoc networks.

Of these referred models, the OperA model by Dignum [15] appears to be particularly relevant as it is intended to support knowledge management between actors participating in open environments. The OperA model provides a flexible way to represent interaction and role enactment since it abstracts from the specific internal representations of the individual agents and separates the modelling of organizational requirements and aims. In OperA, contracts are used to link the different models, and create specific instances that reflect the needs and structure of the current environment and participants. Another relevant contribution is the work proposed by Kamara et al. [24]. Even though their architecture may be applicable to agents involved in virtual enterprise formation, its formal model appears not yet to have been completed. The work by Fasli [18] is interesting to study when further specifying the relationship between roles and obligations of the actors in a virtual breeding environment, but that particular aspect is outside the scope of this chapter.

3.5.2 Agents and Norms in Virtual Organizations

A virtual organization can be described as a set of individuals and institutions that need to coordinate resources and services across institutional boundaries in order to collaborate. In contrast to a virtual enterprise, a virtual organization need not be business-oriented.

Some research efforts on the topic of agents, norms, and virtual organizations have been carried out. For instance, Boella et al. [6] propose a conceptual model of virtual organizations as normative systems, which demonstrates distinctions between local and global authorities, and between local and global norm enforcement policies. In this work, they also describe three kinds of agents that inhabit such a system, referred to as subjects, defenders, and autonomous normative systems. Crucial to their model is that norms are dealt with explicitly, and that system designers should not assume that norms will automatically be obeyed.

In another paper by Boella and van der Torre [5], normative multi-agent systems for secure knowledge management based on access control policies are studied. It is shown how distributed access control is realized by means of local policies of access control systems, and how global policies can enforce the local ones. A contribution is that their framework can be extended to deal not only with policies consisting of regulative rules like obligations, prohibitions and permissions, but also with constitutive rules specifying counts-as relations and institutional facts.

Norman et al. [27] investigate how effective virtual organizations may be rapidly formed for any given purpose, but they exclude norms from their studies. They conclude by pointing out the need for a model that addresses the automation of virtual organization management while emphasizing flexibility and robustness.

Dignum and Dignum [16] present a framework for the support of virtual organizations based on agent societies in which norms are treated as a way for such societies to cope with social order. They propose the use of institutions to implement, and monitor the behaviours of agents within the virtual organization. An important contribution is that institutions can make the organizational goals and norms explicit, and warrant their fulfilment by providing facilitation roles and controlled interaction protocols.

The work by Boella and van der Torre is general and comprehensive and one of their models was included in the model comparison (see Section 3.4). Another relevant piece of their work is presented in more detail in the next subsection. The other articles mentioned above address various aspects of normative multi-agent systems, but they appear to exclude either norms or the society perspective.

3.5.3 Virtual Communities of Agents

Boella and van der Torre [3] investigate the use of design policies composed by prohibitions, permissions, and authorizations for virtual communities of agents on a computational grid. This work partly builds on the work discussed in the previous section, i.e., by Boella and van der Torre [2]. They define a virtual community as a large, multi-institutional group of individuals who use a set of rules, i.e., a policy, to specify how to share their resources. In a virtual community, agents can play both the role of resource consumers, and the role of resource providers. Resource providers retain the control of their resources, and they specify in local policies the conditions for use of their resources thereby giving rise to a third role, authorization, in their model.

In virtual communities, a single set of agents (A), where each can play one or more roles, is defined so that each agent of the agent set can play three roles:

1. *Resource consumer*, denoted as $c(a_i)$, is an agent who manipulates a resource by means of some action. It can access resources to achieve its goals, is subject to norms regulating security, prohibitions and permissions, and also endowed with authorizations to access resources.
2. *Resource provider*, denoted as $p(a_i)$, can provide access to the resources it owns. This is referred to as the normative role, since it can issue norms, i.e., prohibitions and permissions, about the access of a resource, and enforce their respect by means of sanctions, and delegate the power to authorize resource consumers.
3. *Authority*, denoted as $u(a_i)$, can declare resource consumers authorized when they are requested to do so. They know that their declarations are considered authorizations by the resource providers since they have been delegated the power to authorize resource consumers on behalf of resource providers.

Prohibitions and *permissions* are specified in terms of goals and desires of the bearer of the norm and of the normative role. A prohibition is defined as a goal of resource providers whereas a permission is the behavior, which is not considered by a provider as a violation, and thus is not sanctioned. A third concept, *authorization*, is a belief of a provider, which appears as a condition in some permission it issued. Thereby, an authorization has a meaning only if it appears among the conditions of a permission. These concepts are then supplemented with two concepts, namely *violation* and *sanction*. The agent holding the normative role, i.e., the resource provider, can decide if some action is to be regarded as a violation. The possibility to punish violations by means of some sanction is among the preconditions for creating a prohibition. Sanctions provide motivation to fulfill the norms since it is not possible to assume that all agents are cooperative, and that they respect the norms. Thereby, a sanction is an action negatively affecting an agent, i.e., the agent desires the absence of the sanction.

As in the NS model, societal aspects, e.g., agent and environment owner, are excluded in this model. The primary objective of their work was, however, to study the design of policies for agent-based virtual communities, which of course can be useful (in terms of security and coordination) also in the ABE model.

3.5.4 Norms in Electronic Institutions

Electronic institutions are created to establish a social order that can enable successful interactions among heterogeneous and autonomous agents, and they are many times described as business-like environments of collaborating agents [10]. Often, electronic institutions are comprised by legal obligations, and the like.

The work by Cardoso and Oliveira revolves around contract negotiation within (agent-supported) virtual enterprises and electronic institutions. For instance, they identify requirements of virtual enterprise contracts, and develop a normative framework for contract validation and enforcement [8]. In a paper on institutional reality and norms [9], they add surveillance functionality to the electronic institutions that can monitor the agents' ability to comply with norms. They view norms as means used to regulate an existing environment, and to define contracts that make the agents' commitments to each other explicit. Moreover, Cardoso et al. [10] provide a virtual normative environment that assists and regulates the creation and operation of virtual organizations through contract-related services. However, their contributions are mainly in the area of formal models for contract specification, and they do not explicitly address the structures and the elements of norm-governed agent societies.

3.6 Discussion

As can be seen in the Appendix, the analysed models are in fact rather similar. When concepts are missing, e.g., agent communication language (as with NMAS or NS), this may not be out of principle, but because of the scope of the work. However, in one aspect there is a significant difference – the normative models exclude the notion of agent and environment owners, whereas the society-oriented do not. If norms are to be enforced successfully, these concepts are of a crucial nature since they can impact, e.g., the agents' willingness to follow and comply with obligations, permissions, and prohibitions. However, there are a number of issues that need further study, e.g.:

- Investigate what precise types of norms (obligations, permissions, prohibitions, etc.), and on what levels (general or specific) to include in the formal models of ABE.
- Explore the possibility to include the owner in norm-autonomous entities in the NMAS and NS models, so that these models can be applied also to societies populated by agents that are not norm-autonomous.

With respect to the ABE model, the inclusion of the following entities should be further investigated:

- A gate keeper functionality as is done in the semi-open version of the AS model.
- Defender and promoter functionality as is done in the NMAS model.
- Environment owner as in the AS model.

These suggestions are important steps in the analysis of concepts and software that supports the formation and operation of normative and agent-supported virtual

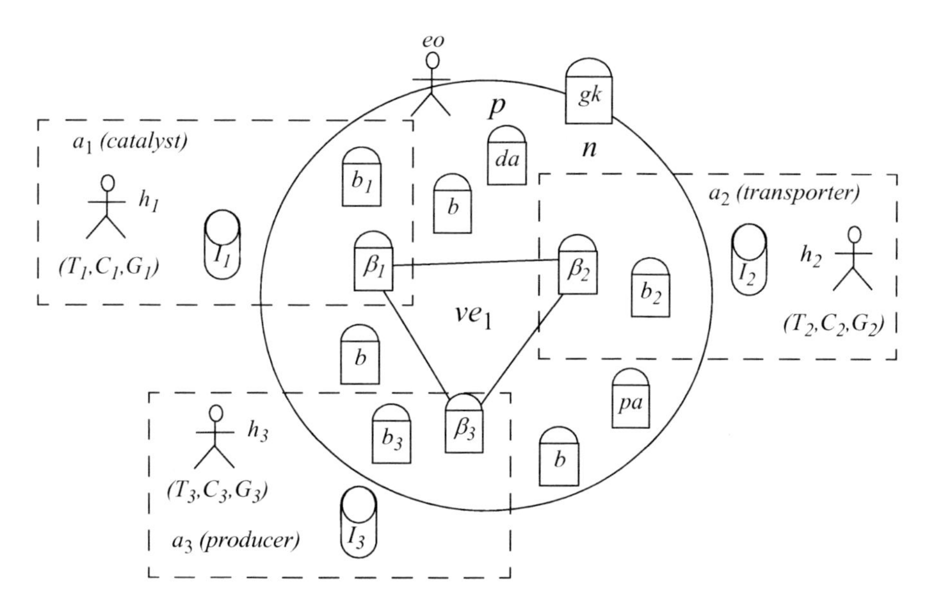

Fig. 3.2. A revised example of a breeding environment where norms (*n*), promoter agents (*pa*), defender agents (*da*), environment owner (*eo*), and gate-keeper (*gk*) are also included

enterprises within breeding environments. A proposal for an updated ABE model is illustrated in Figure 3.2.

3.7 Concluding Remarks

In this paper, we investigated how norm-governed behavior within agent societies can be achieved in the context of virtual enterprises. For this purpose, a comparative analysis of a number of formal models describing artificial societies or normative systems haven been undertaken. A general observation is that the models reviewed are not entirely concordant with each other. For instance, completely norm-autonomous agents are assumed in the norm-focused frameworks, but in the society-oriented model the notion of an agent owner is specified. In the norm-focused models, entities like agent communication language and the physical environment are often not regarded. Moreover, norm-enhancing mechanisms are only included in two models (ABE and NMAS). Based on these findings, we have discussed how model alignment can foster both areas, i.e., modeling of agent societies and normative systems, in general, and how the inclusion of various missing entities may improve the formal models in particular.

With respect to the ABE model, we can conclude that it can be enriched by, for instance, the introduction and definition of norms (both on general and specific levels) that regulate the interaction between agents, and that these types of norms should be accompanied by the specification of goals. On norm enforcement, some opportunities for improvements in the ABE model are to include gate-keeper, as well as, defender and promoter functionality in order to ensure norm compliance and punishments of norm violations. It may also be beneficial to regard the artificial society as a community, i.e., a social structure in which the members can share a common culture and a set of values, in approaching norm-governed virtual enterprises.

References

[1] Artikis, A., Pitt, J.: A Formal Model of Open Agent Societies. In: Proceedings of the 5th International Conference on Autonomous Agents (2001)

[2] Boella, G., van der Torre, L.: Regulative and Constitutive Norms in Normative Multiagent Systems. In: Proceedings of 9th International Conference on the Principles of Knowledge Representation and Reasoning (American Association for Artificial Intelligence) (2004)

[3] Boella, G., van der Torre, L.: Permission and authorization in policies for virtual communities of agents. In: Proceedings of Agents and P2P Computing Workshop at AAMAS 2004 (2004)

[4] Boella, G., van der Torre, L.: Normative multiagent systems. In: Proceedings of Trust in Agent Societies Workshop at AAMAS 2004 (2004)

[5] Boella, G., van der Torre, L.: Security policies for sharing knowledge in virtual communities. IEEE Transactions on Systems, Man and Cybernetics - Part A 36(3), 439–450 (2006)

[6] Boella, G., Hulstijn, J., van der Torre, L.: Virtual organizations as normative multiagent systems. In: Proceedings of the Hawaiian International Conference on Systems Science (2005)

[7] Camarinha-Matos, L.M., Afsarmanesh, H.: Collaborative Networks: A New Scientific Discipline. Journal of Intelligent Manufacturing 16(4-5), 439–452 (2005)

[8] Cardoso, H.L., Oliveira, E.: Virtual Enterprise Normative Framework within Electronic Institutions. In: Gleizes, M.-P., Omicini, A., Zambonelli, F. (eds.) ESAW 2004. LNCS (LNAI), vol. 3451, pp. 14–32. Springer, Heidelberg (2005)

[9] Cardoso, H.L., Oliveira, E.: Institutional Reality and Norms: Specifying and Monitoring Agent Organizations. International Journal of Cooperative Information Systems 16(1), 67–95 (2006)

[10] Cardoso, H.L., Rocha, A.P., Oliveira, E.: Virtual Organization Support through Electronic Institutions and Normative Multi-Agent Systems. In: Handbook of Research on Nature-Inspired Computing for Economics and Management, pp. 786–804. Idea Group Publishing (2007)

[11] Conte, R., Paolucci, M.: Responsibility for Societies of Agents. Journal of Artificial Societies and Social Simulation 7(4) (2004)

[12] D'Atri, A., Motro, A.: VirtuE: a Formal Model of Virtual Enterprises for Information Markets. Journal of Intelligent Information Systems 30(1), 33–53 (2007)

[13] Davidsson, P.: Categories of Artificial Societies. In: Omicini, A., Petta, P., Tolksdorf, R. (eds.) ESAW 2001. LNCS (LNAI), vol. 2203, p. 1. Springer, Heidelberg (2002)

[14] Davidsson, P., Johansson, S.: On the Potential of Norm-Governed Behavior in Different Categories of Artificial Societies. Computational and Mathematical Organization Theory 12(2-3), 169–180 (2006)

[15] Dignum, V.: A model for organizational interaction: based on agents, founded in logic. PhD thesis, Institute of Information and Computing Sciences, Utrecht University, NL (2004)

[16] Dignum, V., Dignum, F.: Towards an agent-based infrastructure to support virtual organizations. In: Proceedings of the International Conference on Infrastructures for Virtual Enterprises (2002)

[17] Dignum, V., Vázquez-Salceda, J., Dignum, F.P.M.: OMNI: Introducing Social Structure, Norms and Ontologies into Agent Organizations. In: Bordini, R.H., Dastani, M., Dix, J., El Fallah Seghrouchni, A. (eds.) PROMAS 2004. LNCS (LNAI), vol. 3346, pp. 181–198. Springer, Heidelberg (2005)

[18] Fasli, M.: On Commitments, Roles, and Obligations. In: Dunin-Keplicz, B., Nawarecki, E. (eds.) CEEMAS 2001. LNCS (LNAI), vol. 2296, pp. 93–102. Springer, Heidelberg (2002)

[19] Fasli, M.: Agent Technology for e-Commerce. John Wiley & Sons, Ltd., Chichester (2007)

[20] Grossi, D., Dignum, F., Dastani, M., Royakkers, L.: Foundations of Organizational Structures in Multiagent Systems. In: Proceedings of the Fourth International Joint Conference on Autonomous Agents and Multiagent Systems (AAMAS) (2005)

[21] Grossi, D., Dignum, F.P.M., Dignum, V., Dastani, M., Royakkers, L.M.M.: Structural Aspects of the Evaluation of Agent Organizations. In: Noriega, P., Vázquez-Salceda, J., Boella, G., Boissier, O., Dignum, V., Fornara, N., Matson, E. (eds.) COIN 2006. LNCS (LNAI), vol. 4386, pp. 3–18. Springer, Heidelberg (2007)

[22] Jacobsson, A., Davidsson, P.: A Formal Analysis of Virtual Enterprise Creation and Operation. In: Król, D., Nguyen, N.T. (eds.) Intelligence Integration in Distributed Knowledge Management. Idea Group Publishing (to appear in, 2008)

[23] Johansson, S.J.: On Coordination in Multi-Agent Systems. Ph.D. Dissertation Series No. 05/02, Department of Software Engineering and Computer Science, Blekinge Institute of Technology, Sweden (2002)

[24] Kamara, L., Pitt, J., Sergot, M.: Towards Norm-Governed Self-Organizing Networks. In: Proceedings of the AISB 2005 Symposium on Normative Multiagent Systems (2005)

[25] López, Y., López, F., Luck, M., d'Inverno, M.: A Normative Framework for Agent-Based Systems. Computational and Mathematical Organization Theory 12(2-3), 227–250 (2006)

[26] Luck, M., McBurney, P., Preist, C.: Agent Technology: Enabling Next Generation Computing (A Roadmap for Agent Based Computing), AgentLink (2003)

[27] Norman, T.J., et al.: Agent-based formation of virtual organizations. Knowledge-Based Systems 17(2-4), 103–111 (2004)

[28] Putnik, G.D., Cunha, M.M.: Virtual Enterprise Integration. Idea Group Publishing, Hershey (2005)

Appendix: A Concise Version of the Model Comparison, Part 1

	AS	ABE	NMAS	NS
Definition of Agent	A software entity that typically acts on the behalf of a person or an institution	-----	Normative agent: - a set of goals - a set of capabilities - a set of motivations (preferences) - a set of beliefs - ability to rank the goals according to preferences - a set of norms - a set of intended norms - a set of rejected norms	Human or artificial. An agent is defined as: -Beliefs (B) -Desires (D) -Goals (G) -Decision variables (X) -Agent description variables (AD) -A priority relation ($\geq$)
Agents	A set of agents	A set of agents (*B*) and for each active virtual enterprise a set of agents (β)	A set of normative agents (*members*)	A set of agents (A)
Agent roles	A set of roles that the agents can play	The roles of initiator and potential participant are implicitly assumed	With respect to a norm, an agent can play either the *addressee* or the *beneficiary*	A set of roles, *R*, that the agents can play
Agent owners	A set of owners (of the agents)	A set of actors (*A*) which have human representatives (*H*)	-----	-----
Agent owner roles	-----	A set of roles (*R*)	-----	-----
Definition of norm	Constraints on the agent communication language, on the agent behavior that results from the social roles they occupy, and on the agent behavior in general	-----	Components are: -Normative goals -Addressee agents -Beneficiary agent -Context -Exception -Have not (complied with norms) -Immunity -Rewards -Punishments	Norms are either *regulative* (defined as obligations, prohibitions, and permissions) or *constitutive* (defined as institutional facts)

A Concise Version of the Model Comparison, Part 2

	AS	ABE	NMAS	NS
Norms	A set of constraints	A set of obligations for each role that the actors can play (O^{role}) in a virtual enterprise. Thus this is not on the agent level	Four sets of norms: *general norms*, *legislation norms*, *enforcement norms*, and *reward norms* (which in turn may be obligations, prohibitions, social commitments or social codes)	Four sets of norms: obligations, permissions, prohibitions and institutional facts
Norm-enhancing mechanisms	In semi-open societies there are a Gate-keeper agent (regulates the entering and leaving from the community)	-----	*Defenders* (agents that are responsible for the application of punishments when norms are violated) *Promoters* (agents that monitor compliance with norms) Legislators (agents that define norms)	-----
Communication language	A communication language	*l* and λ_i is the agent communication language used by the agents *B* and β	-----	-----
The physical environment/ infrastructure	The environment (computation and/or communication infrastructure)	A set of communication infrastructures needed for formation and collaboration (*CI*)	-----	-----
Environment owners	An environment owner	-----	-----	-----
Goals	-----	Each virtual enterprise has a set of goals (G_{VE}) and each actor has a set of goals (G_{actor})	Each agent has a set of goals (*goals*)	Each agent has a set of beliefs (B_a), desires (D_a), and goals (G_a)

4

e-JABAT – An Implementation of the Web-Based A-Team

Dariusz Barbucha, Ireneusz Czarnowski, Piotr Jędrzejowicz, Ewa Ratajczak-Ropel, and Izabela Wierzbowska

Department of Information Systems, Gdynia Maritime University
Morska 83, 81-225 Gdynia, Poland
{barbucha,irek,pj,ewra,iza}@am.gdynia.pl

Abstract. The chapter proposes a middleware called JABAT (JADE-Based A-Team). JABAT allows to design and implement A-Team architectures for solving combinatorial and selected non-linear optimization problems. The JABAT is intended to become a first step towards next generation A-Teams which are fully Internet accessible, portable, scalable and in conformity with the FIPA standards. From the user point of view JABAT is the web-based application, in the paper refereed to as the e-JABAT.

Keywords: JABAT, A-Team, optimization, computionally hard problems, multi-agents systems.

4.1 Introduction

Formerly, research into systems composed of multiple agents was carried out under the banner of Distributed Artificial Intelligence, which has historically been divided into two main fields: Distributed Problem Solving and Multi-Agent Systems. Recently, the term ,,multi-agent systems" has a more general meaning, and is rather used to refer to all types of systems composed of multiple autonomous components [19].

The field of autonomous agents and multi-agent systems is a rapidly expanding area of research and development. It is based on many ideas originating from such disciplines as artificial intelligence, distributed computing, object-oriented systems and software engineering.

Last years, a number of significant advances have been made in both the design and implementation of autonomous agents. A number of applications of agent technology is growing systematically. Nowadays agent technology is used to solve real-world problems in a range of industrial and commercial applications.

One of the successful approaches to agent-based optimization is the concept of an asynchronous team (A-Team), originally introduced by Talukdar [36].

This chapter contains an overview of the Web-based implementation of the A-Team idea called JABAT (JADE-Based A-Team). JABAT is the middleware supporting the construction of the dedicated A-Team architectures used for solving a variety of computationally hard optimization problems. From the user point

N.T. Nguyen, L.C. Jain (Eds.): Intel. Agents in the Evol. of Web & Appl., SCI 167, pp. 57–86.
springerlink.com

of view, JABAT is a web application, refereed to as the e-JABAT, which provides the opportunity to solve instances of optimization problems.

The chapter includes a short review of the A-Team concept, overview of the JABAT functionality and a description of main JABAT components including JABAT engine and web-based user interface. The chapter also includes examples of using JABAT to obtain solutions to several difficult optimization problems, including the resource constrained project scheduling problem, the vehicle routing problem, the artificial neural network training problem and the traveling salesman problem.

4.2 A-Team Concept

A-Team is a multi agent architecture, which has been proposed in [36] and [37]. It has been shown that the A-Team framework enables users to easily combine disparate problem solving strategies, each in the form of an agent, and enables these agents to cooperate to evolve diverse and high quality solutions [33]. Acording to [36] an asynchronous team is a collection of software agents that cooperate to solve a problem by dynamically evolving a population of solutions. As [33] observed agents cooperate by sharing access to populations of candidate solutions. Each agent works to create, modify or remove solutions from a population. The quality of the solutions gradually evolves over time as improved solutions are added and poor solutions are removed. Cooperation between agents emerges as one agent works on the solutions produced by another. Within an A-Team, agents are autonomous and asynchronous. Each agent encapsulates a particular problem-solving method along with the methods to decide when to work, what to work on and how often to work.

A-Team architecture could be classified as a software multi-agent system that is used to create software assistant agents [28]. According to [2] an asynchronous team (A-Team) is a network of agents (workers) and memories (repositories for the results of work). It is possible to design A-Teams to be effective in solving difficult computational problems. The main design issues are structure of the network and the complement of agents.

The ground principal of asynchronous teams rests on combining algorithms, which alone could be inept for the task, into effective problem-solving organizations [37]. Talukdar [38] proposed the grammar to provide a mean for constructing asynchronous teams that might be used in solving a given instance of a family of off-line problems. In other words, the grammar constructively defines the space that must be searched if an asynchronous team that is good at solving the given problem-instance is to be found. The primitives of the grammar are:

- Sharable memories, each dedicated to a member of the family-of-problems, and designed to contain a population of trial-solutions to its problem.
- Operators for modifying trial-solutions.
- Selectors for picking trial-solutions.
- Schedulers for determining when selectors and operators are to work.

A-Team architecture offers several advantages. Rachlin et al. [33] mention modularity, suitability for distributed environments and robustness. Tweedale et al. [41] outlines an abridged history of agents as a guide for the reader to understand the trends and directions of future agent design. This description includes how agent technologies have developed using increasingly sophisticated techniques. It also indicates the transition of formal programming languages into object-oriented programming and how this transition facilitated a corresponding shift from scripted agents (bots) to agent-oriented designs which is best exemplified by A-Teams.

A-Teams have proven to be successful in addressing hard optimization problems where no dominant algorithm exists. Some reported applications include:

- TSP problem [35], [34].
- Railroad routing problem [40].
- Collision avoidance in robotics [21].
- Paper-mill scheduling [5].
- Diagnosis of faults in power systems [10].
- Planning and scheduling in manufacturing [29].
- Computational intelligence methods (genetic algorithms, neural networks, fuzzy controllers) on single machines and clusters of workstations [30].
- Flow optimization of railroad traffic [8].
- Bicriteria combinatorial optimization problem and its applications to the design of communication networks for distributed controllers [9].
- Automatic insertion of electronic components [32].
- Parallel and asynchronous approach to give near-optimal solutions to the non-fixed point-to-point connection problem [11].
- Job-shop scheduling [1].

It should be noticed that several related agent-based architectures have been recently proposed. One of them is the blackboard architecture enabling cooperation among agents, called knowledge sources, by sharing access to a common memory and allowing agents to work on different parts of the problem [15]. The key difference is that A-Teams do not have a central scheduler responsible for sequencing agent invocations.

Co-evolutionary multi-agent systems (CoEMAS) [42], [14] allow co-evolution of several species of agents. CoEMAS can be applied, for example, to multi-objective optimization and multi-modal function optimization (*niching co-evolutionary multi-agent system NCoEMAS*). In CoEMAS several (usually two) different species co-evolve. One of them represents solutions. The goal of the second species is to cooperate (or compete) with the first one in order to force the population of solutions to locate Pareto frontier or proportionally populate and stably maintain niches in multi-modal domain.

Some similarities to A-Team characterize the Decision Making Library (Agent-DMSL) [16] a highly customizable decision making library built using Java 1.4 programming language.

A-Teams belong to two broader classes of systems which are cooperative multi-agent and peer-to-peer (P2P) systems. In cooperative multiagent systems agents work toward achieving some common goals, whereas self-interested agents have distinct goals but may still interact to advance their own goals [25]. In P2P systems a large number of autonomous computing nodes (the peers) pool together their resources and rely on each other for *data* and *services* [23].

The literature reported implementations of the A-Team concept include two broad classes of systems: dedicated A-Teams and platforms, environments or shells used as tools for constructing specialized A-Team solutions. Dedicated (or specialized) A-Teams are usually not flexible and can be used for solving particular problem types only. Among example A-Teams of such type one can mention the OPTIMA system for the general component insertion optimization problem [32] or A-Team with a collaboration protocol based on a conditional measure of agent effectiveness designed for flow optimization of railroad traffic [8]. Majority of applications listed in this section have been using a specialized A-Teams architectures.

Among platforms and environments used to implement A-Team concept some well known include IBM A-Team written in C++ with own configuration language [33] and *Bang 3* - a platform for the development of Multi-Agent Systems (MAS) [30]. Some implementations of A-Team were based on universal tools like Matlab [37]. Some other were written using algorithmic languages like, for example the parallel A-Team of [11] written in C and run under PVM operating system.

The above discussed platforms and environments belong to the first generation of A-Team tools. They are neither scalable nor portable although some might have limited portability or limited scalability. Agents are not in conformity with the FIPA (The Foundation of Intelligent Psychical Agents) standards and there are no interoperability nor Internet accessibility. Migration of agents is either impossible or limited to a single software platform.

To overcome some of the above mentioned deficiencies we propose a solution which is a middleware called JABAT (JADE-Based A-Team). JABAT allows to design and implement A-Team architectures for solving difficult optimization problems. JABAT is intended to become a first step towards the next generation A-Teams which are fully Internet accessible, portable, scalable and in conformity with the FIPA standards. JABAT has been developed as a team work of all authors of this chapter. Earlier version of the system was described in [3].

4.3 Main Features of the JABAT Middleware

JABAT produces solutions to combinatorial optimization problems using a set of optimizing agents, each representing an improvement algorithm.To escape getting trapped into a local optimum the initial population of solutions called individuals is generated. Individuals forming the initial population are, at the following computation stages, improved by independently acting agents, increasing chances for reaching the global optimum through applying the evolutionary computation paradigm. Thus, the process of solving a single task (i.e. a problem instance) in JABAT consists of several steps. At first an initial population of solutions is generated. Then, individuals from the population are improved be optimizing agents. Finally, when the stopping criterion is met, the best solution in the population is taken as the final result.

The following set of features characterizes JABAT functionality:

- The system can in parallel solve instances of several different problems.
- The user, having a list of all algorithms implemented for the given problem, may choose how many and which of them should be used.
- The optimization process can be carried out on many computers. The user can easily add or delete a computer from the system. In both cases JABAT will adapt to the changes, commanding the optimizing agents working within the system to migrate.
- The system is fed in the batch mode - consecutive problems may be stored and solved later, when the system assesses that there is enough resources to undertake new searches.

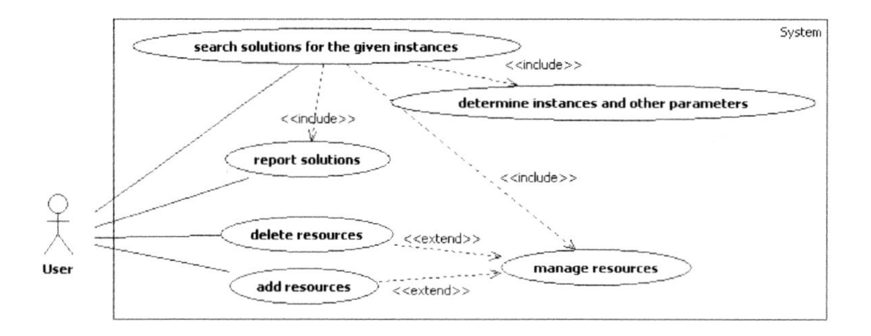

Fig. 4.1. Use case diagram of the functionality of JABAT

From the user point of view JABAT provides mechanisms allowing searching for solutions to the considered problem instances, reporting solutions and managing resources. The general functionality of JABAT is shown on the use case diagram in Fig. 4.1.

Web-based e-JABAT consists of two parts: JABAT engine, responsible for the actual solving of computational task and the web interface, using which the user can upload tasks, set up the required parameters and download the results. e-JABAT architecture is show in Fig. 4.2.

4.4 The JABAT Engine

The JABAT engine is built using JADE (Java Agent Development Framework), a software framework proposed by TILAB [18] for the development and run-time execution of peer-to-peer applications. JADE is based on the agents paradigm in compliance with the FIPA [39] specifications and provides a comprehensive set of system services and agents necessary to implement distributed peer-to peer applications in the fixed and mobile environment. It includes both the

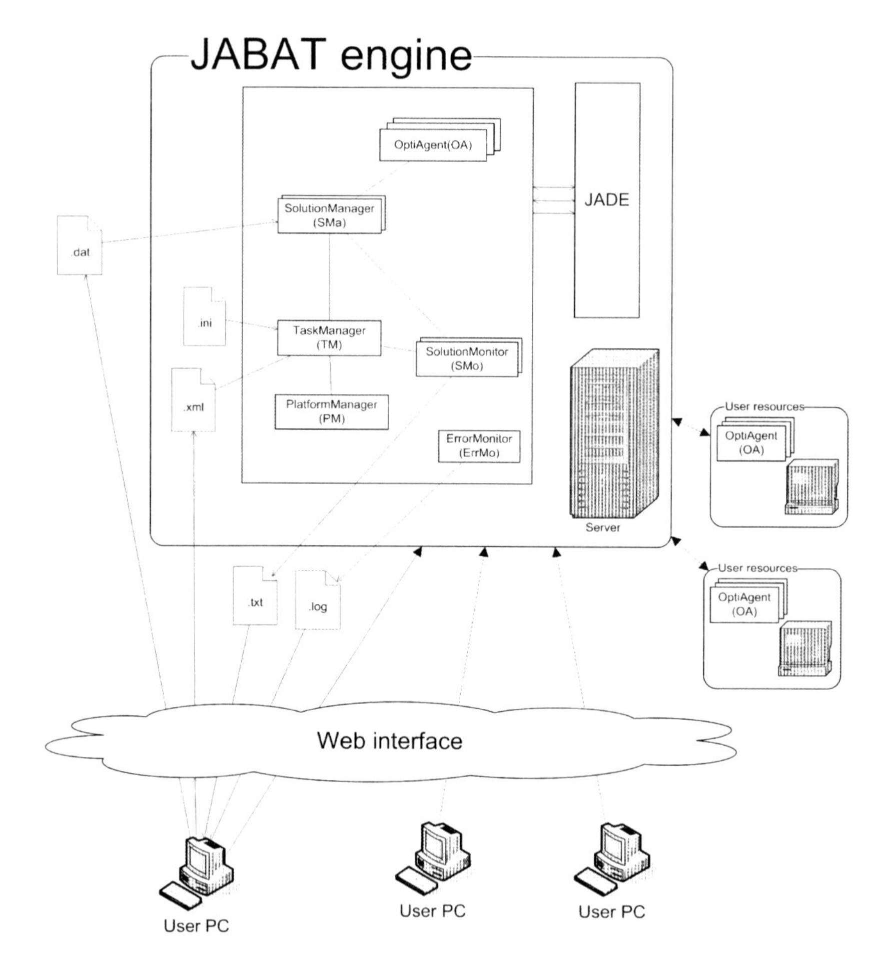

Fig. 4.2. e-JABAT architecture

libraries required to develop application agents and the run-time environment that provides the basic services and that must be running on the device before agents can be activated [6].

JADE as the agent platform includes the AMS (Agent Management System), the DF (Directory Facilitator), and the ACC (Agent Communication Channel). It manages the whole agent life cycle, provides the transport mechanism and interface to send/receive messages to/from other agents and supports debugging, management and monitoring phases with using dedicated graphical tools. It also supports agents migration, complex interaction protocols, messages content creation and management including XML and RDF [6], [7].

JADE platforms have containers to hold agents, not necessarily on the same computer. In JABAT containers placed in different platforms are used to run agents responsible for searching for optimal solutions using pre-defined solution improvement algorithms.

Within the JABAT engine the following types of agents are used:

- *OptiAgents* - representing the solution improvement algorithms,
- *SolutionManagers* - managing the populations of solutions,
- *TaskManager* - responsible for initialising the process of solving of a problem instance,
- *SolutionMonitors* - recording computation results,
- *PlatformManager* - organising the process of migrations between different platforms,
- *ErrorMonitor* - monitoring and reporting unexpected behavior of the system.

Basic functionality ot the JABAT engine realized by the above listed agents are shown on the use case diagram in Fig. 4.3 and described in a detailed manner in the following sections.

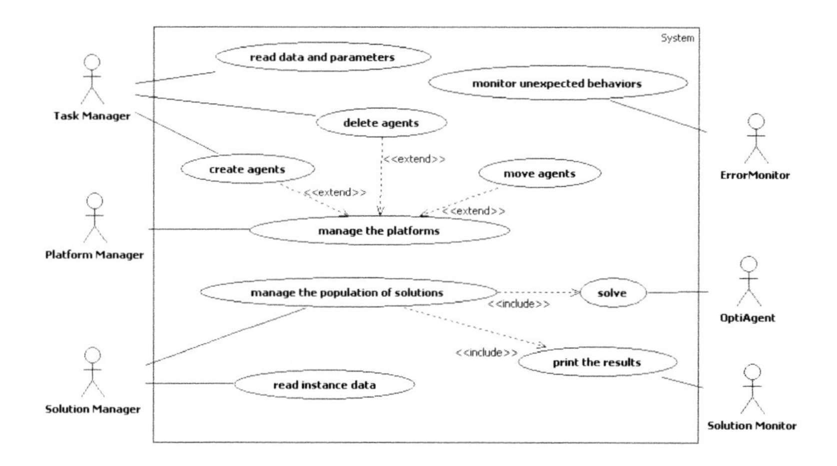

Fig. 4.3. Use case diagram of JABAT from agents' point of view

4.4.1 Agents Responsible for Solving a Task

The most important for the process of solving a task are *OptiAgents* and *SolutionManagers*. They work in parallel and communicate with each other exchanging solutions that are either to be improved when they are sent to *OptiAgents*, or stored back into the common memory when sent to *SolutionManager*.

Each *OptiAgent* is a single improvement algorithm. An *OptiAgent* can communicate with all *SolutionManagers* working with instances of the same problem. An agent sends out the message about its readiness to work. Such message contains information about the number of solutions from the common memmory required to execute the improvement procedure. In response the *SolutionManager* sends the details of the task and appropriate number of solutions. The respective *OptiAgent* processes them and sends back the improved solution or solutions. The process iterates, until some stopping criterion is met.

Each *SolutionManager* is responsible for a single population of solutions created to solve a single instance of the considered problem. Its actions include generation of the initial pool of solutions, sending solutions to the *OptiAgents*, merging improved solutions with the population stored in the common memory and deciding when the whole process of searching for the best solution should be stopped. All these activities are managed in accordance with the population-management strategy that has been defined for the particular problem.

In particular, such a strategy in JABAT defines:

- how the initial population of individuals is created (for example how many solutions it consists of and what method is used to obtain the initial population) - implemented as the *initPopulationOfSolutions()* function,
- how solutions to be sent to optimizing agents are chosen - implemented as the *readSolution()* function,
- how solutions that has been received from optimizing agents are merged with the population - implemented as the *addSolution()* function,
- how the process of searching stops (after a predefined number of iteration, after reaching given solution, or when calculations do not improve current best solution for some length time, etc.) - implemented as the *stop()* function.

Apart from the above *SolutionManager* is also responsible for sending periodical reports on the state of computations to the *SolutionMonitor* that monitors the respective task. The *SolutionMonitor*, in turn, prepares and saves information on the results thus obtained in the report file available to the user.

Fig. 4.4 presents the sequence diagram which shows main messages exchanged between agents in the process of solving the problem instance. After the *SolutionManager* initializes the population of solution, it waits to receive information from any of the available about its readiness to attempt improving a solution (or solutions). *SolutionManager* reads the required number of solutions from the common memory and sends them to the respective *OptiAgent*, which attempt to improve them with their inbuilt improvement algorithms. Next, the improved solution is sent back to *SolutionManager* and added to the common memory. *SolutionManager* may also sent the solution to *SolutionMonitor*, which generates a report on the computation progress and results. Then, again, the *OptiAgent* sends a message informing about its readiness.

If during the process of solving an instance the stopping criterion is met, the respective *SolutionManager* and *SolutionMonitor* are deleted from the system. If there are more instances of the considered problem, *OptiAgents* continue to solve them, otherwise optimizing agents are deleted too.

Throughout the whole lifecycle of *OptiAgents*, the agents may be moved from one container to another, by request of the *PlatformManager*. Also, the agents may be copied, after new resources are added to the system, thus increasing the efficiency of the ongoing optimization process.

Fig. 4.5 presents the state transition diagram for *OptiAgents*. An *OptiAgent* transits between three possible states. After being created the agent is in the

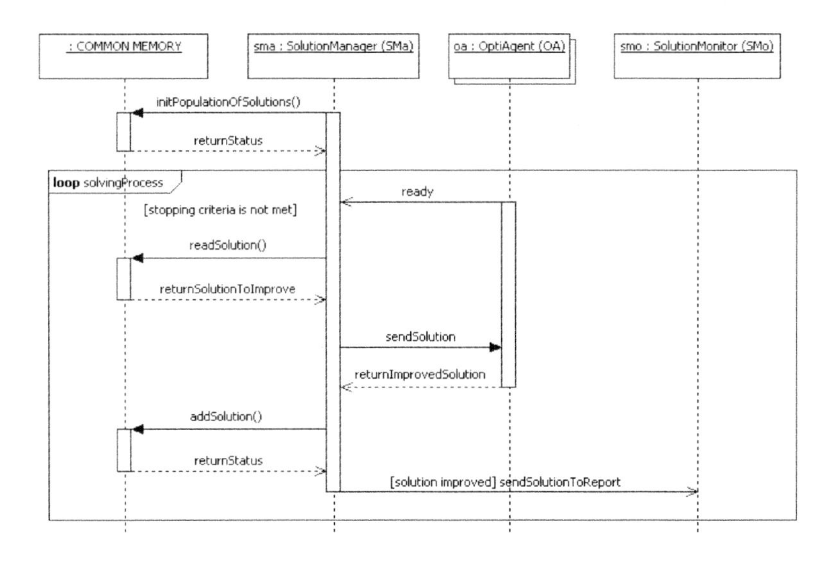

Fig. 4.4. UML sequence diagram of the process of solving an instance of a problem

Ready state and sends messages about its readiness for solving/improving. Once it has received a message from the *SolutionManager* with some solutions to be improved, the agent changes into *Solving/Improving solution* state. After having returned improved solutions to the *SolutionManager* it turns into the *Ready* state again. From both states the agent can be moved to another platform. While being moved, the agent state is changed into the *Migration* state. After the migration process is completed it returns to it's original state (*Ready* or *Solving/Improving solution*).

4.4.2 Agents Responsible for Managing Parallel Searches

JABAT can deal with several searches conducted in parallel, and the process is administered mainly by the *TaskManager* agent. During computations the *TaskManager* is run first. It is responsible for creating other agents and reading all needed parameters from the so called initial directory and from the configuration file where global system parameters are stored.

At first the *TaskManager* creates a single *PlatformManager*. The *PlatformManager* manages optimizing agents and system platforms. It can move optimizing agents among platforms and create their copies to improve the computation efficiency. More information on *PlatformManager* is available in the subsection 4.4.3.

When *TaskManager* finds a request to solve an instance of the problem it creates agents (*SolutionManager*, *SolutionMonitor*, *ErrorMonitor* and *OptiAgents*) designated to this particular task instance. *TaskManager* may initialize the

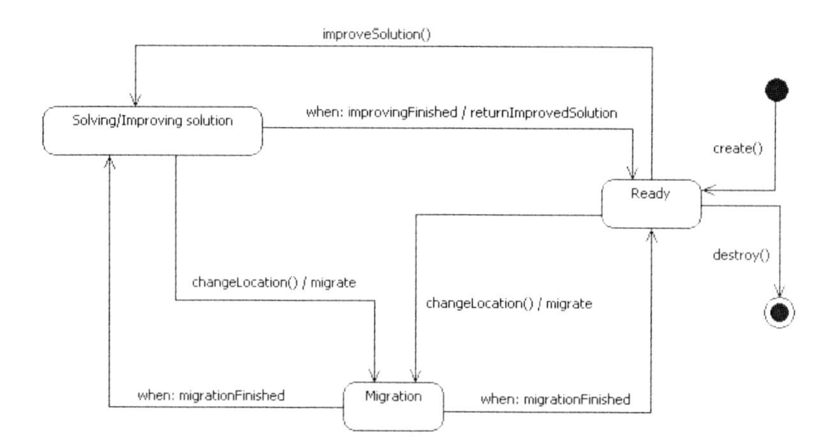

Fig. 4.5. UML state transition diagram of the *OptiAgent*

process of solving the next instance task before the previous search has stopped if there is any such an instance waiting and if the global system setting allows for that (i.e. if the number of tasks that can be solved in parallel has not been exceeded).

The parameters of instances for tasks to be solved are stored as the XML files in the initial directory. These files are read by the *TaskManager* in the order defined by the *first-in-first-out* rule. The example of data stored as the XML file with the set of instance descriptions for the resource constrained project scheduling problem (RCPSP) is shown in *Example* 1. For each problem the respective XML file contains information on the problem name, one or more instance data, list of optimizing agents that should be run in order to solve the tasks, the name of the selected strategy and additional options for this strategy. For each *OptiAgent* the minimal and maximal number of copies may be set. Parameters from each XML file are used by *TaskManager* to create a number of agents: *SolutionMonitors*, *SolutionManagers* and optimizing agents. For each problem instance the *TaskManager* creates one *SolutionMonitor*, which monitors solutions of this instance.

Example 1. Example of XML data for solving an instance of RCPSP

```
<?xml version="1.0" standalone="no" ?>
<!DOCTYPE boundle SYSTEM "http://153.19.114.101/jabat/boundle.dtd">
<boundle>
<problemname>RCPSP</problemname>
 <solutionmanager>
  <dataclass>problems.RCPSP</dataclass>
   <strategy>
```

```
      <class>system.strategies.BASIC</class>
      <option>
        <name>populationSize</name>
        <value>100</value>
      </option>
      <option>
        <name>stopMinutes</name>
        <value>5</value>
      </option>
    </strategy>
  </solutionmanager>
  <datapath>\
   <url></url>
   <dir>\RCPSP\data\mm_j10\</dir>
     <file>j1.mm</file>
     <file>j2.mm</file>
  </datapath>
  <optiagent min=1 max=5>rcpsp.LocalSearch</optiagent>
  <optiagent min=1 max=1>rcpsp.TabuSearch</optiagent>
</boundle>
```

The further behaviour of *TaskManager* depends on whether there are other instances waiting and whether the maximum number of opimization agents has been achieved. The number of agent running at the same time depends on the number of accessible resources. If it is possible to create new agents, *TaskManager* initializes the process of solving the next instance from the same file if there is one, or checks the initial directory for next files. Such checking is done periodically. If running new agents is not possible, *TaskManager* waits for a message from any running *SolutionManager* indicating that the system completed searching for the solution to the current instance. Then, the *SolutionManager* may be destroyed together with other agents that has been created with a view to solve the instance at hand. *TaskManager*, however, may be created and destroyed only by the user.

Main feature of JABAT is its independence from the problem definition. Hence, to create a *SolutionManager* two pieces of information are needed: the name of the class that represents the task and the name of the class replacement strategy selected by the user.

Replacement strategies in JABAT define:

- how the initial population of solutions is created (for example how many solutions it consists of),
- how solutions are chosen to be sent to optimizing agents,
- how solutions that has been received from optimizing agents are merged with the population of solutions, when the process of searching stops.

There are several different predefined strategies in the system to choose from. A simple strategy can, for example, draw a random solution and, when an improved solution is received, replace the worst solution in the population with it. A more sophisticated strategy *StrategyDiv* introduces recursion: it allows for division of the task into smaller tasks, solving them and then merging the results to obtain solutions to the initial task. The strategy may be used for these problems for which specific methods of *Task* and *Solution* are defined that are responsible for dividing and merging.

The user may choose a strategy and furthermore may optionally parameterize it with some options from a predefined set of options. In *Example* 1, *BASIC* is supposed to use not the default size of the population of solutions but the size specified by the *populationSize*.

Apart from the above, *TaskManager* is also responsible for deleting all the agents that are designated to the task (i.e. the current instance), that has been solved. The deletion takes place after *SolutionManager* informs that a stopping criterion for this task has been met.

Deleting an agent may also be caused by the command of *ErrorMonitor*. After an error (or exception) occurs in the work of any agent, it sends an appropriate message to *ErrorMonitor*. On receiving such a message, *ErrorMonitor* tries to close the process of solving the erroneous task in a safe way, i.e. it decides on killing the required number of software agents.

4.4.3 Agent Responsible for Managing the JABAT Platform

JABAT, using the functionality of JADE, makes it possible for optimizing agents to migrate or clone to containers on other computers that have joined the main platform. By using of mobile agents the system allows for decentralization of computation processes. This results in a more effective use of the available resources and reduction of the computation time.

The *PlatformManager* manages optimizing agents and system platforms. It can move optimizing agents among platforms and create (or delete) their copies to improve efficiency of the computations. The *PlatformManager* work is based on a few simple rules:

- the number of *OptiAgents* cannot exceed the maximum number and cannot be smaller than the minimum number of *OptiAgents*, as is specified by the user,
- if JABAT has been activated on a single computer with main container, then all *OptiAgents* would be also placed on this one computer,
- if JABAT has been activated on multiple computers, with main container placed on one computer and the remote joined containers placed on other computers, then *OptiAgents* are moved from the main container to outside containers to distribute the workload evenly.

Thus, the main container is supposed to host the managing agents including the *PlatformManager*, *SolutionManagers* and *TaskManager*, monitoring agents (*SolutionMonitor*, *ErrorMonitor*) while the outside containers host only

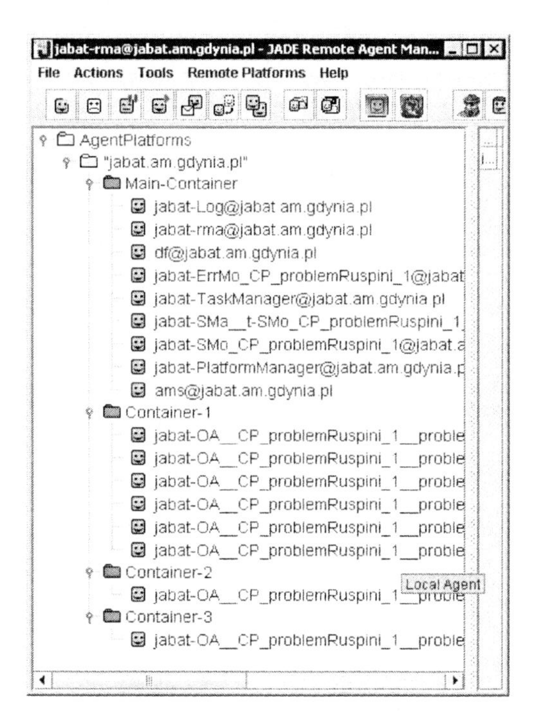

Fig. 4.6. Example of agents working in the main and other containers

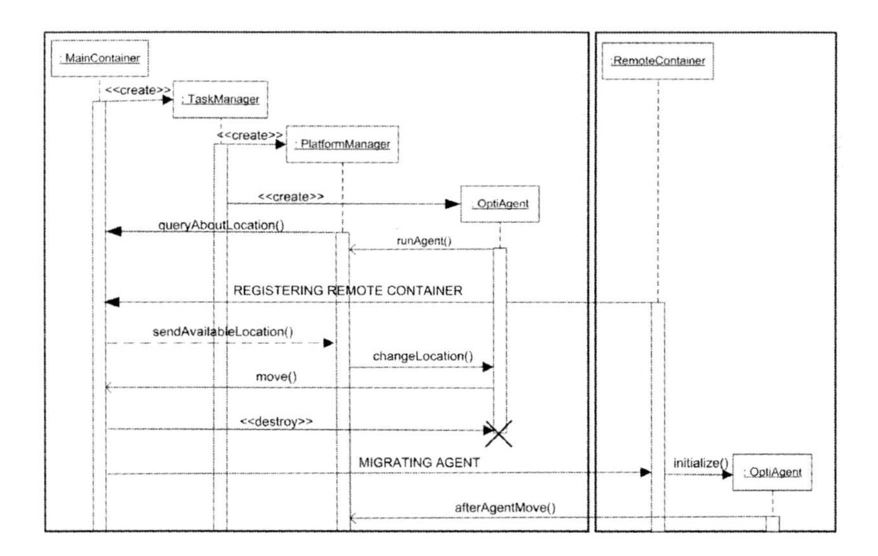

Fig. 4.7. UML sequence diagram of agents mobility in JABAT

OptiAgents. In Fig. 4.6 a view of the main container with all the managing agents activated and a number of containers, in which some optimizing agents work is shown.

PlatformManager does the following:

- manages the number of containers available,
- destroys agents indicated by *TaskManager*,
- registers in which containers *OptiAgents* are working, sends requests to change location to the optimizing agents,
- decides whether and which *OptiAgents* should be cloned or destroyed (basing on the analysis of the available resources).

Fig. 4.7 shows the UML sequence diagram of agents mobility in JABAT.

4.5 The Web Interface of e-JABAT

JABAT is accessible through the web interface, in which most of the original functionality is available to users from all over the world. The web-accessible version of JABAT system with its interface is referred to as e-JABAT. A working copy of e-JABAT can be found at the address http://jabat.wpit.am.gdynia.pl.

4.5.1 Web Interface Architecture

The users obtain access to JABAT engine through the interface that has been created with the use of Java Server Faces and Facelets technologies. All the data required to initialize the interface, like for example the list and descriptions of problems that are available for solving, are read from a set of xml files, that can be easily updated in case of introducing new problems or new optimizing agents into the system.

The interface allows the user to specify the task and to provide some additional information on how the task solving process should be carried out. The task uploaded by the user is saved in the directory from which it can be later read by the JABAT engine. The information given by the user during the process of uploding the task are written in an XML file that is stored in the area called input directory from which JABAT can read it.

4.5.2 Solving Tasks

To solve a task within the system, a user has to register at the e-JABAT website. Registered users obtain access to the part of the website in which tasks can be uploaded for solving. The uploaded tasks are sequenced and solved in the order of uploading. They may be solved in parallel (even instances of different problems). The working copy of the environment allows for parallel solving of two tasks. After the task has been solved the user can download the results saved in a text file in the user's space.

Thus, the user can:

- Upload a file containing a task to be solved,
- Observe the status of all tasks he uploaded (waiting, being solved, solved),
- Observe the logfiles in which some additional information on the process of solving may be found,
- Download the logfiles and the files with solutions,
- Delete the uploaded tasks.

The user who wants to upload a task to the system must also:

- Choose from the available list the problem that the task (i.e. the considered instance) belongs to. At present four different problems have been implemented in JABAT and are available to the users of the web interface. These are: the resource constrained project scheduling problem with single and multiple mode (RCPSP, MRCPSP), the clustering problem (CP), the Euclidean planar traveling salesman problem (TSP) and the vehicle routing problem (VRP). For each of these problems the format of the file containing the task to be solved is specified and published in the website.
- Choose which optimizing agents should be involved in the process of searching for the solution. Each of the optimizing agents within the system represents different optimization algorithm. For the resource constrained project scheduling problem there are simple local search algorithm, algorithm based on the simple evolutionary crossover operator and the tabu search algorithm. For the traveling salesman problem there are implementation of the Lin-Kerninghan algorithm and a simple evolutionary algorithm with the crossover operator. For the vehicle routing problem there are 2-optimum algorithm operating on a single route, interchange local optimization method, a simple evolutionary algorithm with the crossover operator and a local search algorithm. Finally, for the clustering problem there are random local search, hill-climbing local search and tabu search algorithms. For each of these algorithms a short description is available at the JABAT website.
- Define (optionally) for each optimizing agent the minimum and the maximum number of identical agents that may run simultaneously within the system. The system will initially use the minimal specified number of agents and then the number will be automatically increased if there is enough computational resources available,
- Choose the strategy from the list of available strategies and optionally define a set of options for this strategy, for example the size of the initial population or the length of the time interval after which the search for better solutions stops.

Fig. 4.8 presents the task upload screen, where the user's choices are shown.

While solving a task its owner (i.e. the user) can watch two files. The first contains information on the process of searching for the solution. This includes messages on errors encountered, for example when the user uploads a file that is not compatible with the required format of the task for the given problem. Information in the second file (the report file) includes the best solution obtained

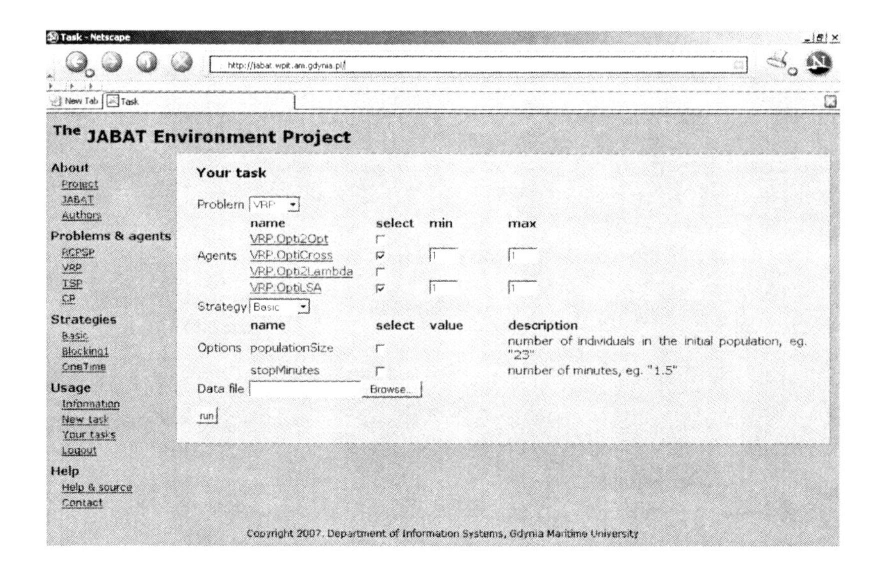

Fig. 4.8. The task uploading screen

so far, average value of the solution from the common memory, the actual running time and the time when the best solution was last reached. The file is created after the initial population has been generated and then the next set of data is appended to the file every time the best solution in the population changes. The final results are added to the content of the file when the stopping criterion has been met and the system is closing down all the processes connected with the task.

The report on the process of searching for the best solution may be later analyzed by the user. It can be easily read into a spreadsheet and converted into a summary report with the use of the pivot table.

4.6 Flexibility of e-Jabat

4.6.1 Solving New Problem with e-JABAT

JABAT has been designed in such a way, that it can be easily extended to solving new problems or solving them with new algorithms. The main idea is to reduce the amount of work of the programmer who wants to solve new problems or wishes to introduce new ways of representing tasks or solutions, new optimizing algorithms or, finally, new population-management strategies. e-JABAT makes it possible to focus only on defining these new elements (Fig. 4.9), while the processes of communication and search for the best solution management will still work.

In order to solve an instance of a new problem not yet implemented within the e-JABAT, classes inheriting from *Task* and *Solution* need to be defined. Apart

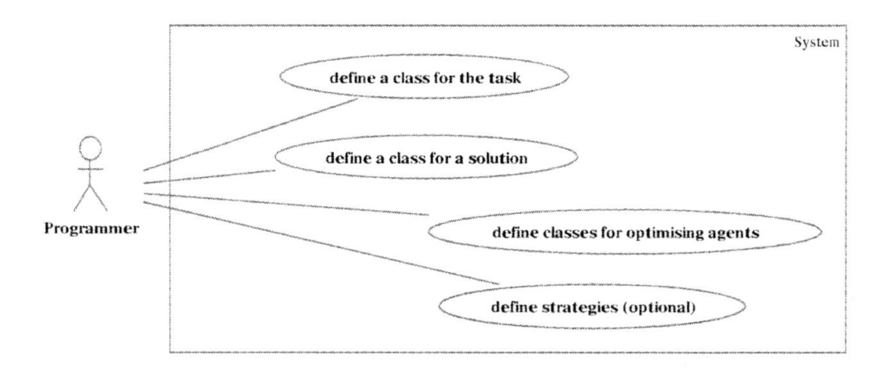

Fig. 4.9. Use case diagram of JABAT from the programmer's point of view

from defining constructors and properties specific to such a problem, several inherited methods should be overridden. These include, for example, *createSolu tion*() creating the initial solution (drawn at random or empty), function *readDataFile*() reading the instance data from a file and function *ontology*() returning the JADE ontology of the class. An ontology defines how a class may be transformed into a text message and how the text message is used to construct the class. *Solution* should also override some methods, for example the method that calculates the fitness (i.e. quality) of the solution, the method that compares the solution with other solutions and again the method returning the ontology of the class. In addition to *Task* and *Solution* at least one optimizing algorithm should be implemented. It can be done by simply creating a class inheriting from *OptiAgent* with overriden method *Optimize*().

It is also possible (optionally) to easily create own population-management strategies, which require defining the class inheriting from one of the predefined strategies already available in the system and to adjust it by overriding the existing methods. Strategies work on population of solution called individuals of the class *Solution* and thus are independent from the problem definition or implementation details of particular solution and can be used with solutions of all types.

4.6.2 Adding/Deleting Resources

JABAT makes it possible for optimizing agents to migrate or clone to other computers. By using mobile agents the system offers decentralization of computations resulting in a more effective use of available resources and reduction of the computation time.

Thus, the user has an additional ability to add his computer or computers to the JABAT platform. To do so, the user has to download and install on his computer JAVA, version 1.4 and JADE, version 3.3. Then, with a system command the user may launch a JADE container on current host and attach it to JABAT running on jabat.wpit.am.gdynia.pl.

4.7 Implementation and Performance Evaluation

This section contains several examples of using JABAT to solve instances of difficult optimization problems. There are also results of several computational experiment carried out to evaluate JABAT performance. In particular, the reported experiments aimed at evaluating:

- quality of results,
- computation time and computation speed-up,
- influence of the population-management strategy on results,
- influence of the number of optimizing agents run in parallel on the quality of results and computation time.

4.7.1 Evaluation of Results Obtained Using JABAT Implementations

Resource Constrained Project Scheduling Problem. The resource-constrained project scheduling problem (RCPSP) consists of a set of n activities, where each activity has to be processed without interruption to complete the project. The dummy activities 1 and n represent the beginning and the end of the project. The duration of the activity j, $j = 1, \ldots, n$ is denoted by d_j where $d_1 = d_n = 0$. There are r renewable resource types. The availability of each resource type k in each time period is r_k units, $k = 1, \ldots, r$. Each activity j requires r_{jk} units of resource k during each period of its duration where $r_{1k} = r_{nk} = 0$, $k = 1, \ldots, r$. All parameters are non-negative integers. There are precedence relations of the finish-start type with a zero parameter value defined between activities. In other words, activity i precedes activity j if j cannot start until i has been completed. The objective is to find the schedule S of activities starting times $[s_1, \ldots, s_n]$, where $s_1 = 0$ and resource constraints are satisfied, such that the schedule duration $T(S) = s_n$ is minimized.

To solve the RCPSP in JABAT four types of optimizing agents representing different optimization algorithms and heuristics have been implemented: tabu search, local search, crossover and path relinking algorithms. The *OptiTSA* agent represents the tabu search algorithm where neighborhood is determined by two activities exchange. The *OptiLSA* represents the local search algorithm which finds the local optimum by moving each activity to all possible places in the solution. Both algorithm are described in detail in [20]. The *OptiCH* represents the crossover heuristic where two initial solutions are crossed and improved by the local search algorithm until a better solution is found or all crossover points are checked. The *OptiPRA* represents the path relinking algorithm where a path is built between two solutions. Each solution from the path is improved by the local search procedure and the best solution found is remembered.

The experiment involving benchmark instances of RCPSP from PSPLIB [22] has been conducted. The instances contain 30, 60, 90 and 120 activities. The population of solutions consisted of 50 elements. The solutions in the initial population were generated randomly. The computation for each problem instance

Table 4.1. Results as mean relative errors form optimal or best known solutions obtained by JABAT implementation for RCPSP for instances from PSPLIB

OptiAgent set	n=30	n=60	n=90	n=120
first	0.17%	0.97%	1.24%	1.98%
second	0.24%	1.14%	1.69%	2.10%

was interrupted after 1 minute. Two sets of optimizing agent were considered. The first included one optimizing agent of each type (first *OptiAgent* set), the second included additionally one *OptiTSA* (second *OptiAgent* set). The computation results were evaluated in terms of the mean relative error calculated as the deviation from the optimal (30 activities) or best known solution. The computational experiment was carried on 12 computers with 1.7 GHz processor where 3 instances are evaluated simultaneously. The results are presented in Table 4.1.

Vehicle Routing Problem. The vehicle routing problem (VRP) can be stated as the problem of determining optimal routes through a set of locations (customers) and defined on a directed graph $G = (V, E)$, where $V = \{0, 1, \ldots, n\}$ is the set of nodes and E is a set of edges. Node 0 is a central depot with NV identical vehicles of capacity W. Each other node $i \in V - \{0\}$ denotes customer with a non-negative demand d_i. Each link $(i, j) \in E$ denotes the shortest path from customer i to j and is described by the cost c_{ij} of travel from i to j by path $(i, j = 1 \ldots, n)$ $(c_{ij} = c_{ji})$. The goal is to find vehicle routes such that each route starts and ends at the depot, each customer is serviced exactly once by a single vehicle and total load on any vehicle associated with a given route does not exceed vehicle capacity in order to minimize total cost of travel (or travel distance).

Table 4.2. Results obtained by JABAT implementation for VRP for selected instances from ORLibrary

Problem	Number of customers	Best known solution	JABAT (VRP)	Error
vrpnc1	50	524.61	524.61	0.00%
vrpnc2	75	835.26	846.91	1.40%
vrpnc3	100	826.14	839.60	1.63%
vrpnc4	150	1028.42	1057.46	2.82%
vrpnc5	199	1291.45	1335.01	3.37%
vrpnc11	120	1042.11	1044.58	0.24%
vrpnc12	100	819.56	820.64	0.13%

In case of JABAT implementation for solving VRP instances, individuals are represented as permutations of N customers, which in process of solving are divided into feasible segments (routes) through applying some heuristics. Four types of optimizing agents have been implemented using various heuristics and local search procedures. These are: *Opti2Opt* - an agent implementing the $2-opt$ local search algorithm, in which for all routes first two edges are removed and next remaining edges are reconnected in all possible ways, *OptiStringCross* - a crossing heuristic in which parts of two randomly chosen routes are exchanged by crossing two edges, *Opti2Lambda* - an implementation of the local search algorithm based on λ - interchange local optimization method and *Opti2LambdaC* - an implementation of local search algorithm which try to remove nodes (customers) situated relatively far from the centroid of the given route and next insert them to the other route.

The experiment involved 7 known benchmark instances of the vehicle routing problem from the OR-Library benchmark set [31]. The selected instances contain 50 to 199 customers and there are only capacity restriction.

Table 4.2 shows the best results obtained by the suggested JABAT implementation for VRP, the best known solutions reported by [24] and relative error from the best known solution.

ANN Training. The artificial neural network training problem requires finding values of weights in the considered ANN such that the network quality in terms of its performance is maximized. JABAT implementation for solving the ANN training problem is based on the population learning algorithm developed by the authors. The earlier implementation of the algorithm consisted of several improvement procedures run sequentially (for particulars see [12]). The main feature of the population learning approach is increasing complexity of the improvement algorithms applied to the population of solutions and decreasing number of individuals as the computation progresses. The agent-based JABAT version is using identical improvement procedures as in the original sequential algorithm, each represented by a different agent type. Within the JABAT-based artificial neural network training system there are seven agent types representing seven different improvement procedures. These are *OptiSM* - the agent which is an implementation of standard mutation, *OptiLSA* - the agent in which the local search algorithm is implemented, *OptiNUM* - the agent representing the non-uniform mutation, *OptiGM* and *OptiGAH* agents representing respectively the gradient mutation and the gradient adjustment heuristics. The last two agents represent the single point crossover heuristic - *OptiSPCH*, and the arithmetic crossover heuristic - *OptiACH*. Within the above described set of agents the crossover agents take care of information exchange and diversification while the remaining agents are used to directly improve the fitness of individuals drawn from the common memory. The details of each of the respective improvement algorithms can be found in [12] and [13].

The experiment involved training of the MLP type artificial neural networks aimed at solving benchmark datasets including instances of four well known classification problems - Cleveland heart disease (303 instances, 13 attributes, 2 classes), credit approaval (690, 15, 2), Wisconsin breast cancer (699, 9, 2) and sonar problem (208, 60, 2). The respective datasets have been taken from [27].

Table 4.3. Single platform JABAT performance versus the best reported (* Source for the best reported with respect to cases of BP and RBF: http://www.phys.uni.torun.pl/kmk/projects/datasets-stat.html)

Problem	ANN structure	Training error	Accuracy(%)			
			JABAT	MLP+PLA	MLP+BP	RBF
Sonar	60-6-1	0.182	88.1	-	83.5*	83.6*
Credit	15-15-1	0.118	85.1	86.6	82.1	85.5*
Cancer	9-9-1	0.021	96.8	96.6	96.7*	95.9*
Heart	13-13-1	0.109	85.7	86.5	76.4	84.0*

Each benchmarking problem has been solved 30 times and the reported values of the quality measures have been averaged over all runs. The quality measure in all cases was the correct classification ratio calculated using the 10-cross-validation approach. The common memory size in JABAT was set to 50 individuals. All optimizing agents, except the crossover ones, have been allowed to continue iterating until an improvement has been achieved or until 100 iterations have been performed.

The results obtained using the proposed JABAT-based training algorithm are shown in Table 4.3. The structure of the respective ANN are shown in the column 2 and the value of training error in column 3 of the discussed table. The table contains also correct classification ratios produced by the ANN trained using JABAT-based training algorithm, as well as by the original population learning algorithm - PLA as reported in [12] and back propagation algorithm - BP. There are also the respective ratios obtained by the radial basis function classifiers (RBF).

Euclidean Planar Traveling Salesman Problem. The Euclidean planar traveling salesman problem is a particular case of the TSP. Given n cities in the plane and their Euclidean distances, the problem is to find the shortest TSP-tour, i.e. a closed path visiting each of the n cities exactly once.

To solve an instance of the Euclidean planar TSP two algorithms have been used to create agents *OptiKL* and *OptiCross*. The algorithm in *OptiKL* is a java implementation of the Lin-Kerninghan algorithm [17], the other is a crossover algorithm. The above architecture has been experimentally tested and several instances from TSPLIB [26] have been successfully solved. Some results obtained are presented in Table 4.4.

TSP is one of the problems, for which some ready and efficient algorithms may be found that have been exploited and optimized for many years. Such algorithms also can be used within JABAT. To prove that, the LKH algorithm for solving the traveling salesman problem has been downloaded from the website of its author [26] and has been used as the algorithm for one of *OptiAgents* in JABAT - *OptiLKH*.

Table 4.4. The results obtained for selected problems from TSPLIB

Problem	Solution	Optimal solution	Error
eil51	428	426	0.47%
berlin52	7542	7542	0.00%
kroa100	21424	21282	0.67%
eil101	645	629	2.57%
bier127	119834	118282	1.31%

LKH algorithm is an effective implementation of the Lin-Kernighan heuristic written in C++ and described in the report [17]. The algorithm has been compiled as DLL library and accessed from the agent through calling the optimizing function from this library. To make the cooperation of the agent and the library possible, the function must be provided that describe the task to be solved and must return the solution. Since the data structures in the LKH and in JABAT are different, also the code for converting the data was written.

Thus, it has been shown, that JABAT may make use of algorithms that have been created for the standalone programs, without the need to rewrite the code of these algorithms.

4.7.2 Evaluation of the Computation Speed-Up Factor

The speed-up factor analysis has been performed using the JABAT implementation of the population learning algorithm for the ANN training problem. The results reported in Table 4.3 have been used to establish the critical value of the training error for each problem type, which, in turn, has been used to evaluate the computational speed-up factor resulting from adding additional computers and enabling agents migration. The respective critical (that is the required) values have been set at the following levels: 0.25 for sonar problem, 0.13 - credit, 0.035 - cancer and 0.12 - heart.

In the experiment the training procedure of the respective ANN classifier has been terminated as soon as the classifier has been able to reach the respective critical value of the training error. In each run computation times were registered. The computations have been carried on several PC computers with Pentium IV 1.7 GHz processors and 256 MB RAM, connected within a local area network.

Dependency between the speed-up factor and the number of containers (computers) used by JABAT is shown in Fig. 4.10 (see the left box). Mean speed-up factor can be estimated as 1.6 to 2.3 for, respectively, 2 to 6 computers. The above results have been achieved without cloning of agents, that is with only 7 optimizing agents distributed among platforms. Increasing the number of copies can still improve the computation speed. This is shown in Fig. 4.10 (see the right box) where the speed-up factor depending on the number of computers and the allowed number of agent clones is shown. The results are shown for 0, 1, 2, 3 and 4 additional clones allowed for each agent type, corresponding, respectively, to 7, 14, 21, 28 and 32 optimizing agents distributed among available platforms.

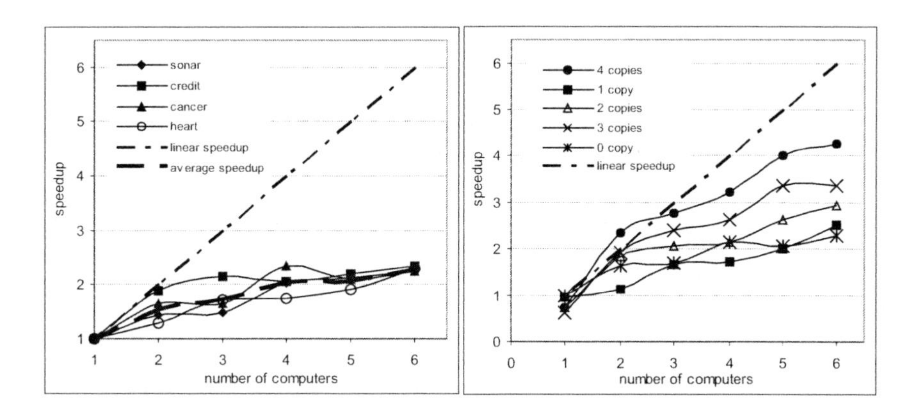

Fig. 4.10. Relations between the speed-up factor and the number of computers used (left box) and between the speed-up factor and the number of computers with different numbers of agent clones allowed - data averaged over all problem types (right box)

Data obtained from the experiment have been used to perform the two-ways analysis of variance with a view to analyze the effect of two qualitative factors (the number of copies of each agent type and the number of computers used) on one dependent variable - computation time. The following null hypotheses have been tested:

- The number of copies of each agent type does not influence computation time.
- The number of computers used does not influence computation time.
- There is no interaction of the above factors.

It has been established that at the significance level of 0.05 all the above null hypotheses should be rejected. Additionally, the *post hoc* Tukey test has shown that there are no significant differences between average computation speed-up achieved while increasing the number of computers from 3 to 4 as compared with the speed-up achieved with the increase from 5 to 6 computers. In all of the remaining cases average speed-ups achieved statistically differ.

4.7.3 Influence of Replacement Strategy on Results

As it was mentioned earlier in order to use JABAT for solving an optimization problem several elements need to be defined. In particular, these include choosing a strategy for managing the population of solutions called individuals. The computational experiment, discussed in this section, aimed at evaluating how the population-management strategy mix influences computation results. It has been evaluated based on nine methods available as replacement strategies (four for *readSolution()* and five for *addSolution()*):

– *RandomRead* - random solution is chosen from the population,

- *Worst* - the worst solution from the population is selected,
- *RandomBlocking* - randomly chosen solution sent to optimizing agents is blocked for a period,
- *Select* - a solution is selected using the proportional selection method and next sent for improvement to an optimizing agent or agents,
- *RandomAdd* - a randomly chosen solution from the current population is replaced by the received solution actually received from an optimizing agent,
- *RandomNotBest* - a randomly chosen solution (except the best one) from the current population is replaced by the solution actually received from an optimizing agent,
- *Worse* - a randomly chosen solution from the current population worse than the actually received from an optimizing agent is replaced by it,
- *Worst* - the worst solution from the current population is replaced by the solution actually received from an optimizing agent,
- *WorstBlocking* - the worst solution from the current population is removed and a newly, randomly generated one is added providing the last consecutive five solutions received from the optimizing agents has not improved the quality of any already obtained solutions within the population of solutions.

By combining possible settings of read and add operations, one obtains twenty strategies, which are summarized in Table 4.5. It contains the name of strategy and *readSolution()* and *addSolution()* methods used in managing a population of solutions.

The influence of the population-management strategies on computation results has been validated using the JABAT implementation for VRP, described in section 4.7.1. The computational experiment has been carried out based on 7 test problems from the *OR-Library*. All 20 strategies presented in Table 4.5 were tested. Each of the 7 instances of the problem was run for each possible strategy, in total giving 140 (7x20) test problems. Moreover, each test problem was repeatedly executed five times and mean results from these runs were recorded.

The proposed approach was evaluated using the mean relative error (MRE) from the best known solution as the criterion. The experiment results are shown in Table 4.6 containing mean relative errors averaged over all runs for each tested strategy and each problem. Together with the problem names the number of customers is given in the brackets. The table shows also the average value (AVG) of errors for each strategy, as well as the average (AVG), minimum (MIN) and maximum (MAX) value of MRE for each problem separately.

By looking at the columns of Table 4.6 corresponding to the instances of the analyzed problem and trying to determine how a strategy of the population-management influences the quality of solutions, one can see that although there are some differences between quality of solutions obtained for particular instances there are no significant differences between results obtained for each instance using different strategies. Taking into account both, reading and adding operations, for all instances the best mix for the considered set of instances seems to be the *RandomBlocking* (randomly choosing a solution with blocking mechanism) together with the *RandomNotBest* (replacing randomly chosen solution except

Table 4.5. Replacement strategies available in JABAT implementation

Strategy	*readSolution()*	*addSolution()*
strategy01	*RandomRead*	*RandomAdd*
strategy02	*RandomRead*	*RandomNotBest*
strategy03	*RandomRead*	*Worse*
strategy04	*RandomRead*	*Worst*
strategy05	*RandomRead*	*WorstBlocking*
strategy06	*Worst*	*RandomAdd*
strategy07	*Worst*	*RandomNotBest*
strategy08	*Worst*	*Worse*
strategy09	*Worst*	*Worst*
strategy10	*Worst*	*WorstBlocking*
strategy11	*RandomBlocking*	*RandomAdd*
strategy12	*RandomBlocking*	*RandomNotBest*
strategy13	*RandomBlocking*	*Worse*
strategy14	*RandomBlocking*	*Worst*
strategy15	*RandomBlocking*	*WorstBlocking*
strategy16	*Select*	*RandomAdd*
strategy17	*Select*	*RandomNotBest*
strategy18	*Select*	*Worse*
strategy19	*Select*	*Worst*
strategy20	*Select*	*WorstBlocking*

the best one). It is represented as strategy12 in the experiment and provided the minimum value of MRE.

Detailed description of the experiment as well as the analysis of the obtained results may be found in [4].

4.7.4 Influence of the Number of Optimizing Agents Used on the Quality of Results and Computation Time

As it was mentioned to use JABAT for solving an optimization problem several elements need to be defined including the number of optimizing agents. The computational experiment discussed in this section aimed at evaluating the influence of the number of optimizing agents on the computation results and time. The experiment was based on the JABAT implementation for the resource constrained project scheduling problem.

Four optimizing agents described in 4.7.1 were used: *OptiTSA* (T), *OptiLSA* (L), *OptiCH* (C) and *OptiPRA* (P). The number of each agent type was set to 1 and 2, and the number of computers used during computation was constant. The experiment consisted of solving all instances with 30 activities from the benchmark data set. Each problem instance was calculated 4 times for each combination of the number of agents. The initial set of agents included 4 agents - one optimizing agent of each kind - 1T, 1L, 1C, 1P. Next, the number of agents was increased by 1 to obtain all possible combinations with 5 and 6 agents. The

Table 4.6. The mean relative error (MRE) in % from the best known solution obtained by the JABAT implementation for VRP for selected instances from ORLibrary

	PROBLEM							
STRATEGY	vrpnc1 (50)	vrpnc2 (75)	vrpnc3 (100)	vrpnc4 (150)	vrpnc5 (199)	vrpnc11 (120)	vrpnc12 (100)	AVG
strategy01	0.24%	9.45%	6.27%	8.09%	5.32%	2.38%	1.93%	4.81%
strategy02	0.40%	5.73%	6.07%	8.09%	8.40%	0.41%	1.15%	4.32%
strategy03	0.48%	9.36%	6.17%	8.09%	10.11%	0.43%	2.36%	5.28%
strategy04	0.34%	9.29%	6.27%	8.09%	8.25%	3.03%	5.23%	5.78%
strategy05	0.09%	9.46%	6.17%	8.09%	4.66%	2.86%	2.36%	4.81%
strategy06	1.73%	6.99%	6.01%	8.09%	9.48%	2.86%	2.36%	5.36%
strategy07	2.30%	6.77%	4.70%	7.04%	6.88%	2.86%	2.29%	4.69%
strategy08	2.21%	9.37%	5.63%	8.09%	7.74%	2.86%	2.36%	5.47%
strategy09	1.53%	5.79%	4.73%	8.09%	7.54%	2.86%	2.36%	4.70%
strategy10	2.01%	9.29%	6.17%	8.09%	5.96%	2.86%	5.72%	5.73%
strategy11	1.62%	9.15%	5.28%	7.17%	4.94%	1.40%	1.19%	4.39%
strategy12	0.57%	8.23%	6.16%	8.16%	4.91%	1.42%	0.63%	4.30%
strategy13	0.66%	8.00%	5.42%	8.09%	7.31%	0.82%	1.47%	4.54%
strategy14	0.44%	9.45%	6.27%	8.09%	4.67%	6.88%	2.36%	5.45%
strategy15	0.01%	9.28%	6.16%	8.09%	6.07%	2.86%	2.36%	4.98%
strategy16	0.72%	9.29%	6.17%	8.09%	5.24%	1.90%	1.93%	4.76%
strategy17	0.25%	9.46%	6.27%	8.35%	4.63%	2.38%	2.36%	4.81%
strategy18	0.75%	9.46%	6.17%	8.09%	4.34%	2.86%	2.36%	4.86%
strategy19	0.09%	9.28%	6.27%	8.09%	4.77%	2.86%	6.10%	5.35%
strategy20	0.00%	9.29%	6.17%	8.39%	6.78%	2.38%	2.36%	5.05%
AVG	0.82%	8.62%	5.93%	8.02%	6.40%	2.46%	2.56%	4.97%
MIN	0.00%	1.40%	1.63%	2.82%	3.37%	0.24%	0.13%	
MAX	4.59%	9.77%	8.91%	9.60%	10.11%	22.94%	21.10%	

Table 4.7. Results obtained by JABAT implementation for RCPSP for instances from PSPLIB where n=30

Set of optimization agents	Mean relative error	Average computation time [s]
1T, 1L, 1C, 1P	0.24%	16.3
2T, 1L, 1C, 1P	0.17%*	33.0
1T, 2L, 1C, 1P	0.21%	27.0*
1T, 1L, 2C, 1P	0.22%	28.1
1T, 1L, 1C, 2P	0.21%	27.7
2T, 2L, 1C, 1P	0.25%	26.7*
2T, 1L, 2C, 1P	0.23%*	28.3
2T, 1L, 1C, 2P	0.23%*	28.3
1T, 2L, 2C, 1P	0.26%	29.0
1T, 2L, 1C, 2P	0.26%	28.0
1T, 1L, 2C, 2P	0.24%	30.0

respective results averaged over all instances and runs are presented in Table 4.7. The best results for 4, 5 and 6 optimizing agents are marked with the star.

The results of this experiment show that even increasing the agent number by 1 (that is using 5 agents instead of 4) results in significantly better results, especially if the additional agent is a more sophisticated one (here it was the tabu search agent). Unfortunately, with the addition of a single agent computation time grew twice. Increasing further the number of agents up to 6 did not significantly improve the quality of solutions, or even resulted in decreasing it. However computation time in case of using 5 and 6 optimizing agents was fairly stable.

4.8 Conclusion

The goal of the research presented in this paper was to propose a middleware environment allowing Internet accessibility and supporting development of the A-Team systems solving effectively difficult optimization problems. The solution described in the paper has achieved this goal. Some of the advantages of the JABAT have been inherited from JADE. Among them the most important seems to be JABAT ability to simplify the development of the distributed A-Teams composed of autonomous entities that need to communicate and collaborate in order to achieve the working of the entire system. The software framework that hides all complexity of the distributed architecture plus the set of predefined objects are made available to users, who can focus just on the logic of the A-Team application and effectiveness of optimization algorithms rather than on middleware issues, such as discovering and contacting the entities of the system. It is believed that the proposed approach has resulted in achieving Internet accessible, scalable, flexible, efficient, robust, adaptive and stable A-Team architectures. Hence, JABAT can be considered as a step towards next generation A-Team solutions.

During the test and verification stages JADE-A-Team has been used to implement several A-Team architectures dealing with well known combinatorial and non-linear optimization problems. Functionality, ease of use and scalability of the approach have been confirmed.

References

1. Aydin, M.E., Fogarty, T.C.: Teams of Autonomous Agents for Job-shop Scheduling Problems: An Experimental Study. Journal of Intelligent Manufacturing 15(4), 455–462 (2004)
2. Baerentzen, L., Avila, P., Talukdar, S.: Learning Network Designs for Asynchronous Teams, MAAMAW, pp. 177–196 (1997)
3. Barbucha, D., Czarnowski, I., Jędrzejowicz, P., Ratajczak, E., Wierzbowska, I.: An Implementation of the JADE-base A-Team Environment. International Transactions on Systems Science and Applications 3(4), 319–328 (2008)
4. Barbucha, D., Jędrzejowicz, P.: An Agent-Based Approach to Vehicle Routing Problem. International Journal of Applied Mathematics and Computer Sciences 4(1), 18–23 (2007)
5. Bassak, G.: Decision Support Trims a Paper Company's Costs. IBM Research Magazine 2 (1996)
6. Bellifemine, F., Caire, G., Poggi, A., Rimassa, G.: JADE. A White Paper, Exp. 3(3), 6–20 (2003)
7. Bellifemine, F., Poggi, A., Rimassa, G.: JADE - A FIPA-Compliant Agent Framework. In: Proceedings of PAAM 1999, London, pp. 97–108 (1999)
8. Blum, J., Eskandarian, A.: Enhancing Intelligent Agent Collaboration for Flow Optimization of Railroad Traffic. Transportation Research Part A 36, 919–930 (2002)
9. Camponogara, E., Talukdar, S.: Designing Communication Networks for Distributed Control Agents. European Journal of Operational Research 153, 544–563 (2004)
10. Chen, C.L., Talukdar, S.N.: Causal Nets for Fault Diagnosis. In: 4th International Conference on Expert Systems Application to Power Systems, Melbourne, Australia, Jan 4-8 (1993)
11. Correa, R., Gomes, F.C., Oliveira, C., Pardalos, P.M.: A Parallel Implementation of an Asynchronous Team to the Point-to-Point Connection Problem. Parallel Computing 29, 447–466 (2003)
12. Czarnowski, I., Jędrzejowicz, P.: An Approach to Artificial Neural Network Training. In: Bramer, M., Preece, A., Coenen, F. (eds.) Research and Development in Intelligent Systems, vol. XIX, pp. 149–162. Springer, London (2003)
13. Czarnowski, I., Jędrzejowicz, P.: Implementation and Performance Evaluation of the Agent-Based Algorithm for ANN Training. In: Nguyen, N.T., Grzech, A., Howlett, R.J., Jain, L.C. (eds.) KES-AMSTA 2007. LNCS (LNAI), vol. 4496, pp. 131–140. Springer, Heidelberg (2007)
14. Dreżewski, R.: A Model of Co-evolution in Multi-agent System. In: Mařík, V., Müller, J.P., Pěchouček, M. (eds.) CEEMAS 2003. LNCS (LNAI), vol. 2691, pp. 314–323. Springer, Heidelberg (2003)
15. Erman, L.D., Hayes-Roth, F., Lesser, V.R., Reddy, D.R.: The Hearsay-ii Speech-Understanding System: Integrating Knowledge to Resolve Uncertainty. In: Webber, B.L., Nilsson, N.J. (eds.) Readings in Artificial Intelligence, pp. 349–389. Kaufmann, Los Altos (1981)

16. Far, B.H.: Modeling and Implementation of Software Agents Decision Making. In: Proceedings of the Third IEEE International Conference on Cognitive Informatics (ICCI 2004) (2004)
17. Helsgaun, K.: An Effective Implementation of the Lin-Kernighan Traveling Salesman Heuristic. European Journal of Operational Research 126(1), 106–130 (2000)
18. Jade - Java Agent Development Framework, http://jade.tilab.com/
19. Jennings, N.R., Sycara, K., Wooldridge, M.: A Roadmap of Agent Research and Development. Autonomous Agents and Multi-Agent Systems 1, 7–38 (1998)
20. Jędrzejowicz, P., Ratajczak, E.: Population Learning Algorithm for Resource-Constrained Project Scheduling. In: Pearson, D.W., Steele, N.C., Albrecht, R.F. (eds.) Artificial Neural Nets and Genetic Algorithms, pp. 223–228. Springer, Wien (2003)
21. Kao, J.H., Hemmerle, J.S., Prinz, F.B.: Collision Avoidance Using Asynchronous Teams. IEEE International Conference on Robotics and Automation 2, 1093–1100 (1996)
22. Kolisch, R., Sprecher, A.: PSPLIB - A project scheduling problem library. European Journal of Operational Research 96, 205–216 (1996)
23. Koubarakis, M.: Multi-agent Systems and Peer-to-Peer Computing: Methods, Systems, and Challenges. In: Klusch, M., Omicini, A., Ossowski, S., Laamanen, H. (eds.) CIA 2003. LNCS (LNAI), vol. 2782, pp. 46–61. Springer, Heidelberg (2003)
24. Laporte, G., Gendreau, M., Potvin, J., Semet, F.: Classical and Modern Heuristics for the Vehicle Routing Problem. International Transactions in Operational Research 7, 285–300 (2000)
25. Lesser, R.: Cooperative Multiagent Systems: A Personal View of the State of the Art. IEEE Transactions on Knowledge and Data Engeneering 11(1), 133–142 (1999)
26. LKH version 2.0, http://www.akira.ruc.dk/~keld/research/LKH/
27. Merz, C.J., Murphy, P.M.: UCI Repository of Machine Learning Databases. Department of Information and Computer Science, University of California, Irvine, CA (1998), http://www.ics.uci.edu/~mlearn/MLRepository.html
28. Muller, J.: The Right Agent (Architecture) to do the Right Thing, in Intelligent Agents V. In: Proceedings of the Fifth International Workshop on Agent Theories, Architectures, and Languages, ATAL (1998)
29. Murthy, S., Rachlin, J., Akkiraju, R., Wu, F.: Agent-Based Cooperative Scheduling. In: Constraints and Agents, Technical Report WS 1997-97-05, AAAI Press, Menlo Park (1997)
30. Neruda, R., Krusina, P., Kudova, P., Rydvan, P., Beuster, G.: Bang 3: A Computational Multi-Agent System. In: Proceedings of the IEEE/WIC/ACM International Conference on Intelligent Agent Technology, IAT 2004 (2004)
31. ORLibrary, http://people.brunel.ac.uk/~mastjjb/jeb/orlib/vrpinfo.html
32. Rabak, C.S., Sichman, J.S.: Using A-Teams to Optimize Automatic Insertion of Electronic Components. Advanced Engineering Informatics 17, 95–106 (2003)
33. Rachlin, J., Goodwin, R., Murthy, S., Akkiraju, R., Wu, F., Kumaran, S., Das, R.: A-Teams: An Agent Architecture for Optimization and Decision-Support. In: Rao, A.S., Singh, M.P., Müller, J.P. (eds.) ATAL 1998. LNCS (LNAI), vol. 1555, pp. 261–276. Springer, Heidelberg (1999)
34. de Souza, P.: Asynchronous Organizations for Multi-Algorithm Problems. Ph.D. Thesis, Carnegie Mellon University, USA (1993)
35. Talukdar, S.N., de Souza, P.: Scale Efficient Organizations. In: Proceedings of the 1992 IEEE International Conference on Systems, Man, and Cybernetics, Chicago, Illinois (1992)

36. Talukdar, S.N., de Souza, P., Murthy, S.: Organizations for Computer-Based Agents. Engineering Intelligent Systems 1(2) (1993)
37. Talukdar, S., Baerentzen, L., Gove, A., de Souza, P.: Asynchronous Teams: Cooperation Schemes for Autonomous Agents. Journal of Heuristics 4, 295–320 (1998)
38. Talukdar, S.N.: Collaboration Rules for Autonomous Software Agents Decision Support Systems 24, 269–278 (1999)
39. The Foundation for Intelligent Physical Agents, `http://www.fipa.org/`
40. Tsen, C.K.: Solving Train Scheduling Problems Using A-Teams, Ph.D. Thesis, Carnegie Mellon University, USA (1995)
41. Tweedale, J., Ichalkaranje, N., Sioutis, C., Jarvis, B., Consoli, A., Phillips-Wren, G.: Innovations in Multi-agent Systems. Journal of Network and Computer Applications 30, 1089–1115 (2007)
42. Yong, C.H., Miikkulainen, R.: Cooperative Coevolution of Multi-agent Systems. Technical Report AI 2001-287, Department of Computer Sciences, University of Texas at Austin (2001)

5

Adaptive and Intelligent Agents Applied in the Taking of Decisions Inside of a Web-Based Education System

Alejandro Canales Cruz and Rubén Peredo Valderrama

Computer Science Research Center of National Polytechnic Institute, Mexico City, Mexico
{acc, peredo}@cic.ipn.mx

Abstract. A new Agents and Components Oriented Architecture for the development of adaptive and intelligent Web-Based Education systems is presented. This architecture is divided into client and server parts to facilitate adaptability in various configurations such as online, offline and mobile scenarios. This architecture shows an implementation of adaptability and intelligent technology in the development of authoring and evaluation tools, which are oriented to offer application level interoperability under the philosophy of Web Services.

5.1 Introduction

Web-Based Education (WBE) is currently an important research and development area and it has opened new ways of learning for many people. A basic problem faced by the learning community is how to produce and deliver content online learning experiences, being able to suitable for the particular requirements of each individual. But at the same time, they have to be flexible and available for being tailored and used by a wide community of developers and students, respectively.

To respond to these situations an adaptive and intelligent Multi-Agent System (MAS) is presented. This MAS is applied for development of WBE systems and it considers the diversity of requirements and provides the needed functionalities based on the facilities of the Web.

Wherefore, the purpose of this chapter is to show a new architecture for development of WBE systems based on the IEEE LTSA (Learning Technology System Architecture) specification [1]. In order to achieve this goal, this chapter as organized as follows: In section 5.2, the IEEE LTSA overview is presented; whereas in section 5.3, the components system layer is presented, these components are based on the pattern of the Intelligent Reusable Learning Components Oriented Object (IRLCOO), developed by Peredo et al [2]. IRLCOO are a special type of Sharable Content Object (SCO) according to the Sharable Content Object Reference Model (SCORM). SCORM is used to create reusable and interoperable learning content [3]. IRLCOO are labeled with RDF [4] and XML [5] to be used for the rule-based inference engine known as JENA [6], and JOSEKI server for to implement a semantic platform [7].

Afterwards, in section 5.4 an adaptive and intelligent MAS is showed. MAS contains all the logical for taking of the decisions in order to deliver adaptive and intelligent teaching–learning experiences suitable for the particular requirements of

N.T. Nguyen, L.C. Jain (Eds.): Intel. Agents in the Evol. of Web & Appli., SCI 167, pp. 87–112.
springerlink.com

each individual. This MAS is a set of standard specifications supporting inter-agent communication and key middleware services. All the communications is realized across the entire FIPA2000 standards [8] and they are represented through an agent-specific extension of UML, known as AUML or Agent.

In section 5.5 and 5.6 the investigation projects called SiDeC (authoring content) and Evaluation System are respectively depicted. Besides, these systems are based on IRLCOO, Semantic Web and MAS. These tools are oriented to offer application level interoperability under the philosophy of WS (Web Services). Other used technologies for the implementation are AJAX (Asynchronous JavaScript And XML) [9], this is used for communication between components and LMS's API (Application Programming Interface), JADE (Java Agent DEvelopment Framework) [10], Struts implements the design pattern MVC (Model-View-Controller) [11], Servlets, JSPs and JavaBeans implements other functionalities of the system under the model MVC, DOM (Document Object Model) and XSLT (Extensible Stylesheet Language Transformations) for XML (eXtensible Markup Language) persistence [5], Hibernate for object/relational persistence and query service [12]. The section 5.7 contains a reference list of resources relevant to the topic.

Finally, in Section 5.8 we conclude with some comments and evaluation critics about this work.

5.2 Agents and Components Oriented Architecture

Between the key issues of domain engineering is the aim for developing software reusable components [13]. Thus, components are widely seen by software engineers as a main technology to address the "software crisis."

The Agent-Oriented Programming (AOP) is a relatively new software paradigm that brings concepts from the theories of artificial intelligence into the mainstream realm of distributed systems. AOP essentially models an application as a collection of components called agents that are characterized by, among other things, autonomy, proactivity and an ability to communicate. Being autonomous they can independently carry out complex, and often long-term tasks. Being proactive they can take the initiative to perform a given task even without an explicit stimulus from a user. Being communicative they can interact with other entities to assist with achieving their own and others' goals [10].

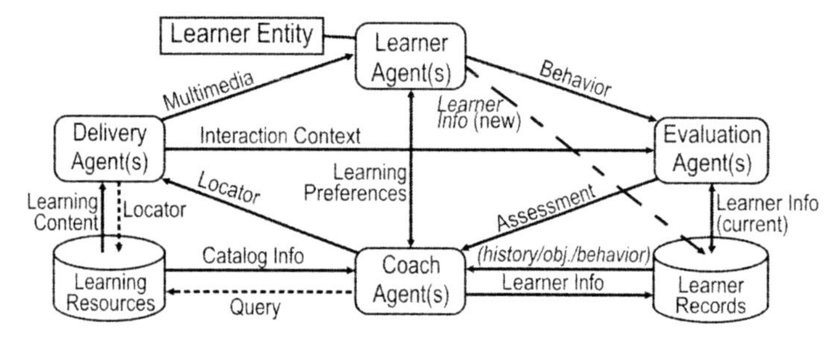

Fig. 5.1. Agents and Components Oriented Architecture

Our Agents and Components Oriented Architecture is based on layer 3 of IEEE 1484 LTSA specification. This architecture provides a framework to understand and applied the reusability, interoperability and portability for Learning Content Management System (LCMS) [1]. The layer 3 is depicted in Fig. 5.1, and consists in four processes manage by agents: learner entity, evaluation, coach, and delivery process; two stores: learner records and learning resources; and fourteen information workflows.

The layer 3 of IEEE 1484 LTSA specification was modify to supply processes to adapt to the learners' needs in intelligent form. Initially, the coach process has been divided in two subprocesses: coach and virtual coach. The reason is because we considered that this process has to adapt to the learners' individual needs in a quick way during the learning process. For this, some decisions over sequence, activities, examples, etc., can be made manually for the coach but in others cases these decisions can be made automatically for the virtual coach.

Briefly, the overall operation has the following form:

(1) Learner Entity. The learning styles, strategies, methods, etc., are negotiated among the learner and other stakeholders and are communicated as learning preferences.

(2, *new proposal*) The learner information (behaviour inside the course, e.g., trajectory, times, nomadicity, command voice, etc.) is stored in the learner records.

(3) The learner is observed and evaluated in the context of multimedia interactions.

(4) The evaluation produces assessments and/or learner information (current).

(5) The learner information (keyboard clicks, mouse clicks, voice response, choices, written responses, etc., all over learner's evaluation) is stored in the learner history database.

(6) The coach reviews the learner's assessment and learner information, such as preferences, past performance history, and, possibly, future learning objectives.

(7, *new proposal*) The virtual coach reviews the learner's behaviour and learner information, and automatic and smartly he makes dynamic modifications on the course sequence (personalized to learner's needs) based on the learning process design.

(8) The coach/virtual coach searches the learning resources, via query and catalog info, for appropriate learning content.

(9) The coach/virtual coach extracts the locators (e.g., URLs) from the available catalog info and passes the locators to the delivery process, e.g., a lesson plan or pointers to content.

(10) The delivery process extracts the learning content and the learner information from the learning resources and the learner records respectively, based on locators, and transforms the learning content components to an interactive and adaptive multimedia presentation to the learner.

It is important to point out that the task 2 and 7 are new and they were incorporated to the IEEE 1484 LTSA standard.

In section 5.3, the Agents and Components Oriented Architecture from learning components viewpoint is described.

5.2.1 Layer of the Architecture

Fig. 5.2 illustrates an overview of how Agents and Components Oriented Architecture is integrated. In general, the software architecture is divided into four layers: application, agents & components, database, and server layers. The application layer includes an administration system, which is the ADL platform, to allow system administrators, instructors, and learners to manage learner records and curriculum. On the left side below the administration system, asynchronized ours systems are incorporated. Thus, structure authoring systems are separated from learning content.

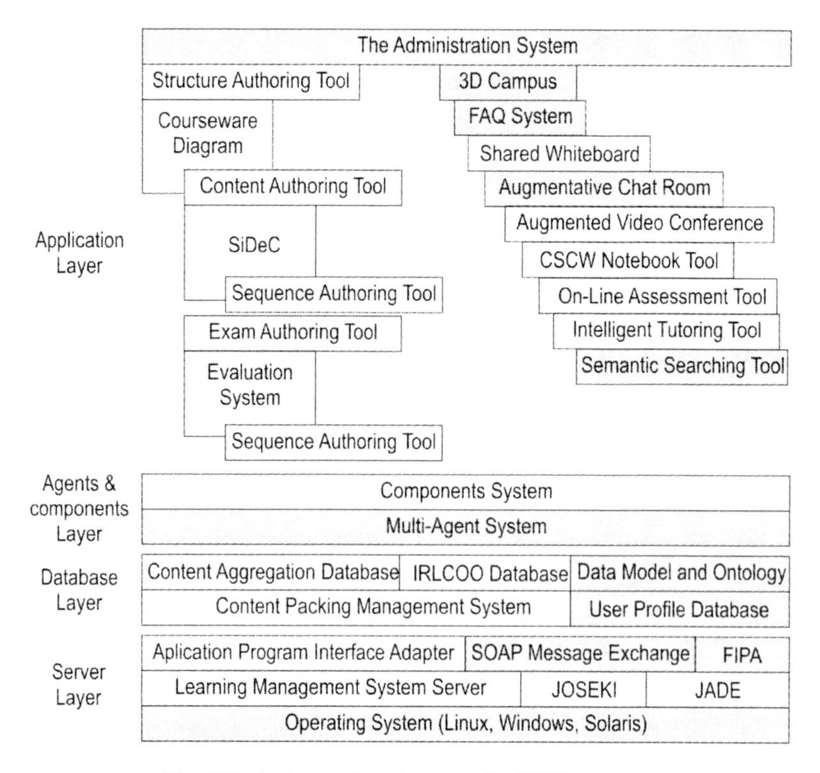

Fig. 5.2. An integral architecture for WBE systems

5.2.2 Adaptation in Web-Based Education Systems

The term "adaptation" is an important aspect in WBE systems. It has been shown that the application of adaptation can provide better learning environments in such systems. There are two forms of WBE systems development for supporting the learners' individual needs:

- Those that allow the learner to change certain system parameters and adapt their behaviour accordingly are called adaptable.
- Those that adapt to the learners' needs in intelligent form and automatically based on the system's conjecture about are called adaptive.

In this work, we are centered in the development of adaptive and intelligent WBE systems. The constant change of these systems requires taking into account the following aspects:

- The adaptation with respect to current domain competence level of the learner.
- The suitability with respect to domain content.
- The adaptation with respect to the context in which the information is being presented.

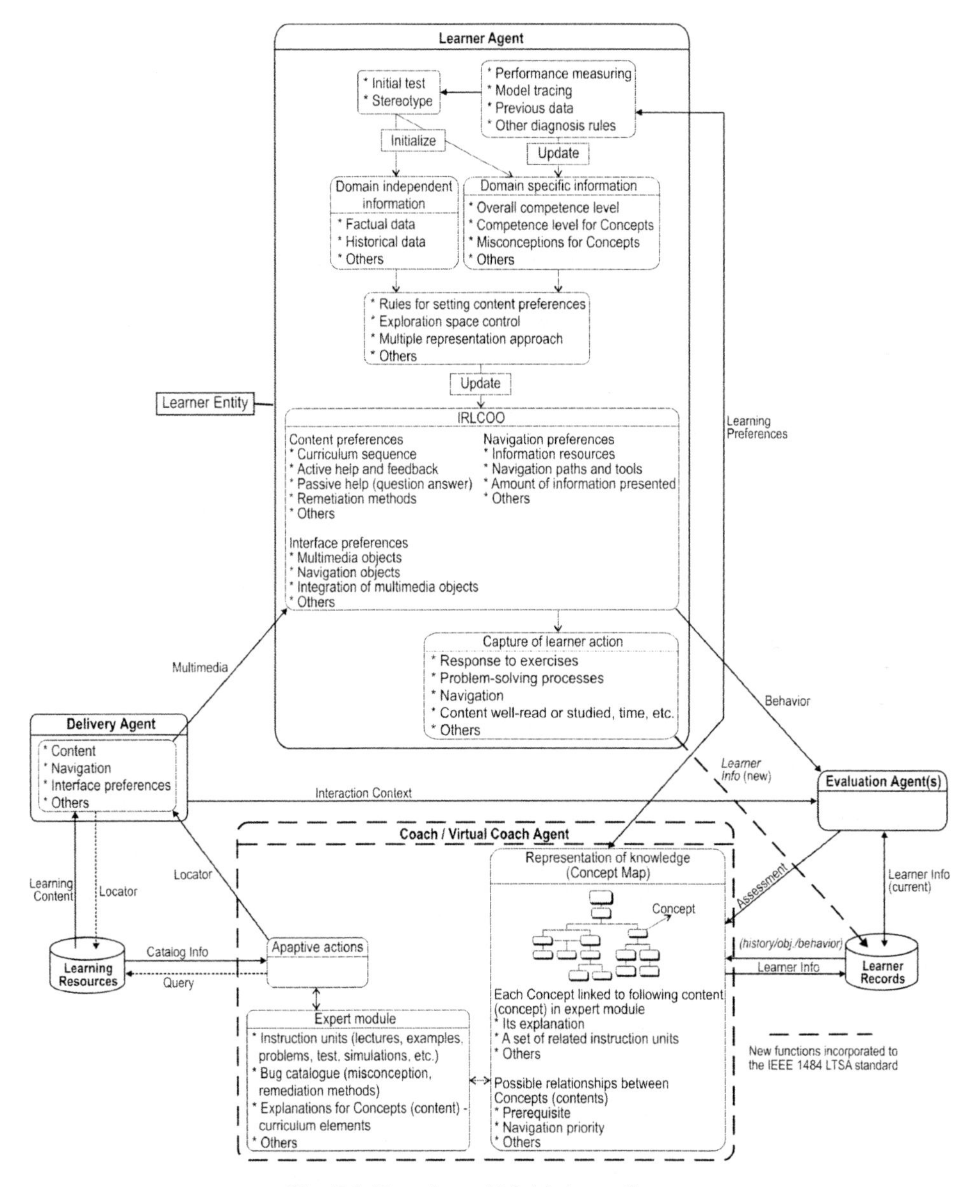

Fig. 5.3. Top view of Multi-Agent System

To cover these aspects, we propose to develop a new architecture, which captures the interactions of learners with the system to extract information about their competence level for various domain concepts and tasks represented in the system.

According to the potential and the criteria of the dynamism in the WBE systems, we are developing a suitable way to capture interactions over the Internet and to provide a continuous interaction pattern for a given learner (Learner process agent, see Fig. 5.1). In WBE systems, the interactions between client and server normally take place using Hypertext Transfer Protocol (HTTP). HTTP is a stateless protocol, which makes it difficult to track the learner progress and hence to analyze the mental processes of the learner. However, using a judicious mixture of LMS API, Web Services, Ajax [9], Hibernate [12], Struts Framework [11] and Semantic Web Platform, we can be more precise about browsing behaviour. Part of the new architecture resides on the server and part on the user's machine.

5.2.3 Learner Entity

The Learner Entity allows adaptive behaviour of the system by providing information about the Learner agent. The Learner agent processes granular information about learner's competence level for various tasks represented in the system (depicted in Fig. 5.1). The Learner agent is used by the system to:

- support adaptive navigation guidance - based on prioritized successors and learner's needs,
- support context based on previous learning components,
- support dynamic messaging and feedback, e.g. navigation, learning content, current context and progression.

Fig. 5.3 shows the summary of Learner agent which provides adaptation to the learners in intelligent form and automatically based on Virtual Coach agent.

5.3 Intelligent Reusable Learning Components Oriented Object

IRLCOO were developed with ActionScript 3 [14]. ActionScript was reengineered from top to bottom as a true object-oriented programming (OOP) language; using reusable design patterns, which are ideal way to solve common problems [15]. ActionScript has evolved from a few statements in Flash to a full fledged Internet programming language in the latest release of Flash and Flex. The idea of design patterns is to take a patterns set and solve recurrent problems. The founding principles of design patterns are laid down in the Gang of Four's canon [16]. Two essential principles are:

- Program to an interface, not an implementation.
- Favor object composition over class inheritance.

ActionScript enables the design of client components for multimedia content [17]. At Run-Time, the components load media objects and offer a programmable and adaptive environment to the learner's needs, through of the composite pattern. The composite pattern provides a robust solution to build complex systems that are made up of several smaller components. The system is made by components that may be individual media objects or containers that represent collections of media objects. The composite pattern streamlines the building and manipulation of complex media objects that are composed of several related pieces. The complex media objects are built as hierarchical trees, the structure components can be individual components (primitives or indivisible objects) or composite components that hold a collection of other components, and they allow to the clients to treat both individual components and composite components the same way, simplifying the interface [17]. This pattern has particular utility in ActionScript, allowing easily building and manipulating complex media objects.

Flash already has Smart Clips for the learning elements denominated Learning Interactions (LI). The first aim was to generate a multimedia library of IRLCOO for WBE systems with the purpose to separate the content from the navigation. Thus, the components use different levels of code inside the Flash Player (FP). With this structure, it is possible to generate specialized components which are small, reusable, and suitable to integrate them inside a component container at Run-Time. The Object Oriented paradigm allows passing from component IRLC to IRLCOO. With these facilities, IRLCOO are tailored to the learner's needs. In addition, the IRLCOO platform owns certain communication functionalities inside of the Application Programming Interface (API) developed for the LMS/MAS (shown in section 5.1), dynamic load of Assets, and Assets composition. The middleware for the components uses different frameworks as: Ajax [9], Hibernate [12], Struts [11], etc.

IRLCOO are meta-labeled with the purpose of complete a similar function as the product bar codes, which are used to identify the products and to determine certain

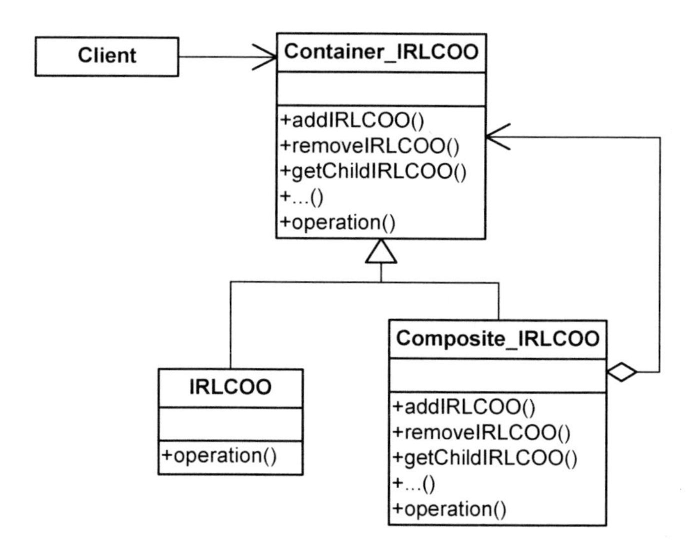

Fig. 5.4. Class diagram: Composite pattern of IRLCOO

characteristics specify of themselves. This contrast is made using the meta-labeled SCORM and Resource Description Framework (RDF-XML) [4], which allows enabling certain grade inferences on the materials by means of the Semantic Web Platform.

5.3.1 Composite Pattern

The nodes that contain other components are composite_IRLCOO components. The IRLCOO nodes are indivisible or primitive components, which cannot have any children. Each IRLCOO node is a child of a composite_IRLCOO node and each composite_IRLCOO node can have multiple children, including other composite_IRLCOO nodes. As shown in the class diagram in Fig. 5.4, the composite pattern provides a common interface to deal with both composite_IRLCOO and IRLCOO nodes.

5.3.2 Persistent Layer - Hibernate

Hibernate is an ambitious project that aims to be a complete solution to the problem of managing persistent data in Java. It mediates the application's interaction with a relational database, leaving the developer free to concentrate on the business problem at hand. Hibernate is a non-intrusive solution. Almost all applications require persistent data. Persistence is one of the fundamental concepts in application development. If an information system did not preserve data entered by users when the host machine was powered off, the system would be of little practical use. Object persistence means that individual objects can outlive the application process; they can be saved to a data store and be recreated at a later point in time. Object/relational mapping (ORM) is the name given to automated solutions to the mismatch problem. The applications built with ORM middleware can be expected to be cheaper, more performance, less vendor-specific, and more able to cope with changes to the internal object or underlying SQL schema [12]. In our case, Hibernate is used in the subsystems of login, XML, Upload, SiDeC, Evaluation System, FAQ, Chat, etc. The Fig. 5.5 shows the class HIBERNATE.INTERPRETE for our FAQ (Frequently Asked Questions) subsystem.

The following code shows as obtaining the user's ID using Hibernate:

```
MANAGER.beginTransaction();
Criteria crit = MANAGER.getSession().createCriteria
  (BEANS.Users.class);
crit.add(Restrictions.eq("nick",nick));
crit.add(Restrictions.eq("clave",password));
MANAGER.commitTransaction();
if(crit.list().size()==1){
  Users User = (Users) crit.list().iterator().next();
  Answer=User.getId_user();}
MANAGER.closeSession();
MANAGER.rollbackTransaction();
return Answer;
```

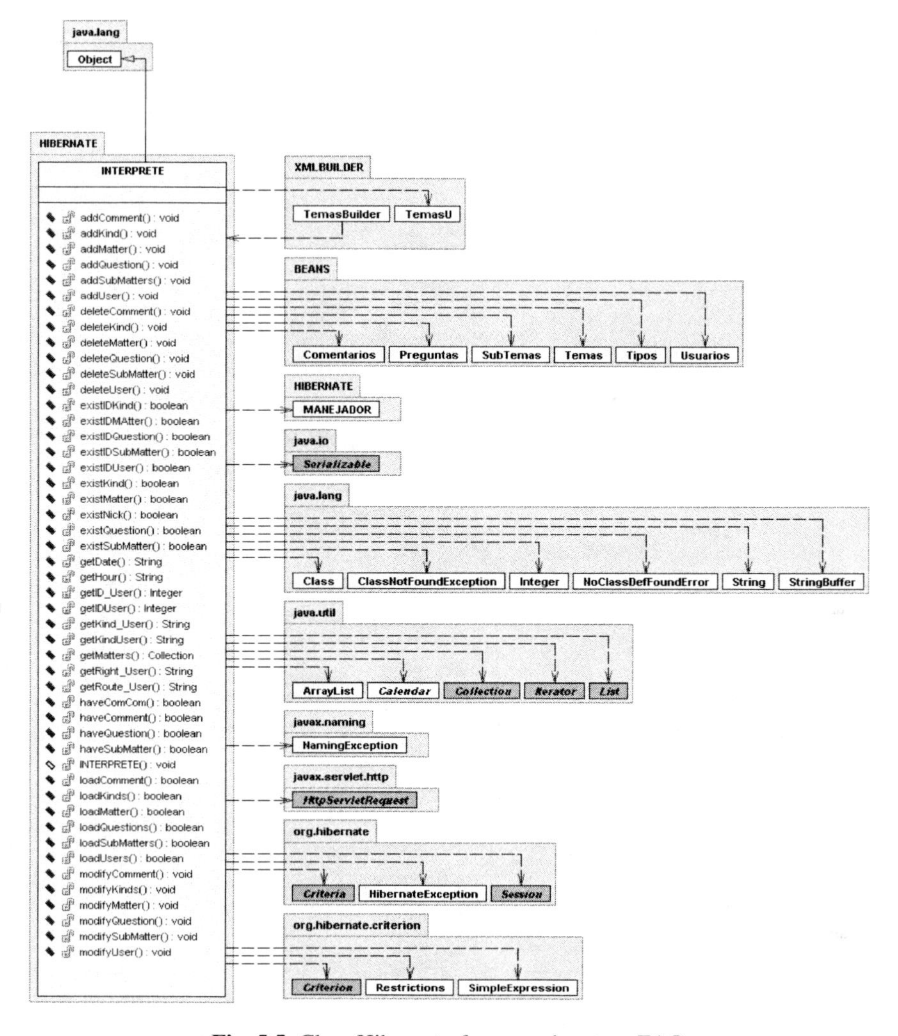

Fig. 5.5. Class Hibernate for our subsystem FAQ

5.3.3 Web Services

The WS have emerged and are becoming the interoperable ubiquitous technology. WS are a key of the next generation of World Wide Web —the Semantic Web, in which the content is not only human-readable, but also machine-readable. The other driving force behind WS is enabling business-to-business communication. Companies can share data, and integrate heterogeneous data and services— in one conglomerate Service Oriented Architecture (SOA) [18].

WS enable remote procedure calls and asynchronous messaging. WS are generally implemented as XML messages over the HTTP protocol. There are numerous standards surrounding this definition. The principal goal of WS is interoperability. The formation of the open industry organization, Web Services Interoperability

Organization (WS-I) demonstrates the importance of interoperability. WS are an implementation of independent platform for the remote procedure calls. An XML encoded request is sent to the server and an XMLencoded response is returned. Normally WS are sent over the HTTP protocol via the HTTP POST operation. XML is posted to the Web server, and XML is returned from the HTTP POST operation. While the underlying RPC protocol itself is in XML, the application data passed to and from the server is also encoded in XML—and can be any valid XML described by XML Schema.

The WS-I Basic Profile is a set of implementation directives for WS interoperability. The implementation of WS in Java 6 is the Java APIs for XML WS (JAX-WS), conforms to the WS-I Basic Profile, version 1.1. Although the WS-I Basic Profile references other specifications and protocols, it does not actually define any. The Simple Object Access Protocol (SOAP) is used for the WS message format and the Web Services Description Language (WSDL) is used to describe what is in a particular service's messages.

ActionScript 3.0 adds the component WebServiceConnector and WebServices classes to connect to WS from the IRLCOO. The WebServiceConnector component enables the access to remote methods offered by a LMS through SOAP protocol. This gives to WS the ability to accept parameters and return a result to the script, in other words, it is possible to access and join data between public or own WS and the IRLCOO. It is possible to reduce the programming time, since a simple instance of the WebServiceConnector component, which is used to make multiple calls to the same functionality within the LMS. The components discover and invoke WS using SOAP and UDDI, via middleware and a JUDDI server. Placing a Run-Time layer between a WS client and server dramatically increases the options for writing smarter, more dynamic clients. Reducing the dependence necessity inside the clients. It is only necessary to use different instances for each one of the different functionalities. WS can be unloaded using the component and deployed within an IRLCOO. The next code shows a request from the learning content to the middleware, requesting the WS notes. The answer (URL) is used to call the WS from IRLCOO:

```
// Inquire WS from a server JUDDI
Inquire.Url =
 "http://148.204.45.65:8080/juddi/inquiry";
FindService fs = new FindService();
fs.Names.Add("es", "Notes");
ServiceList sl = fs.Send();

// Calling WS from a client IRLCOO
import mx.services.WebService;
var oListener:Object = new Object();
oListener.click = function(oEvent:Object):Void {
 var sWSDLURL:String = FindWebService;
 var wsNotes:WebService = new WebService(sWSDLURL);
 var oCallback:Object =
  wsNotes.getNotes(cmi.core.student_id,cmi.core.lesson_
  location);
 oCallback.onResult = function(nNote):Void {
  ctiNote.text = nNote; }; };
cbtSubmit.label = "Submit";
cbtSubmit.addEventListener("click", oListener);
```

The WebService object acts as a local reference to a remote web service. When it creates a new WebService object, the WSDL file that defines the web service gets downloaded, parsed, and placed in the object. Then, it can call the methods of the web service directly on the WebService object and handle any callbacks from the web service. When the WSDL has been successfully processed and the WebService object is ready, the WebService.onLoad callback is invoked. If there is a problem loading the WSDL, the WebService.onFault callback is invoked. When it is necessary to call a method on a WebService object, the return value is a callback object. The object type of the callback returned from all web service methods is PendingCall. These objects are normally constructed automatically as a result of the webServiceObject.webServiceMethodName() method that was called. These objects are not the result of the WebService call, which occurs later. Instead, the PendingCall object represents the call in progress. When the WebService operation finishes executing (usually several seconds after a method is called), the various PendingCall data fields are filled in, and the PendingCall.onResult or PendingCall.onFault callback you provide is called.

5.4 Adaptive and Intelligent Multi-Agent System

The construction of our adaptive and intelligence MAS is carried out with Jadex. Jadex is a software framework for the creation of goal-oriented agents following the belief-desire-intention (BDI) reasoning engine that allows for programming intelligent software agents in XML and Java [19]. The reasoning engine is very flexible and can be used on-top of different middleware infrastructures such as JADE. Jadex aims to build up a rational agent layer that sits on top of a middleware agent infrastructure and allows for intelligent agent construction using software engineering foundations.

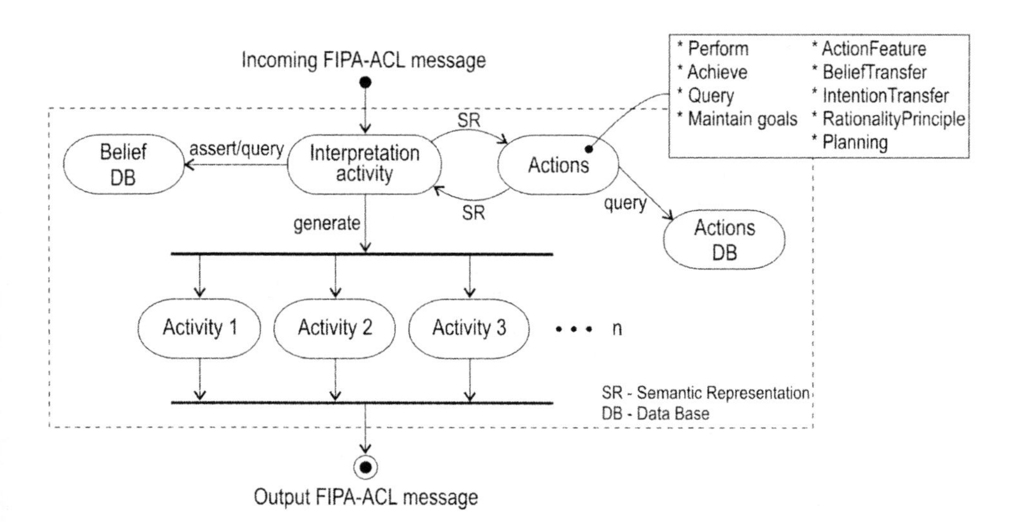

Fig. 5.6. Jadex abstract architecture for Learner agent

Moreover, Jadex allows developing MAS addressed by the Foundation for Intelligent Physical Agents (FIPA) specifications.

The FIPA has defined a standard agent communication language, namely FIPA-ACL [8]. This language has the advantage that is relies not only on a syntactic definition, but also on a semantic form. In other words, FIPA-ACL standard formally specifies a precise meaning for each communication primitive in the language.

According with Fig. 5.1, our main MAS is conformed by learner entity, evaluation, coach, and delivery agents. In this case, it is focused on Learner agent. In Fig. 5.6 an overview of the abstract Jadex architecture is presented. Viewed from the outside, the Learner agent is a black box capable of sending and receiving messages. The main activity of Learner agent consists of interpreting the received FIPA-ACL messages. This interpretation activity can be refined into two main functions: the first one produces some sense about the input message, while the second one consumes this sense and updates the agent's respective activities and beliefs.

5.4.1 Beliefs

Beliefs represent the state of the agent's knowledge about the world, itself and other agents [19]. For example, Performance measuring, Model tracing, Previous data, etc. (see Fig. 5.3). The belief representation in Jadex allows arbitrary Java objects to be stored instead of relying on a logic-based representation. This facilitates integration with our Web-Based Applications (SiDeC and Evaluation System), e.g. classes generated by ontology modelling tools and our database mapping layers can be directly reused. Objects are stored as named facts (called beliefs) or named sets of facts (called belief sets). Using the belief names, the 'belief data base' can be manipulated by setting, adding or removing facts. In addition, a more declarative way of accessing beliefs and belief sets is provided by Object Query Language (OQL) - like queries.

5.4.2 Goals

Goals are the motivational force driving an agent's activities. They come in different flavours allowing various attitudes of an agent to be expressed. Activities consist of performing some arbitrarily complex courses of action [19]. These actions include the communicative actions defined by FIPA-ACL, such as inform or request as inform or request, as well domain-specific actions.

One of the core concepts used by the interpretation activity is the Semantic Representation (SR), which represents a produced or consumed sense. Jadex currently supports four application-relevant goal types: perform, achieve, query and maintain goals. Next, an example of activities based on Fig. 5.3 (Learner Agent) is described.

First, the "ActionFeature" produces several semantic representations (SRs) from a received message that represent the semantic features of the corresponding communicative action. The FIPA-ACL standard defines two semantic features for each communicative action: the feasibility precondition and the rational effect. The former mainly gives rise to a SR starting the precondition was necessarily true before the communicative action was performed. The latter gives rise to an SR (usually called "intentional effect") stating the sending agent intends the rational effect to become true. In our case, the Model tracing represents the intention of a definition action, and the Domain specific information denotes the corresponding rational effect. These SRs feed the other activities, leading to the agent's reaction.

The "BeliefTransfer" manages the adoption of beliefs suggested by other agents. It applies to any SR stating an external agent intends the interpreting agent to believe a fact, and produces a new SR stating the interpreting agent actually believes this fact. For our example, the Model tracing results from applying the "BeliefTransfer" to the SR Navigation priority.

Similarly, the "IntentionTransfer" manages the adoption of other agents' intentions. It applies to any SR stating an external agent has an intention, and produces a new SR stating the interpreting agent actually has the same intention. This application may be customized to specify the expected cooperative attitude for the interpreting agent in terms of the intentions to adopt and the external agents to cooperate with. In our case, the SR of Domain specific information results from applying the "IntentionTransfer" to Model tracing, Initial test and Stereotype.

Finally, the last three activities manage the planning capabilities of the interpreting agent by creating proper activities in order to satisfy her/his intentions.

5.4.3 Plans

Means–end reasoning is performed with the objective of determining suitable plans for pursuing goals or handling other kinds of events such as messages or belief changes.

The "Perform" tries to directly perform an intended action (for example, Content preferences in Fig. 5.3). The "RationalityPrinciple" searches the agent's base of actions for an action whose rational effect matches a given intention (for example, Delivery agent is necessary to Content preferences). Lastly, the "Planning" uses an external planner to find an (arbitrarily complex) action plan whose performance brings about the input intention.

5.4.4 Capabilities

A capability results from the packaging of specific functionality into a module with precisely defined interfaces [20]. An agent can be composed of an arbitrary number of capabilities that themselves may include any number of subcapabilities. Jadex contains several generic plans and predefined capabilities in the package *jadex.planlib*. Basic platform features can be accessed by using the Agent Management System (AMS) and Directory Facilitator (DF) capabilities. The AMS capability offers goals for agent management such as creating new agents or destroying existing ones, whereas the DF capability can be used for accessing yellow pages services such as registering agents or searching specific services via goals.

5.4.5 Implementing MAS

Implementing our MAS with Jadex is simple. It actually consists of implementing the cooperative and domain-specific agent features rather than analyzing in detail and coding all possible messages and interaction protocols. MAS programming is carried out following three main tasks.

The first task consists of implementing the domain-specific actions - these are part of the MAS. In the scenario of Fig. 5.1, the only four domain specifications that have been implemented are "put-on" and "take-off".

The second task consists in coding the agent's belief base management to handle domain-specific facts.

Finally, the last task consists of customizing the cooperation principles of the agent, including the "BeliefTransfer" and "IntentionTransfer" activities.

In most cases, a significant part of the code developed to implement an agent is domain-specific. Hence, it can be reused by other agents in this domain without any additional development.

5.5 SiDeC

To facilitate the development of learning content, it was built an authoring system called SiDeC (Sistema de Desarrollo de eCursos - eCourses Development System). SiDeC is a system based on Agents and Components Oriented Architecture to facilitate the authoring content to the tutors who are not willing for handling multimedia applications. In addition, the Structure and Package of content multimedia is achieved by the use of IRLCOO, as the lowest level of content granularity.

SiDeC is used to construct Web-based courseware from the stored IRLCOO (Learning Resources, see Fig. 5.1), besides enhancing the courseware with various authoring tools. Developers choose one of the SiDeC lesson templates and specify the desired components to be used in each item. At this moment, the SiDeC lesson templates are based on the cognitive theory of Conceptual Maps (CM), but in the future we will consider others theories such as: Based-Problems Learning (BPL), the cases method, etc.

The inclusion of CM, as cognitive elements, obeys to the instructional design pattern for the development of the courses. Thus, the courses do not only have theoretical or practical questions, but rather they include a mental model about individual thought process. CM is a schema to structure concepts with the purpose of helping the learners to maximize the knowledge acquisition. A CM is a graphical technique used during the teaching-learning process, among other forms as instructional and learning strategy, and as schematic resource or navigation map [21].

SiDeC has a metadata tool for supporting the generation of XML files for IRLCOO to provide on-line courses. This courseware captures learners' metrics with the purpose to tailor their learning experiences (through Learner agent). Furthermore, the IRLCOO offer a friendly interface and flexible functionality. These deliverables are compliance with the specifications of the IRLCOO and with learning items of SCORM 1.2 Models (Content Aggregation, Sequencing and Navigation, and Run Time Environment) [3]. Metadata represent the specific description of the component and its contents, such as: title, description, keywords, learning objectives, item type, and rights of use. The metadata tool provides templates for entering metadata and storing each component in the SiDeC or another IMS/IEEE standard repository.

SiDeC proposes a course structure based on the idea of a compound learning item as a collection of IRLCOO and Reusable Learning Component Oriented Object (RLCOO). These atoms are grouped together to teach a common task based on a single learning objective, as is depicted in Fig. 5.7. IRLCOO are an elementary atomic piece of learning that is built upon a single learning objective. Each IRLCOO can be classified as: concept, fact, process or procedure. The IRLCOO provide the information of learner's behaviour within the course, e.g., trajectory, times, and assessments. This information is processed by Learner agent and stored in the learner history database (learner records).

On the other hand, a RLCOO is an atomic piece of information that is built upon single information object. It may contain up to seven different content items, such as: overview, introduction, importance, objectives, prerequisites, scenario, and outline. Each RLCOO is an elementary no divisible piece of information that is built upon single information object.

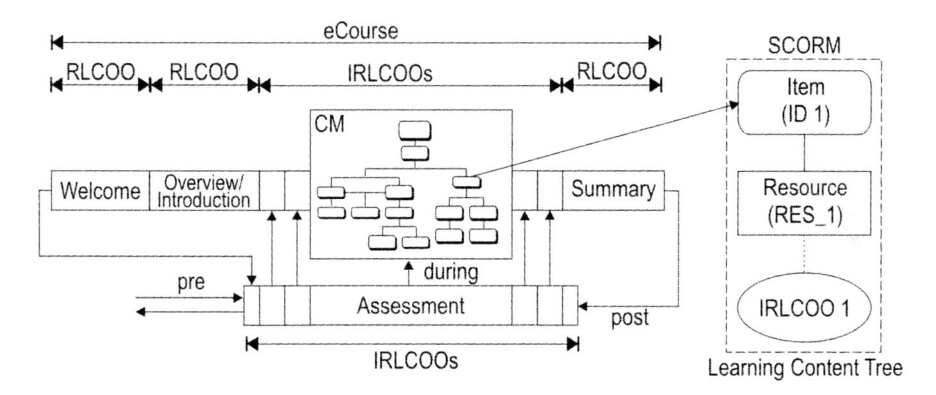

Fig. 5.7. Learning content generated for the SiDeC

In Fig. 5.7, the SiDeC implements the CM as a navigation map or instructional and learning strategy allowing to the learner to interact with content objects along the learning experiences. These experiences follow an instructional-teaching strategy. These kinds of strategies carry out modifications of the learning content structure. Such modifications are done by the designer of the learning experience with the objective of provide significant learning and to teach the learners how to think [22]. The learning content can be interpreted in a Learning Content Tree. A Learning Content Tree is a conceptual structure of learning activities managed by the Delivery process agent for each learner. The Tree representation is just a different way for presenting content structure and navigation. This information is found in the manifest that is defined into the SCORM Content Aggregation Model (CAM) [3]. According with Fig. 5.7, the next fragment code shows how the course structure is organized as *imsmanifest.xml*.

```
<manifest>
  <organizations>
     <organization>
        <item>
          <item identifier="ID1" identifierref=
          "RES_1">
             <adlnav:presentation>
```

```
                    <adlnav:navigationInterface>
                       <adlnav:hideLMSUI>previous
                          </adlnav:hideLMSUI>
                       <adlnav:hideLMSUI>continue
                          </adlnav:hideLMSUI>
                    </adlnav:navigationInterface>
                 </adlnav:presentation>
                 <imsss:sequencing>
                    <imsss:controlMode choice="false"
                       flow="true"/>
                    <imsss:rollupRules
                       rollupObjectiveSatisfied="false"/>
                 </imsss:sequencing>
              </item>
            <item>   ...   </item>
        </organization>
      </organizations>
       <resources>
          <resource identifier="RES_1">   ...
       </resources>
          ...
   </manifest>
```

5.5.1 Communication between IRLCOO and LMS

Our communication model uses an asynchronous mode in Run-Time Environment (RTE) and joins to LMS communication API of ADL [3], AJAX (Asynchronous JavaScript And XML) [9] and Struts Framework [11] for its implementation. The LMS communication API of ADL consists of a collection of standard methods to let the Client to communicate with the LMS.

AJAX is a way of developing Web applications to create interactive applications that it is executed in client side, in other words, the Web browser maintains the asynchronous communication with the server in backstage. This way it is possible to carry out changes in the same page without necessity of reload it. This increases the interaction speed.

On the other hand, the Struts Framework is a tool for Web application development under the Java MVC (Model-View-Controller) architecture; with this Framework is defined the independent implementation of the Model (business object), the View (interface with the user or another system) and the Controller (controller of the application workflow). This Framework provides the advantage of maintainability, performance (tags pooling, caching, etc.), and reusability (contains tools for the field validation that it is executed in client or server sides). The browser-based communication model is depicted in Fig. 5.8.

According to Fig. 5.8, the communication model starts: (1) when an IRLCOO generates an event. (2) Form the browser interface is made a JavaScript call to the function *FileName_DoFSCommand(command,args)*, which handles all the *FSCommand* messages from IRLCOO, LMS communication API, and AJAX and Struts methods. Next, a fragment of this code is showed:

```
function FileName_DoFSCommand(command,args)  {
  doInitialize();
  doSetValue(name,value); // i.e. (StudentName,name)
  doEnding();
callAjaxStruts();  }
```

The communication with the LMS starts when: (I) the standard methods call to the Communication Adapter (written in JavaScript). (II) The communication adapter implements the bidirectional communication ADL´s API between the Client and the LMS. (III) The LMS realizes the query-response handling and the business logic, i.e., the access to the database. The purpose of establishing a common data model is to

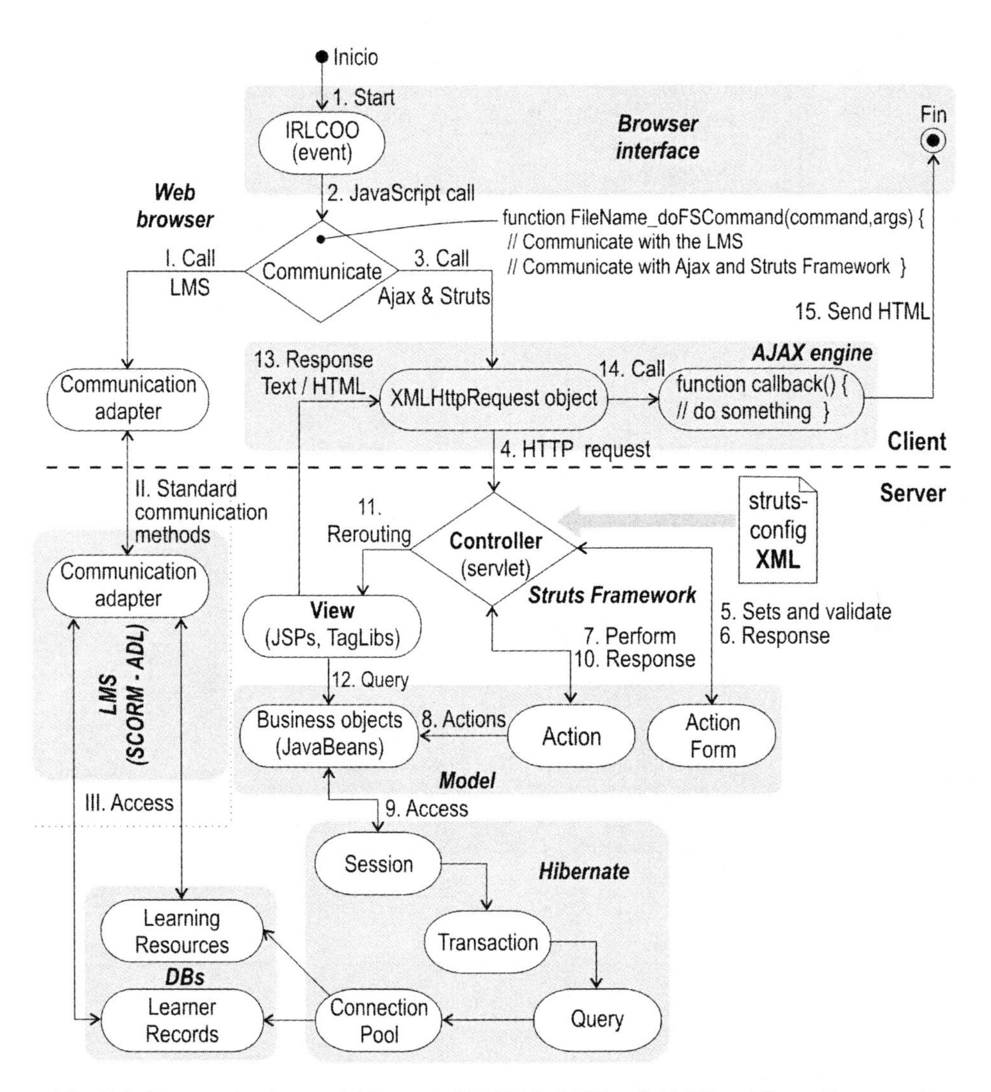

Fig. 5.8. Communication model between IRLCOO, LMS and AJAX and Struts Framework

make sure that a defined set of information about content can be tracked by different LMS environments. If, for example, it is determined that tracking a student's score is a general requirement, then it is necessary to establish a common way for content to report scores and for LMS environments to process such information (for interoperability and reuse of such content). If every chunk of content used its own unique scoring representation learning management systems would not know how to receive, store or process such information. These definitions are derived from the IEEE Standard for Computer-Managed Instruction document (P1484.11), which originated within the Aviation Industry CBT Committee (AICC) [23]. As an example, a code fragment to get the student's name and their score is showed:

```
function doGetValue(name) {
  loadPage();
  var studentName= "!";
  var lmsStudentName=fscommand("LMSGetValue",
     "cmi.core.student_name");
  if(lmsStudentName != " ") {
    studentName=""+lmsStudentName + "!"; }
  fscommand("doContinue","completed");
  ...
  if(lmsInialized == true) {
  var lmsStudentScore= fscommand("LMSGetValue",
     "cmi.core.score.raw" + score); } }
```

The communication with AJAX and Struts Framework begins when AJAX-Struts method is called. (3) An instance of the *XMLHttpRequest* object is created. Using the *open()* method, the call is set up, the URL is set along with the desired HTTP method, typically GET or POST. The request is actually triggered via a call to the *send()* method. This code might look something like this:

```
function callAjaxStruts ()   {
  createXMLHttpRequest();
  var url = "register.do?";
  var urlString = createURLString();
  xmlHttp.open("POST",url,true);
  xmlHttp.onreadystatechange = processStateChange;
  xmlHttp.setRequestHeader("Content-Type",
    "application/x-www-form-urlencoded;");
  xmlHttp.send(urlString);   }
```

(4) A request is made to the server, this might be a call to a servlet or any server-side technique. (5) The Controller is a servlet (called ActionServlet), which coordinates all applications activities, such as: register of user data, (6) data validations, and control flow. The Controller is configured through of *struts-config.xml*. (7) The Controller calls to *execute* method of *Action*, it passes to this method the data values and the *Action* reviews the characteristic data that correspond to the Model. (8) The business objects (Java Beans) realize the handling of the model, (9) usually a database access. (10) The *Action* sends the response to the Controller. (11) The Controller reroutes and generates the interface for the results to the View (JSPs). (12) The View makes the query to the Business objects based on the correspondent interface. (13)

The request is returned to the browser. The Content-Type is set to text/xml, the *XMLHttpRequest* object can process results only of the text/ html type. In more complex instances, the response might be quite involved and include JavaScript, DOM manipulation, or other technologies. (14) The *XMLHttpRequest* object calls the function *callback()* when the processing returns. This function checks the *readyState* property on the *XMLHttpRequest* object and then looks at the status code returned from the server. (15) Provided everything is as expected, the *callback()* function sends HTML code and it does something interesting on the client, i.e. advanced dynamic sequence.

This communication model provides new wide perspectives for the WBE systems development, because it improves the capabilities of communication, interaction, interoperability, security, and reusability, between different technologies. For example, the LMS communication API allows us to make standard database queries of learners' information such as personal information, scores, assigned courses, trajectory, etc. While the communication with AJAX and Struts Framework provides the capability of modify the learner's trajectory according to variables from the learner records in RTE (advanced dynamic sequence), components management (IRLCOO) – remember that these components are built and programming with ActionScript 3.0 and consuming XML files – then, this model provides the way to write, load, change and erase XML files in the Server side.

5.6 Evaluation System

The Evaluation System (ES) for WBE is based on the components model of IRLCOO. The ES uses the composite pattern as a solution to building complex systems that are made up of smaller components. The components that make up the system are constituted by: containers, components and objects.

Key system functionality is based on the analysis of the learner's profile, which is built during the teaching-learning experiences. The profile is based on metrics that extracted from the learner's behaviour at Run-Time. These measures are stored into the learner records that compose the profile. The generation of new sequences of courses is in function of the results obtained, besides the account of the adaptation level.

The ES combines IRLCOOs, additional metadata, and a Java Agent platform. Also, some technologies of the Artificial Intelligence field are considered in order to recreate a Semantic Web environment. Semantic Web aims for assisting human users to achieve their online activities. Semantic Web offers plenty of advantages, such as: reduction of the complexity for potential developers, standardization of functionalities and attributes, definition of a set of specialized APIs, and deployment of a Semantic Web Platform [24].

All resources have a Universal Resource Identifier (URI). An URI can be a Unified Resource Locator (URL) or some other type of unique identifier. An identifier does not necessarily enable access to a resource. The XML layer is used to define the IRLCOO's metadata SCORM that are used to interchange data over the Web. XML Schema tier corresponds to the language used to define the structure of metadata. RDF level is represented by the language used for describing all information and metadata sorts. RDF Schema layer is carried out by the Framework that provides meaning to the vocabulary implemented. The Ontology tier is devoted to define the semantics for establishing the usage of words and terms in the context of the vocabulary. Logical level corresponds to the reasoning used to establish consistency and correctness of data sets and to infer conclusions that are not explicitly stated. The Proofs layer explains the steps of logical reasoning. The Trust tier provides authentication of identity and evidence of the trustworthiness of data, services and agents.

In resume, the components and operation of the ES are outlined in Fig. 5.9. Basically the Evaluation System is fulfilled through two phases. The first phase is supported by the LMS, and is devoted to present the course and it is structure. All the actions are registered and the presentation of the contents is realized with content IRLCOO. The evaluations are done by evaluating IRLCOO and in some cases by simulators based on IRLCOO. These processes are deployed by the Framework of Servlets/Java Server Pages/JavaBeans and struts.

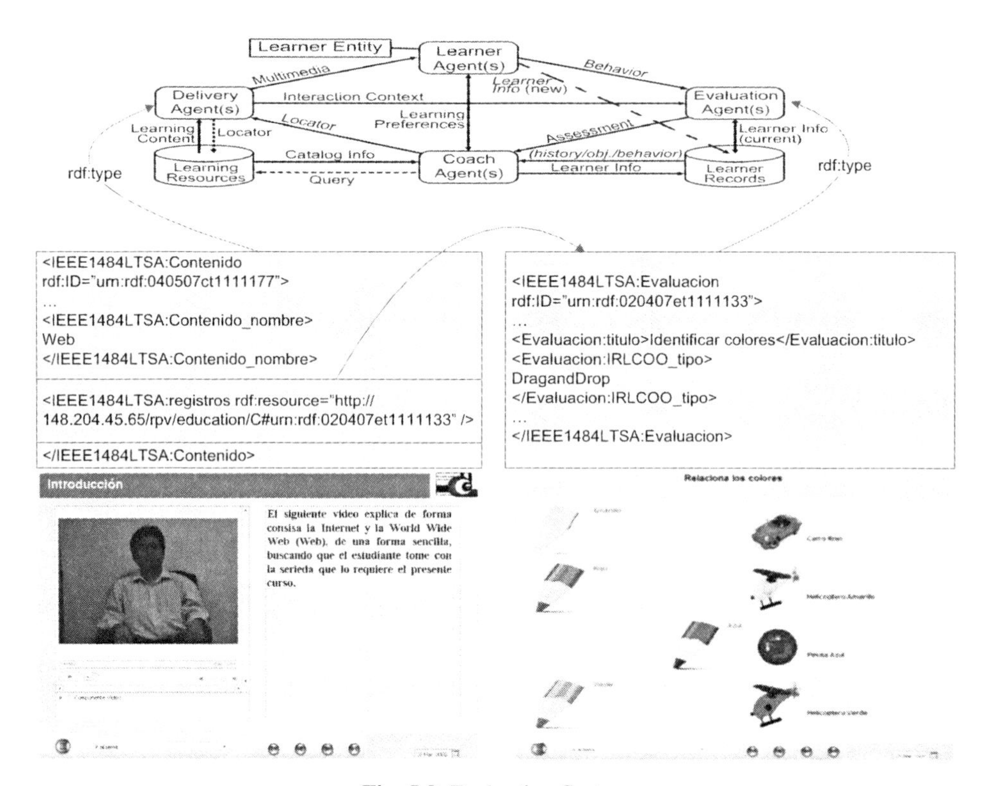

Fig. 5.9. Evaluation System

The second part analyzes the learner's records carried out by the Server based on JADE MAS. This agent platform owns seven agents: Snooper, Buffer, Learner, Evaluation, Delivering, Coach, and Info. The fundamental idea is to automate the learner's analysis through the coach, and to give partial results that can be useful for the learner's final instruction. These agents are implemented as JavaBeans programs, which are embedded in the applications running both at the client and server sides.

The Snooper Agent works as a trigger by means of the INFORM performative, which activates the MAS server's part. This agent is deployed into a Java Server Page (JSP), which uses a JavaBean. During the lesson or once evaluation is finished, the graphical user interface activates the Snooper Agent and sends it the behaviour or evaluation metrics (using Agents Communications Language) to be analyzed at the server-side of the MAS. The Snooper Agent activates the system, whereas the Buffer Agent manages the connection and all the messages from the client. Both tasks are buffered and send them to the Coach Agent. Then the Coach Agent requests to the learner records for the preferences learner: trajectory, results, previous learner monitoring information, etc. The Coach Agents analyzes this information to determine if the learner needs help. If this situation is true, the Coach Agent requests to the learning resources needful via learning content (URLs), and it sends the learning contents (URLs) to the Learner Records. The Delivery Agent via Coach Agent sends the learning content to Evaluation Agents for it is presentation. The Evaluation agent employ the dynamic sequencing to change the course or assessment sequence, and modify the navigation SCORM original via the *imsmanifest.xml*, creating a specific learner sequence customize via a file *learner_n_customize_imsmanifest.xml*. The sequencing is defined for the instruc-tional strategy based on CM and it employs the SCORM Sequencing/Navigation via IRLCOO. Once the necessary information is received, this is represented via one view as feedback, which is constructed dynamically by the rule-based inference engine known as JENA [6] and JOSEKI server, to generate dynamic feedback [7].

5.6.1 Java Agent Development Framework

This work used Java Agent DEvelopment framework (JADE), possibly the most widespread agent oriented middleware in use today [10]. The framework facilitates the development of complete agent based applications by means of a Run-Time Environment (RTE), implementing the life cycle support features required by agents, the core logic of agents themselves, and graphical tools. JADE is a completely distributed middleware system with a flexible infrastructure allowing easy extension by means of add-on modules. JADE is written completely in Java, it benefits from the enormous set of language features and third party libraries, allowing developers to construct JADE SMA with relatively minimal expertise in agent theory. JADE implements the complete Agent Management specification including the key services of Agent Management System (AMS), Directory Facilitator (DF), Message Transport Service (MTS) and Agent Communication Channel (ACC).

The agents execute several behaviours concurrently. It is important to note that the scheduling of behaviours in an agent is not preemptive, but cooperative. This means that when a behaviour is scheduled for execution its *action() method* is called and runs until it returns. Each agent has a message queue where the JADE run time posts messages sent by other agents. Whenever a message is posted in the message queue the receiving agent is notified.

With the basic features of the JADE platform implement real world applications by means of only these is in general terms complex. For our application, it was necessary to use the advanced features of the JADE platform that are presented next. These concern the manipulation of complex content expressions by means of ontologies and the

Semantic Web Platform. The ontology tier is devoted to define the semantics for establishing the usage of words and terms in the context of the vocabulary [24].

The overall architecture of Semantic Web Platform, which includes three basic engine representing different aspects, is provided in Fig. 5.10.

1. The query engine receives queries and answers them by checking the content of the databases that were filled by info agent and inference engine.
2. The database manager is the backbone of the entire systems. It receives facts from the info agent, exchanges facts as input and output with the inference engine, and provide facts to the query engine.
3. The inference engine use facts and ontologies to derive additional factual knowledge that is only provided implicated. It frees knowledge providers from the burden of specifying each fact explicitly.

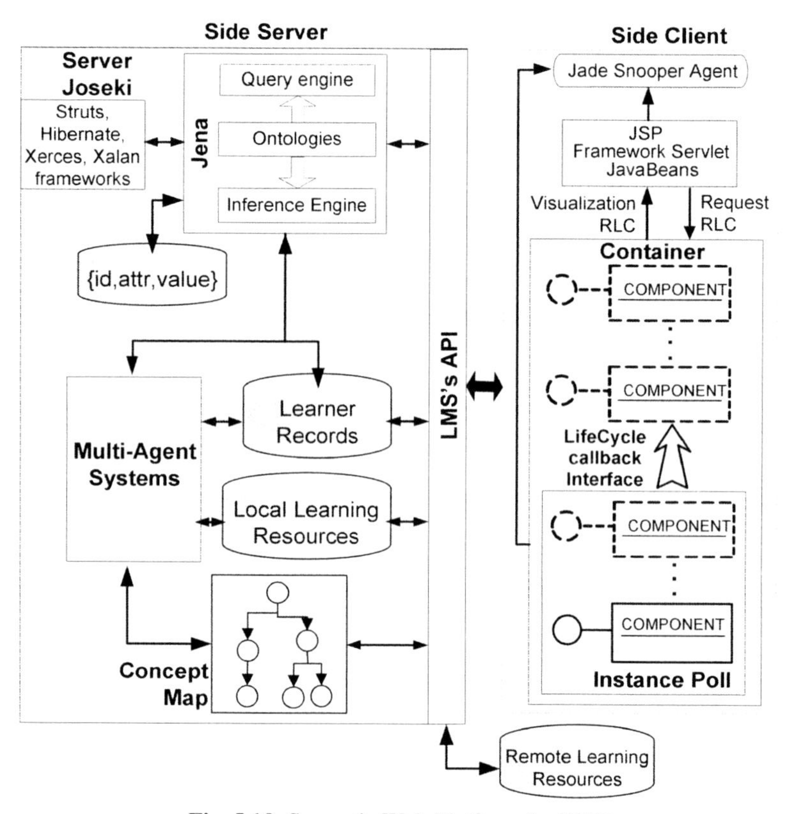

Fig. 5.10. Semantic Web Platform for WBE

The ontologies are the overall structuring principle. The info agent uses them to extracts facts, the inference engine to infer facts, the database manager to structure the database, and query engine to provide help in formulating queries. JENA was selected as the inference engine. It is a Java framework for building Semantic Web applications. It provides a programmatic environment for RDF, RDFS and OWL, SPARQL and includes a rule based inference engine [6]. While JOSEKI was selected as Web API and server. It is an HTTP and SOAP engine supports the SPARQL Protocol and the SPARQL RDF Query language. SPARQL is developed by the W3C RDF Data Access Working Group [7].

5.6.2 Generation of Learning Materials

The components IRLCOO generated by the tutors using SiDeC and Evaluation systems for learning materials are shown in Fig. 5.11. (1) The tutor use an interface (template) to develop the learning content. Basically, these systems generate: XML behaviour files for the IRLCOO components, RDF metadata and SCORM for the learning content. The learning content is structured as dynamic SCO, for later they are used by the LMS of ADL [3]. (2) The learning content is generated by the middleware and the components IRLCOO. Afterwards, it is deployed for the Learner agent. (3) Allowing to capture the metrics through of the Learner agent in RTE and to be adaptive to the learner's true necessities. (4) Then, the LMS takes the control of IRLCOO components. (5) When the learner has completed a learning unit, the components deposit the results in the Learner Records. (6) The Virtual Coach agent

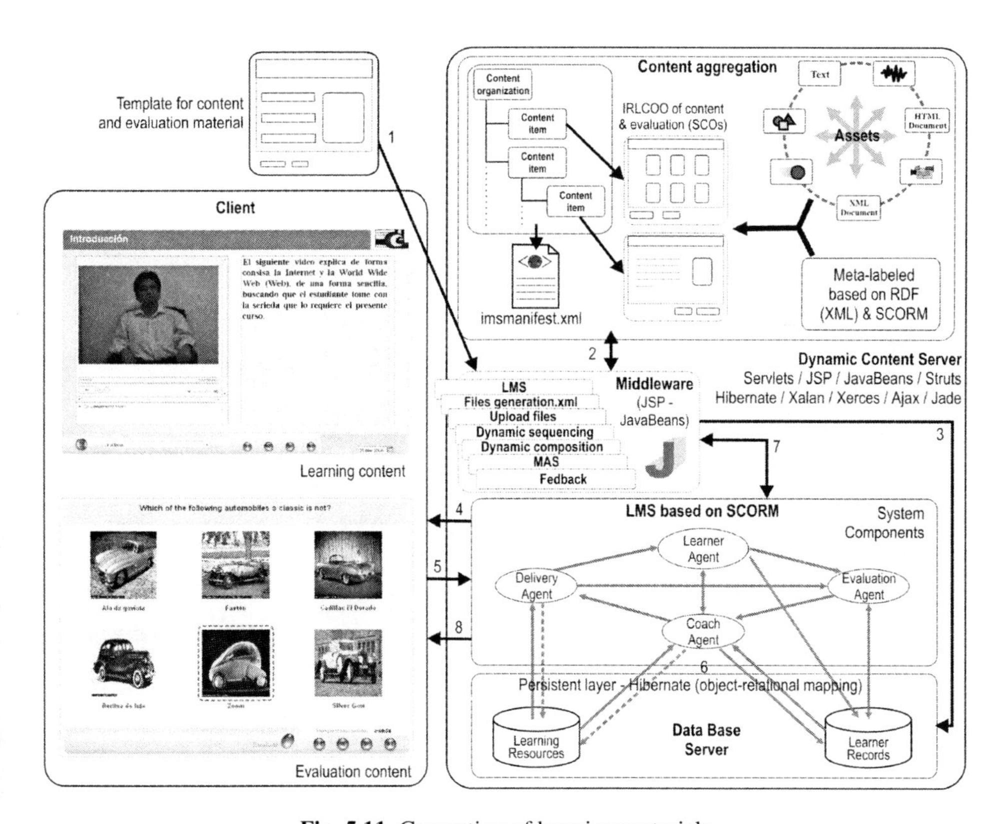

Fig. 5.11. Generation of learning materials

analyzes the results and requests to reconfigure the *learner_n_imsmanifest.xml* file for customize it to the learner's necessities. (7) By means of the middleware, a new learning trajectory for the learner is generated. (8) The new learning content (trajectory) is presented to the Learner.

5.7 Resources

As relevant standard available in the WWW for consulting, it is recommended:

— IEEE 1484.1/D9 LTSA [1].
— The Foundation for Intelligent Physical Agents (FIPA) specifications [8].
— Resource Description Framework (RDF) [4].

Next it is presented a download resources list used for developing our Web-Based Education system:

— SCORM 2004 3rd Edition Sample Run-Time Environment Version 1.0.2 [25].
— JADE (Java Agent DEvelopment Framework) [26], it is necessary register to access to this Web site.
— Jena is a Semantic Web Framework for Java [6].
— Joseki is a SPARQL Server for Jena [7].
— Struts is a Framework for developing Java Web applications [27].
— Xerces Java Parser [28].
— Xalan Java is an XSLT processor for transforming XML documents into HTML [29].

5.8 Conclusions

SiDeC, Evaluation system, FAQ, Chat, etc. were created under the Agents and Components Oriented Architecture for the development of adaptive and intelligent WBE systems. This architecture is integrated by: IRLCOO Components, MAS and Semantic Web Platform. Our approach focuses on: reusability, accessibility, durability, and interoperability of the learning contents, which are built as SCOs using IRLCOO, as the main concept for delivering learning and evaluation content.

Our communication model improves the LMS communication API through of the use of some technologies such as: Ajax, Struts Framework, Hibernate, IRLCOO, WS, Semantic Web, and server JUDDI. This model provides new development capabilities for WBE systems, because it integrates complementary technologies. The different systems were developed under this model to help in the communication and complexity reduction of the learning contents process.

The incorporation of Web Semantic Platform and the modifications to IEEE 1484 LTSA standard help us to create intelligent and adaptive systems (bidirectional communication), according to the learner's true necessities in Run-Time. Everything thanks to the collection of certain learner's metric in an automatic way.

The ADL-Schema manages dynamic sequencing, composition, content separation, and navigation in RTE for development learning and evaluation content. While, our architecture has the same ADL advantages and adds the capacity of generate desk and Web tools using the same learning and evaluation components generated.

Adaptivity in WBE systems has been proved very important, particularly in the distance learning scenarios where direct contact between tutor and learner is not ensured. A survey of various techniques, used in such systems to provide adaptivity, is presented before discussing one such implementation in SiDeC and Evaluation system projects. These projects have a major thrust at providing an integrated system for the management, authoring, delivery and monitoring of such material. They are based on a sound research foundation and are being developed in a manner that provides simple-to-use interfaces in order to reach out to the educators in developing archives of indexed material that may be accessed adaptively by the learners in the same way as they would interact with a couch or tutor.

The Jadex BDI reasoning engine allows the development of rational MAS using mental notions. The engine is specifically designed to build on established software engineering principles and practices as an independent layer that can be flexibly deployed on middleware platforms such as JADE. The Jadex BDI reasoning engine enables the construction of complex real-world applications.

References

1. IEEE 1484.1/D9 LTSA: Draft Standard for Learning Technology - Learning Technology Systems Architecture (LTSA). New York, USA (2001), `http://ieee.ltsc.org/wg1`
2. Peredo, R., Balladares, L., Sheremetov, L.: Development of intelligent reusable learning objects for web-based education systems. Expert Systems with Applications 28(2), 273–283 (2005)
3. ADL, Advanced Distributed Learning Consortium (2008), `http://www.adlnet.org`
4. RDF specification (2005), `http://www.w3.org/RDF/default.htm`
5. XML specifications (2006), `http://www.w3.org/XML/`
6. JENA (2006), `http://jena.sourceforge.net/`
7. JOSEKI server (2006), `http://www.joseki.org/`
8. FIPA XC00061, FIPA ACL message structure specification (2001), `http://www.fipa.org`
9. Grane, D., Pascarello, E., James, D.: Ajax in Action. Manning Publications, Greenwich (2006)
10. Bellifemine, F., Caire, G., Greenwood, D.: Developing Multi-Agent Systems with JADE, 1st edn. John Wiley & Sons Ltd., England (2007)
11. Holmes, J.: Struts: The Complete Reference, California, USA. Mc Graw Hill – Osborne Publications (2004); Edited by Herbert Schild
12. Peak, P., Heudecker, N.: Hibernate Quickly. Manning Publications, Greenwich (2006)
13. Simos, D., Klingler, C., Levine, L., Allemang, D.: Organization Domain Modeling (ODM). Guidebook –Version 2.0., Technical Report, Lockheed Martin Tactical Defense Systems, USA (1996)
14. Adobe (2008), `http://www.adobe.com`

15. Wang, A., Qian, K.: Component-Oriented Programming, pp. 3–5. John Wiley & Sons, Inc., Publication, Georgia (2005)
16. Gamma, E., Helm, R., Johnson, R., Vlissides, J.: Design Patterns Elements of Reusable Object-Oriented Software, 1st edn. Addison-Wesley Pub. Co., Reading (1995)
17. Sanders, W., Cumaranatunge, C.: ActionScript 3.0 Design Patterns. O'Reilly Media, Inc., Publication, Sebastopol (2007)
18. Newcomer, E., Lomow, G.: Understanding SOA with Web Services. Addison Wesley Professional Publication, USA (2004)
19. Pokahr, A., Braubach, l., Lamersdorf, W.: Jadex: A BDI Reasoning Engine. Programming Multi-Agent Systems, pp. 149–174. Kluwer Academic Publishers, Dordrecht (2005)
20. Busetta, P., Howden, N., Ronnquist, R., Hodgson, A.: Structuring BDI Agents in Functional Clusters. In: Jennings, N.R. (ed.) ATAL 1999. LNCS, vol. 1757, pp. 277–289. Springer, Heidelberg (2000)
21. Novak, J., Gowin, B.: Learning How to Learn. Cambridge University Press, Cambridge (1984)
22. Díaz-Barriga, F.: Educational strategies for a significant learning, 2nd edn. Mc Graw Hill Publication, México (2002)
23. IEEE 1484.11.1, Working Draft 14, Draft standard for data model for content object communication
24. Passin, T.: Explorer's Guide to Semantic Web. Manning Publications Co., USA (2004)
25. ADL, SCORM 2004 3rd Edition Sample Run-Time Environment Version 1.0.2 (2008), http://www.adlnet.gov/downloads/DownloadPage.aspx?ID=280
26. JADE, JADE version 3.6 (May 5th, 2008), http://jade.tilab.com/dl.php?file=JADE-all-3.6.zip
27. Struts, Release of Apache Struts Version 1.3.8 (2008), http://struts.apache.org/download.cgi#struts138
28. Xerces, Xerces Java Parser 1.4.4 Release (2005), http://archive.apache.org/dist/xml/xerces-j/Xerces-J-bin.2.6.1.zip
29. Xalan Java, Xalan Java Version 2.7.1, 28. URL: Xerces (2005), http://archive.apache.org/dist/xml/xerces-j/Xerces-J-bin.2.6.1.zip

6

A Resource Discovery Method Based on Multiple Mobile Agents in P2P Systems

Yasushi Kambayashi and Yoshikuni Harada

Department of Computer and Information Engineering,
Nippon Institute of Technology,
4-1 Gakuendai, Miyashiro-cho, Minamisaitama-gun, Saitama 345-8501 Japan
yasushi@nit.ac.jp, c1015367@cstu.nit.ac.jp

Abstract. A peer-to-peer (P2P) system consists of a number of decentralized distributed network nodes that are capable of sharing resources without centralized supervision. Many applications such as IP-phone, contents delivery networks (CDN) and distributed computing adopt P2P technology into their base communication systems. One of the most important functions in P2P system is the location of resources, and it is generally hard to achieve due to the intrinsic nature of P2P, i.e. dynamic reconfiguration of the network. We have proposed and implemented an efficient resource locating method in a pure P2P system based on a multiple agent system. The model of our system is a distributed hash table (DHT)-based P2P system that consists of nodes with DHT (high performance nodes) and nodes without DHT (regular nodes). All the resources as well as resource information are managed by cooperative multiple agents. In order to optimize the behaviors of cooperative multiple agents, we utilize the ant colony optimization (ACO) algorithm that assists mobile agents to migrate toward relatively resource-rich nodes. Quasi-optimally guided migrating multiple agents are expected to find desired resources effectively while reducing communication traffic in the network. Efficient migration is achieved through the clustering of nodes that correlates nodes into a group by logical similarity, and through an indirect communications that are typical of social insects, called *stigmergy*. When an agent finds a resource-rich node, it strengthens the path toward the node so that further efficiency is gained. Strengthening of the route is achieved by pheromone laid down by preceding agents that guides succeeding agents. The numerical experiments through simulation have shown a significant reduction of generated messages.

Keywords: P2P, Multi-agent system, Mobile agent, DHT, Resource discovery, Swarm intelligence, Ant colony optimization.

6.1 Introduction

As the Internet spreads throughout the world, it is used for a variety of human interactions. User interactions across various applications require software that exchanges resources and information within the network community. The traditional client-server model on computer networks barely accommodates the real-world situations such as the advent of video-streaming services. Intensive accesses on a server can easily create a bottle-neck in a network. Peer-to-peer

N.T. Nguyen, L.C. Jain (Eds.): Intel. Agents in the Evol. of Web & Appl., SCI 167, pp. 113–135.
springerlink.com

(P2P) systems can provide a solution to this problem. A P2P system consists of a number of decentralized distributed network nodes that are capable of sharing resources without central servers. Many applications such as IP-phone, contents delivery networks (CDN) and distributed computing adopt P2P technology into their base communication systems. P2P systems provide resource and information exchanges within nodes as peers. A P2P system includes an overlay network where the nodes can interact and share resources with one another. Here, 'resources' means the variety of services that are provided by the network nodes.

One of the most important problems in P2P systems is the location of resources, and it is one of the hardest mechanisms to implement. Napster avoids this problem by using a central server that provides indexing service [1]. Such a server, however, can be the most vulnerable point, where a failure can paralyzes the entire network. Therefore, P2P systems without any central server (pure P2P) are the area of active research in current P2P system developments. We have proposed and implemented an efficient resource location method based on a multiple-agent, pure P2P system [2][3].

In this chapter, we report our experience with a multi-agent system with an enhanced resource location mechanism. The efficiency is gained through a technique inspired by social insects like ants. In order to optimize the behavior of cooperating multiple agents, we integrate the ant colony optimization (ACO) algorithm that assists mobile agents to migrate toward relatively resource-rich nodes. Quasi-optimally guided migrating multiple agents are expected to find desired resources effectively while reducing communication traffic in the network. Efficient migration is achieved through the clustering of nodes into logically similar groups, and through indirect communication that are typical of social insects, called *stigmergy*. When an agent finds a resource-rich node, it strengthens the path toward the node so that further efficiency is gained. Strengthening of the route to a desirable node is achieved by pheromone applied by preceding agents that guides succeeding agents so that they can easily reach that node.

Most current P2P applications use message flooding for resource discovery. Message flooding makes the quantity of messages in the network increase rapidly as the number of nodes in the network increases, and causes saturation easily. It can be said that message flooding creates a problem with scalability. In our previous papers [2] [3], we proposed a method that demonstrated the coordination of multiple agents; the use of a distributed hash table (DHT) and clustering of nodes to enable us to solve the problem of flooding-based resource discovery methods. Through simulation, we have demonstrated the method's effectiveness, i.e. finding the desired contents with a reasonable quantity of messages.

Even though we demonstrated that our multiple-agent based resource discovery system successfully guides the searching agent according to the user's preference, the system did not take into account the difference between the nodes that are altruistic and have many resources and the nodes that are ungenerous and have few resources. Researchers have reported that a large number of nodes in P2P file sharing systems are free-riders [4]. Free-riders are defined as the nodes that have less than 100 files, and queries seldom hit [4] [5]. Therefore sending search agents as well as query messages to such free-riders is futile, and should be avoided. In order to overcome this free-riding problem, we integrate an indirect communication called *stigmergy* into node clustering so that our search agents have a strong tendency to migrate toward altruistic nodes. In this new system, our search agents perform node clustering and are indirectly guided by pheromone to achieve greater search efficiency compared with previous systems.

The structure of this chapter is as follows. The second section describes the background. The third section describes the P2P system we are proposing. Static and mobile multiple agents are effectively working together with stigmergy to find network resources in the P2P system. We also describe the use of a distributed hash table (DHT), clustering and the ant colony optimization (ACO) algorithm that contribute efficiency to resource discovery. The fourth section describes the resource discovery algorithm that uses the multiple agents and the stigmergy to find network resources in the P2P system. The fifth section describes how the system is implemented using an overlay construction tool kit and a mobile agent construction framework. The sixth section demonstrates the efficiency of our algorithm with the results of numerical experiments. The seventh section discusses related works. Finally, the eighth section discusses future work and conclusions.

6.2 Background

The famous pure P2P system, Gnutella, employs message flooding for locating resources [1]. The advantage of such a system is its simplicity, but it is impractical for a large scale network system, because flooding resource discovery messages alone can easily saturate entire networks. In order to solve this problem, the use of a distributed hash table (DHT) is proposed and used [6] [7] [8] [9]. Even though DHT is one of the most promising methods and it certainly provides fast resource lookup ($O(\log n)$ computational complexity) for pure P2P systems, it has the following disadvantages: 1) since the basic mechanism of DHT is mapping keys to nodes, it is hard to find objects based on multiple keys or contents held in objects; 2) it is hard to find multiple nodes that are related to a given key or a set of keys. In other words, DHT based resource lookup methods are too rigid to process flexible and intelligent queries.

In order to mitigate the rigidity of DHT based systems, several P2P systems employ message flooding for object lookup to complement DHT [10]. Since the message flooding causes dense communication traffic, it is not applicable for mobile communication environments where network connections are intermittent.

Since DHT has a serious problem with topology maintenance, the current pure P2P system researches focus on controlled and constrained message flooding techniques. The Gnutella system employs a controlled flooding technique called dynamic querying (DQ), and it is reported that the DQ technique predicts a proper time-to-live (TTL) value to reduce network traffic load [11]. Jian et al proposed an enhanced DQ technique, DQ+, that employs a confidence interval method to provide a safety margin on the estimate of popularity of the searched resources so that it can further reduce network traffic load [12].

On the other hand, systems based on multiple mobile agents are recently popular for various fields [13]. Mobile agents in P2P systems have the following advantages: 1) mobile agents package necessary interactions between nodes and make them local by conveying the necessary processing to the destination where the desired resources reside, 2) mobile agents are asynchronous so that the node that originates the mobile agents can perform completely different tasks or even leave the network temporarily, and 3) mobile agents are autonomous and they can learn about the network as they progress through it [14].

One of the authors has engaged in a project where autonomous agents play major roles in an intelligent robot control system [15] [16] [17]. The mobile agents in the project can bring the necessary functionalities and perform their tasks autonomously, and they have achieved reduction of communications as well as flexible behaviors. Thus, it is natural for us to employ not only static agents but also mobile agents in our P2P system in order to provide flexible search. The mobile agents are expected to reduce the quantity of the query messages.

Thus, we have implemented a resource discovery method that uses mobile multiple agents in a pure P2P system based on DHT. Mobile multiple agents provide much flexibility as well as some intelligence to the DHT base P2P systems. We also integrate clustering of nodes in order to achieve further performance improvement in agent migrations [2] [3].

Furthermore, algorithms that are inspired by behaviors of social insects such as ants to communicate to each other the shortest paths by an indirect communication called *stigmergy* are becoming popular [18] [19]. Upon observing real ants' behaviors, Dorigo et al found that ants exchanged information by laying down a trail of a chemical substance (called pheromone) that is followed by other ants. They adopted this ant strategy, known as ant colony optimization (ACO), to solve various optimization problems such as the traveling salesman problem (TSP) [20] [21] [22] [23] [24]. Schoonderwoerd et al successfully applied ACO to telecommunication (telephone) network load balancing [25] [26].

As a notable application, Di Caro et al proposed a novel approach to adaptive learning in routing tables in communication networks in an application called AntNet, based on ACO [27] [28]. In AntNet, a set of concurrent distributed agents collectively solves the adaptive routing problem. From each network node, mobile agents are asynchronously launched toward randomly selected destination nodes. This happens concurrently with the data traffic at regular intervals. Each agent searches for a minimum cost path joining its source and destination nodes. While moving, the agents collect information about the time length and the congestion status of the followed paths. When the agents arrive at the destination node, they go back to their source nodes through the same paths as they come but in the opposite direction. During the backward travel, the agents update the routing tables in the nodes along the paths with information they have collected during the travel. Thus, they adaptively build routing tables [27] [28]. Di Caro et al successfully showed the ability of artificial ants to optimize the network configuration.

While AntNet's application builds routing tables for a connectionless network such as the Internet, the Anthill framework developed at the University of Bologna [29] [30] is a framework for P2P system. The Anthill framework employs intelligent agents for nodes and resource discovery. Mobile agents called *ants* migrate across the nodes in a P2P network to discover resources and perform distributed tasks. They disseminate information about resources into the network as well as discover the desired resources. The approach is biologically inspired, and evolutionary computations such as genetic algorithms are used for governing the behaviors of mobile agents. Even though it is a framework and does not specify the resource discovery algorithm (making ants intelligent is left for users), the system demonstrates that artificial ants behave well on P2P systems. Thus, it is reasonable for us to integrate the ACO algorithm into our multi-agent system in order to guide search agents toward resource-rich nodes. Integration should provide further efficiency for our multiple-agent based resource discovery system.

6.3 The P2P System

The model of our system is a DHT based P2P system that consists of nodes with DHT (high performance nodes) and nodes without DHT (regular nodes). All the resources as well as resource information are managed by cooperative multiple agents. They are: 1) information agents (IA), 2) search agents (SA), 3) node management agents (NA), and 4) DHT agents (DA). These four agents are the minimum configuration. SAs are the only mobile agents. IAs encapsulate all the interfaces between users and applications that utilize this P2P system so that all the other parts (agents) can be independent from any applications. We separate DA from NA, because only high performance nodes have DHT. When constructing more application-oriented P2P systems, one may add more application-oriented agents. For our purpose, however, we construct only a multiple-agent based framework for P2P systems. The followings are the descriptions of each agent.

1. *Information Agent (IA):* Each node has a static information agent (IA) that manages resource information. IA also interacts with users.
2. *Search Agent (SA):* Upon accepting a user query, the information agent creates a mobile search agent (SA) and dispatches it. Dispatched SA travels through the network to find the requested resources. The SAs also read and write pheromone values as they travel.
3. *Node Management Agent (NA):* Each node has a static node management agent (NA) that has neighbor information based on one or more cluster words and the numbers of resources the neighbors have. An NA has a table that contains the IP addresses of neighbors, correlations with them based on the cluster word, and pheromone values. The pheromone values are determined by the neighbors' correlations and the numbers of resources they have. Each

link to a neighbor has a pheromone value that is initialized at the joining phase and updated through the behaviors of visiting SA. A traveling SA refers to this table in order to determine the node to which it migrates next. The clustering connects related nodes much more tightly; it should increase the possibility of finding the desired resources. The pheromone value increases the attractiveness of those paths to mobile agents.

4. *DHT agent (DA):* DHT agents (DA) construct DHT through cooperation with other DAs that reside on other nodes. Only high-performance nodes have DAs so that we can construct pure P2P systems in a heterogeneous environment. Though current implementation implements Chord [8] as the DHT algorithm, we can replace it with other algorithms just through replacing DAs [6] [7] [9].

In order to improve the efficiency of resource discovery, our method supports clustering of nodes based on cluster words. The cluster words are key words that evaluate the correlations between a node and its neighbors. Clustering makes the logical distance between correlated nodes shorter. Upon joining a node to the network, the user of the system is required to specify key words that represent the resources the node has. The node management agent (NA) then calculates the logical distances (correlations) between it and neighbors. These logical distances are set in the node management table held by NA.

When we join a node to the network, we first decide whether the node has a DHT agent (DA). Then we register the IP address of the joining node to the nearest bootstrap node. The joining node, at the same time, receives the addresses of neighbors, the number of resources and key words that represent resources they have from the bootstrap node and constructs the node management table. NA takes care of this task. The initial pheromone value for each neighbor is calculated the following formula 6.1. a and b are constants and x represents the number of resources the node has and y represents the correlation value calculated by using the cluster words. We will discuss about the cluster words later in this section. The preliminary experiments indicate $a = 1.542$ and $b = 0.000048$ provide reasonably good results.

$$f(x, y) = ae^{bxy} \tag{6.1}$$

Integrating pheromone values is expected to strengthen the connections to resource-rich nodes. Each node usually links to the top four neighboring nodes with high correlation values. At the joining phase, a new node tries to link to nodes with correlation values over forty. If it cannot find such nodes, it gives up temporarily (e.g. for ten minutes), and retries from time to time. We chose the value forty for the threshold correlation value that determines whether a node is linked or not through the results of our preliminary experiments. Choosing a lower value lets search agents move toward free-riders or irrelevant nodes, and choosing a higher value causes search agents not be able to go anywhere. It is an interesting field of research to find the optimal correlation value to connect nodes in a P2P system, but it is out of scope for the current discussion.

The bootstrap nodes are special nodes that are only used by newly joining nodes to get into the P2P system. They are not included in the P2P system, and their sole role is to collect the node information, such as IP address, port number, cluster words and the number of resources of each P2P node that comprises the P2P network. They keep such information in an encrypted format. The bootstrap nodes are only visible to certain special users. The bootstrap nodes do not participate in the P2P network for system security of the P2P network. Addition of authorization and other security features to the bootstrap nodes, and restricting network joining to the nodes authorized by the bootstrap node, are designed to keep malicious users out of the P2P system.

When a node leaves the network, the NA requests that DAs in the neighborhood to erase its entries from their DHT. It also requests other NAs in its neighborhood to erase its entries from their node management tables. If there are any migrating search agents (SAs) in the node, NA makes them leave the node immediately.

6.3.1 DHT

In order to achieve efficiency on discovering resources, we have employed DHT to locate resources directly when explicit keywords are given. DHT is the technology that makes it possible to treat the entire P2P system as if it were one big hash table, and gives the desired resources or the addresses of the resources from related keywords [6] [7] [8] [9]. The most notable advantage of DHT is that it provides $O(\log n)$ computational complexity for finding resources in pure P2P system. Even though it has certain disadvantages as described in the previous section, it is one of the essential technologies for efficient resource discovery systems. We have implemented the algorithm proposed in Chord, one of such technologies, in order to integrate DHT into our agent-based resource discovery system.

In Chord, nodes construct a cyclic list, and the user performs a binary-search-like method to find the desired resources [8]. The reason that we employ the Chord algorithm is that its data structure and algorithm is relatively simple and easy to implement. In Chord, a small amount of routing information is sufficient to implement consistent hashing tables in a distributed environment. Each node has to aware of just its successor node in the cyclic list. Queries for a given keyword are passed around the cyclic list through these successor pointers until it encounters a node that succeeds the key word.

For example, let m be the number of bits of hash value. Each keyword is mapped into this m-bit hash value. Also each node has a m bits node identifier. Each node maintains a routing table with at most m entries and is responsible for only that range of the hash values (of key words) on the cyclic list. One routing table is responsible for nodes that are $successor(x + 2^{i-1})$, where x is the hash value of the node and where $1 \leq i \leq m$.

Thus, each routing table in a node has information about only small number of other nodes, and each node knows more about nodes closely following it on the cyclic list than about nodes farther away. Since the routing table in a node generally does not contain enough information to determine the successors of an

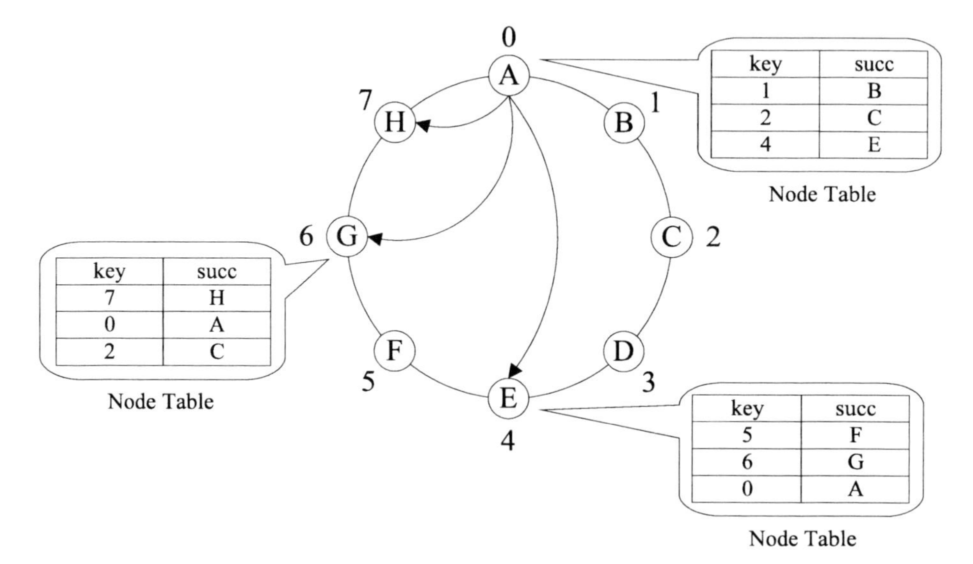

Fig. 6.1. Resource finding in Chord

arbitrary hash value k, it must find a node whose identifier is closer to k. Such a node should know more about the range on the cyclic list in which k resides than the node x does. Thus the node x searches its routing table for the node j whose identifier immediately precedes k, and asks the node j for the node that knows whose identifier is closest to k. By repeating this process, the node with key x learns about nodes with identifiers closer and closer to k.

Fig. 6.1 illustrates an example of the behaviors of the algorithm, where $m = 3$. There are eight nodes ($2^3 = 8$). When node A desires to search a resource with the hash value of a key word 7, it looks up its routing table. We will call this hash value *key*. Since the immediately preceding node with key 7 is node E, it requests node E to find the resource with the key 7. Upon receiving the request, node E looks up its routing table, and finds G is the node that immediately precedes the node with the key 7, and returns the node identifier of node G to node A. Upon receiving the reply, node A sends a query to node G, and receives a reply that node H is the node that is responsible to the resource with key 7. Finally node A can send the request to node H to get the resource. Thus fast resource discovery can be achieved through fast distributed computations of hash functions mapping keys to nodes responsible for them. Of course, one has to update the routing tables when a node joins or leaves the network; a join or leave requires $O(\log^2 n)$ computational complexity.

6.3.2 Clustering

In order to achieve further efficiency in discovering resources, we have employed node clustering by using cluster words. Clustering makes logically related nodes form into a group and shortens their logical distances; thus it is expected to

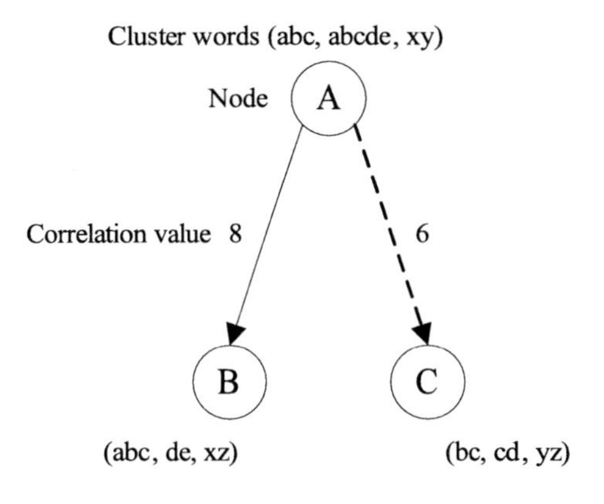

	abc	abcde	xy
abc	3	3	0
de	0	2	0
xz	0	0	0

Correlation value bitween A -B
3+3+2=8

Fig. 6.2. Logical similarity by the cluster words

decrease the number of hops that an SA must migrate. Cluster words are keywords that are used to evaluate the logical similarity between a node and its neighbor nodes. Each node has several keywords that are specified by users when it is created so that each node can evaluate correlations with neighbors and determine any logical similarities.

Fig. 6.2 shows an example of evaluating correlation between two nodes A and B. Node A has three keywords namely, "abc," "abcde," and "xy" and the node B has "abc," "de," "xz." In order to evaluate the correlation, the node makes pairs from each of its keywords and the neighbor's, then the counts the number of matched characters as shown in the table in Fig. 6.2. We can say that the more a pair has matched characters, the more the two nodes are correlated. We define a rule that the contribution to node's correlation value from any two keywords is equal to the number of characters matching in the keyword if and only if one keyword string completely includes the other. That is, for "abc" and "bc" the contribution to the correlation value equals two, but when the two only have some overlapping characters such as "abc" and "bcd", then the contribution to the correlation value is zero. The reason why we set this rule is largely experimental. We need to make some distinction between nodes and we found that counting overlapping characters in the keywords tended to make all correlation values too similar, voiding the utility of the correlation value. In our simulation, this rule works well. We are re-implementing a P2P system on a real computer network, and plan to accumulate more data to improve our clustering methodology.

6.3.3 Ant Colony Optimization

In order to achieve much increased efficiency in discovering resources, we have employed the ant colony optimization algorithm. This optimization technique uses iterative and concurrent simulations that are carried out by a population of artificial agents called *ants*. The artificial ants are expected to generate new solutions to NP-hard combinatorial optimization problems such as the traveling salesman problem. Each ant stochastically performs local search, and builds solutions in an incremental way. Information collected by ants' past simulations is propagated indirectly to other ants, and directs future searches for better solutions. The means of propagating information indirectly is the pheromone.

For example, consider solving the traveling salesman problem (TSP) in the artificial ant colony approach. Each ant builds a solution by using two types of information locally accessible: 1) problem-specific information, i.e. distance among cities, and 2) information added by ants, as pheromone values, during previous iterations of the algorithm. While building a solution, each ant collects information on the problem characteristics and on its own performance, and uses this information to modify the representation of the problem. The modified representation is the new environment shared by all the ants and seen locally by other ants. The representation of the problem is modified in such a way that information contained in past good solutions can be exploited to build new better solutions. This form of indirect communication mediated by the environment is called *stigmergy*, and is typical of social insects [20] [23].

Suppose an ant k (in m ants) is at city i, and N^k is the set of k's unvisited cities. Then, the ant chooses city j to visit next in a probabilistic way as follows:

$$p^k(i,j) = \frac{[\tau(i,j)]^\alpha \, [\eta(i,j)]^\beta}{\sum_{l \in N^k} [\tau(i,l)]^\alpha \, [\eta(i,l)]^\beta} \tag{6.2}$$

Here, $\eta(i,j)$ represents the problem-specific information, and in TSP, it is the reciprocal of the distance between two cities i and j. α and β are non-negative constants, and their values are determined by whether one attaches importance to global information or local information. $\tau(i,j)$ represents the information added by ants, i.e. the pheromone value on the route between two cities i and j. This value is updated in the following rules, when one iteration is done.

$$\tau(i,j) \leftarrow (1-\rho)\tau(i,j) + \sum_{k=1}^{m} \Delta\tau^k(i,j) \tag{6.3}$$

$$\Delta\tau^k(i,j) = \begin{cases} 1/L^k & \text{if } (i,j) \in T^k \\ 0 & \text{otherwise} \end{cases} \tag{6.4}$$

Here, T^k represents the set of all routes from the start city to the goal city, and L^k represents the collective distance from the start city to the goal city. Since the pheromone value put in a specific route is determined by the reciprocal of the total distance, the shorter route has denser pheromone. ρ represents the vapor rate so that newly put pheromone has more influence over old ones. Thus after certain number of iteration, one can expect to obtain quasi-optimal route selection for a traveling salesman.

Upon reflecting on our problem, we can not allow the luxury of such optimal route finding ants, because such probing ants congest the network, and that is contrary to our design. Indeed, our purpose is to reduce redundant query messages produced in flooding-like methods by employing search agents released only on receipt of a query. Therefore an SA in our system adds pheromone to modify the environment for succeeding SAs as well as being guided by the existing pheromone value set by other SAs or itself. While searching resources, when an SA arrives at a node, the NA of that node updates the pheromone value in the node management table as in the following formulae:

$$r \leftarrow r + f(x, y) \tag{6.5}$$

$$p_{ind} \leftarrow p_{ind} + \begin{cases} (1-r)(1-p_{ind}) & \text{if the SA selects the link} \\ -(1-r)p_{ind} & \text{otherwise} \end{cases} \tag{6.6}$$

r is the initial pheromone value calculated by the correlation value and the number of resources the linked node has, as in the formula 6.1. Then, it is updated as SA visit the node where the pheromone value is kept in that node management table as $r \leftarrow r + f(x, y)$. p_{ind} represents each pheromone value corresponding to each link from the current node. When the SA selects one link to migrate, the pheromone value of that link strengthens, and all other pheromone values weaken. Those pheromone values are kept in the node management table in the NAs. The probability s_{ij} that the link that connects node i to node j is selected is calculated as in the following formula:

$$s_{ij} = \frac{p_{ij}}{\sum_{p_{ind} \in P^i} p_{ind}} \tag{6.7}$$

where p_{ij} is the pheromone value of the link from node i to j, and P^i is the set of pheromone values kept in the node management table at node i.

In the previous system, we only employed cluster words to guide SAs, and SAs sometimes migrated toward free-riders [3]. Integrating pheromone values are expected to prevent such futile migrations. Thus SAs are expected to be guided toward nodes with a high correlation value based on the cluster words and resources of the destination nodes.

6.4 Resource Discovery Algorithm

Our resource discovery method is based on DHT and mobile multiple agents augmented by ACO. When a user requests that the information agent (IA) in the current node locate a resource, the user has to specify the lookup keywords and search terminating conditions such as the number of hops, duration time, and the number of found nodes. The user is also required to specify how IA should behave when the dispatched SA does not return due to some accidents. Fig. 6.3 shows interactions of the cooperative agents to locate desired resources. The resource discovery algorithm that the coordinated multiple agents perform is as follows:

1. IA creates an SA for a specific search.
2. The SA requests the NA to select a neighbor to which it should migrate. When the resource name is available, NA demands DA to give the IP address of the node where the resource resides. If the request is ambiguous and NA

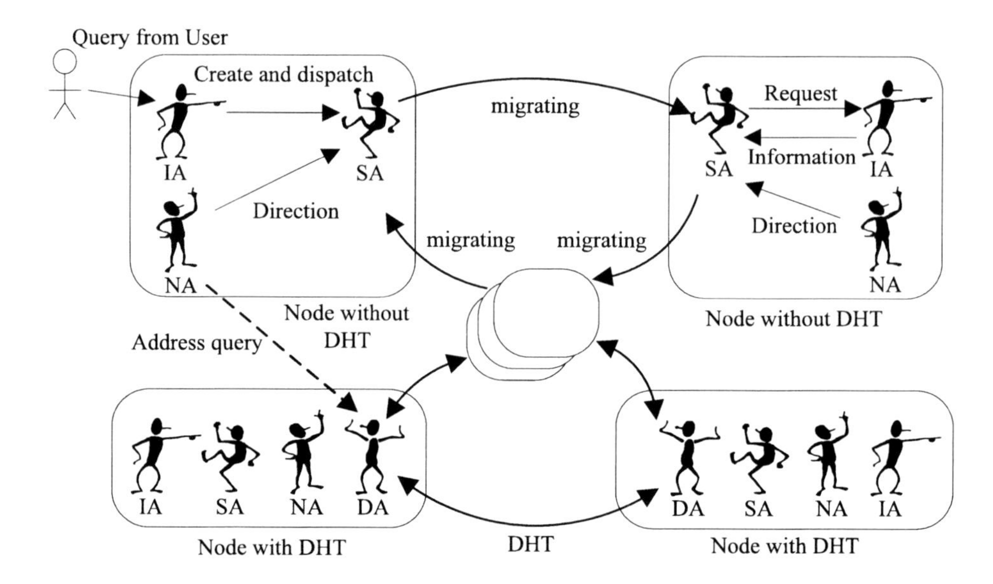

Fig. 6.3. Cooperative agents and resource discovery protocols

cannot tell the exact node to which SA should migrate, then the NA selects a neighbor that is expected to lead toward the desired resource, and gives the SA that address.
3. The SA migrates to the selected neighbor.
4. The SA then interacts with the IA in the arriving node to find whether it has the requested resource.
5. If the node has the resource, the SA stores the resource information and the IP address of the node, then goes to step 6, otherwise repeats at step 2.
6. The SA checks the terminating condition, and if it is satisfied, migrates back to the original node where the SA was created, updating pheromone values on its reverse path.

In step 2, when the NA cannot use DHT due to ambiguity of the resource request, it has to select the neighbor to which the SA migrates. The basic algorithm is to select the logically nearest (the most closely correlated) and resource-rich node. The selection is performed as follows: 1) the NA computes the pheromone values of neighbors with the key words that the SA provides. The computation of pheromone value is performed by the formulae 6.1, 6.5 and 6.6 above; 2) the NA updates the pheromone values in the node management table with the newly computed values; and 3) the NA probabilistically choose one of the neighbors by using the pheromone values of the links to the neighbors. The calculation of the probability is performed by the formula 6.7. Fig. 6.4 illustrates the behaviors of the NA upon request from an SA. Fig. 6.5 shows the flowcharts for the behaviors of the NA.

When the resource discovery request is ambiguous and the IA cannot specify the resource name to use DHT, the SA travels from one node to another to find

Node Management Table

IP address	Port Number	cluster word value	number of resource	pheromone value
64.233.xxx.xxx	50001	40	510	3.48
72.14.xxx.xxx	50010	25	220	1.91
...	...	...	...	...

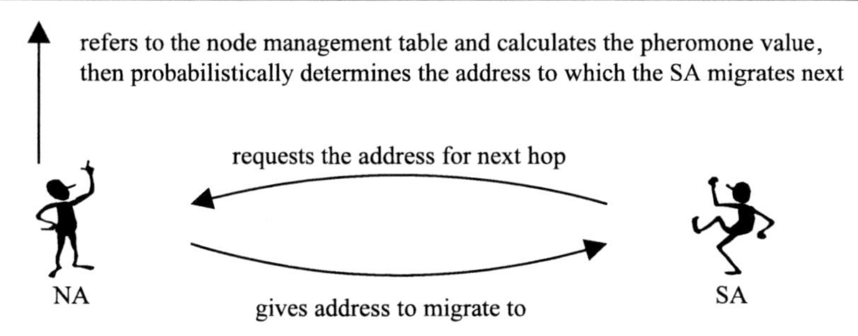

Fig. 6.4. NA updates the node management table periodically and when an SA comes to the node, it probabilistically determines where the SA migrates by using the node management table

the resource name from the resource information in the subsequent node. The NA probabilistically directs the SA to a link that has the strongest pheromone value. The probabilistic selection of the next node is to prevent the NA from sending all incoming SA to the same neighbor, which is not desirable. The NA increases the pheromone value of the link that is selected, and decreases the values of others as in shown in formula 6.6. Even though our ACO does not implement the 'vapor' mechanism included in AntNet's ACO described above, this decrement of pheromone values on all the not-selected links prevents pheromone values from monotonically increasing. Since many SAs move around in a P2P network concurrently, this mechanism provides the same safeguard as 'vapor' and improves routing at the same time.

When a candidate for the resource name is found, the SA returns to the original node and reports it to the IA so that the user can re-issue a new search request. This time, newly created SA may be able to migrate to the desired node with its IP address obtained from DHT. In order to avoid a problem of cyclic migration paths, the node management table is adjusted so that an SA is not sent to the same neighbor.

When an SA finds the right resource that the user wants, it comes back to the node where it is created with the IP address of the target node, but it doe not jump to the node of origin directly. Instead, it goes back to the source node by migrating along the same path but opposite direction just three links before the jump to the node of origin, in order to increase the pheromone values of these three links. Therefore, links around nodes with useful resources automatically have relatively strong pheromones that attract other SAs. This

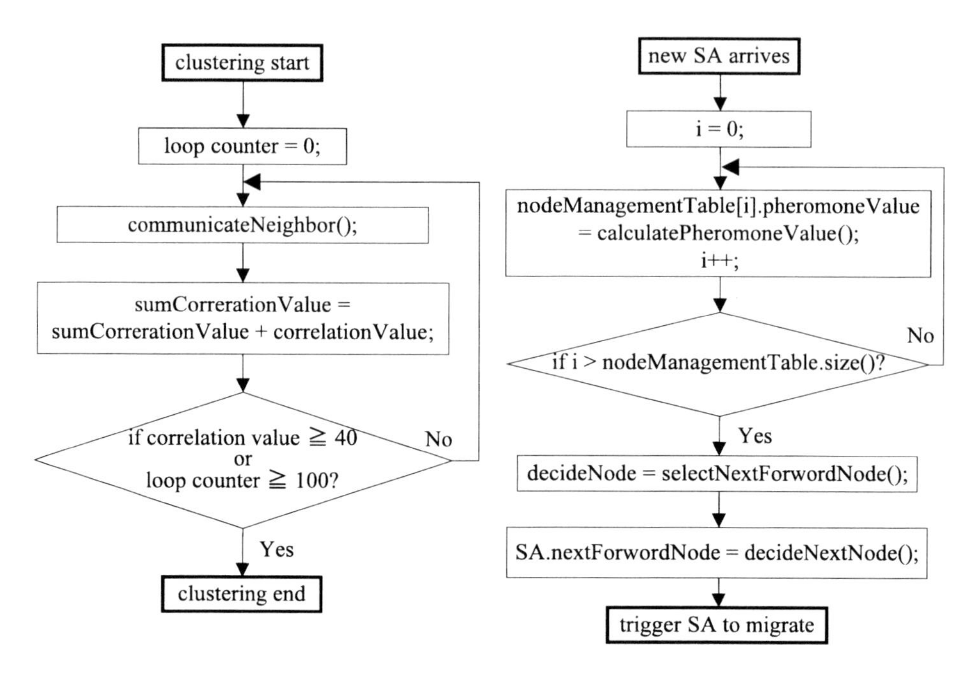

(a) NA periodically performs clustering

(b) When an SA arrives, the NA updates the node management table and then makes the SA migrate to the appropriate neighbor node

Fig. 6.5. Behaviors of NA

mechanism provides additional effectiveness to our resource discovery method. Going backward just three links instead of going back all the way to the node of origin is yet another engineering choice. In AntNet, the agents who find the target go back to their source nodes through the same paths as they come in order to update routing tables. The objective of agents is different, though. Their objective is to build quasi-optimal routing table for the entire network, and ours is to locate specific resources from a specific starting point. On the other hand, we still need to have a system that directs search agents toward nodes that are altruistic and have many resources in ambiguous search. The trade-off point we have found experimentally is to go backward just three links. In the preliminary experiments, that value produces the best result. In real computer networks, the value many vary with the number of participating nodes. That should be yet another research topic.

Since our resource discovery method uses DHT as well as migration of mobile agents, the search agent can migrate to a much farther node directly. Also a node that has the resource information leaves the P2P system during the time that the SA is conducting its search, if the SA gets the resource name before it leaves, the SA can find the desired resource by using DHT. Further, by using DHT, the SA can find multiple nodes that have the desired resource simultaneously. Traditional P2P resource discovery systems using mobile agents lack these advantages.

6.5 Implementation

We have implemented our resource discovery methods by using Overlay Weaver [31] and Agent Space [32], and conducted numerical experiments on the distributed computing environment emulator of Overlay Weaver.

Overlay Weaver is an overlay construction tool kit for the Java language, and provides common APIs for high-level services such as DHT and Multicast. It provides implementations of Kademlia [6], Pastry [7], Chord [8] and Tapestry [9] as routing algorithms, and an emulator for evaluating new algorithms implemented by the user. This emulator can handle several thousand (virtual) nodes and records the number of produced messages and their duration time.

Agent Space is a framework for constructing mobile agents. By using its library, the user can implement a mobile agent environment with the Java language.

We have implemented a P2P network simulator by dispatching an agent system implemented by using Agent Space on the distributed computing environment emulator of Overlay Weaver. The emulator executes our P2P system (a Java application), and controls the system by a specified scenario. Overlay Weaver provides a message-passing mechanism that allows programs to communicate with each other. We have replaced the socket communication mechanism of Agent Space with Overlay Weaver's message-passing mechanism so that we can save memory space and construct larger numbers of nodes. When an agent migrates from one place to another in the original Agent Space, the agent is serialized and then the serialized data is transferred through sockets. In order to replace the socket communication mechanism of Agent Space, we have caused the serialized data to be directly passed to Overlay Weaver's message-passing mechanism. We employed Chord as the DHT algorithm in the distributed computing environment emulator of Overlay Weaver.

In Agent Space, mobile agents are defined as collections of call-back methods, and we then implement interfaces defined in the system. When an agent leaves from a node and arrives at another node, the `leave` and the `arrive` methods are invoked, respectively. In order to kill an agent, the `destroy` method is invoked. The agent also provides service APIs such as the `move` method to migrate an agent and the `invoke` method to communicate to another agent.

The following is an example of SA implementation:

1. Invoke `create` method to create a mobile agent, and specify the search conditions.
2. SA communicates to NA and determines the destination.
3. Invoke `move` method so that the SA can migrate to the specified node.
4. Invoke `leave` method.
5. The agent actually migrates to the specified node.
6. Invoke `arrive` method in the destination node, and the SA communicates to IA in order to receive necessary information.

```
public void requestResourceInfo(Context context) {
  String info = ``'';
  AgentIdentifier[] aids = context.getAgents(``IA'');
  if (aids != null) {
    try {
      Message msg = new Message(``getResourceInfo'');
      msg.setArg(``abc'');
      Object obj = context.invoke(aids[0].msg);
      if (obj != null && obj instanceof String) {
        info = (String)obj;
      }
    } catch (Exception ex) {
      ex.printStackTrace();
    }
  }
}
```

Fig. 6.6. Agent communication

7. Check the terminate conditions; if they are satisfied the SA returns to the original node, otherwise receive the next node address and continue migration.

Step 3 creates a duplication of the SA in the specific node, and step 4 erases the original SA in the original node, so that the SA migrates from a node to the specific destination (step5).

The agents communicate with each other by invoking the `invoke` method. Fig. 6.6 shows a situation where an SA asks IA whether it has a resource named "abc." SA invokes its own `requestResourceInfo` method to communicate with IA in the node where the SA resides, and invokes the `getResourceInfo` method of IA.

6.6 Numerical Experiment

We have conducted numerical experiments to measure the effectiveness of our resource discovery methods. We have compared the number of produced messages by a simple flooding search method and our proposed methods with ambiguous resource names. We have constructed an environment with 4,000 nodes. The simple flooding method employs four message-sending links and makes each query at most seven hops. The number of DHT nodes, which are solely used in our method, is 600. DHT nodes use Chord for their DHT algorithm. Each node has one IA and one NA. We have provided 100 cluster words and each node randomly selects five of them. Each node has ten candidates for SA migration. An SA receives initial information for lookup from IA that interacts with the users, and it moves at most 150 hops. We have randomly chosen five nodes and made IA start a search for randomly located resources.

Using the above framework, we have conducted 10 searches for each resource lookup with three methods, and compared the total number of the transmitted

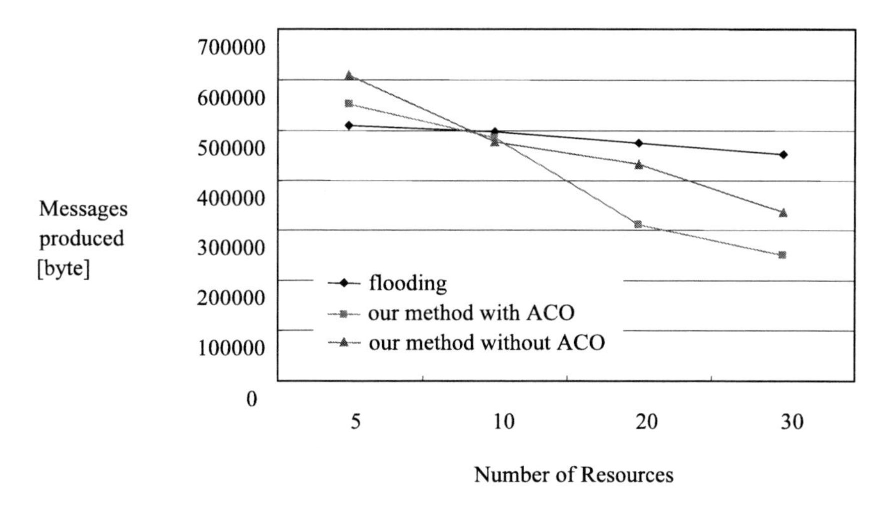

Fig. 6.7. Quantities of produced messages

messages. One is a simple flooding method; another is our method's previous implementation without ACO; and the other is our method's new implementation with ACO. Fig. 6.7 illustrates the results of the experiments for various numbers of resources. The quantity of messages produced by the simple flooding method does not change as the number of resources varies; while the quantity of messages produced by our two methods decreases as the number of resources increases. Even though our two methods display fewer messages produced, our new mobile-agent based search method with ACO is shown to be superior to our previous implementation without ACO. The reason for this is that even though an ACO algorithm produces some redundant messages to adjust pheromone values, it provides more efficiency in searching for resources.

We can observe that, in general, the more resources a network has, the more efficiency our new method with ACO can achieve. We can assume that the number of the generated messages dominates the overall performance, because all the interactions between mobile search agents and local agents are local function calls and therefore have negligible impact on system performance.

6.7 Related Works

The works most closely related to ours are an agent framework for a P2P system developed at the University of Nebraska, Omaha [33], and the ant search algorithm proposed at Institute of Information Science, Academia Sinica [34].

The former framework improves the behaviors of search agents by trails established from previous searches. Similar to our NA, information agents direct search agents by referring them to a node. Unlike our selection, referred nodes are determined by the trails from previously visiting search agents of the same origin, resource name and resource type. It does not rely on routing tables for directing search agents such as is provided by the node management agent in our system. It does not have links with pheromone values. Since there is no trail information initially, the first search agent starts with a blind search. In contrast, our system utilizes cluster words and each link has some initial pheromone value to guide search agents from the beginning. So far, we are not aware of any system that uses not only trail information but also cluster words to compute pheromone values to strengthen or weaken the links of a P2P system. Their agent platform also does not integrate DHT. In contrast, the method described in this chapter integrates DHT for efficient search while suppressing unnecessary communication traffic and accommodating ambiguous searches by using mobile agents.

As in our system, the ant search algorithm proposed at Institute of Information Science, Academia Sinica, employs ACO in order to solve the free-riding problem. As mentioned in the first section, free-riding is a serious problem. Their system, AntSearch, is a new search algorithm that reduces the redundant messages during a query flooding search [34]. In AntSearch, each node maintains a pheromone value that represents its hit rate for a previously received query. Based on the pheromone values, the AntSearch algorithm floods a query only to those nodes that are likely not to be free-riders. Even though the algorithm successfully identifies the free-riders and prevents sending messages to those nodes in order to reduce the message load, its purpose is to reduce the unproductive messages during a query flooding, and does not uses pheromone to direct ant-like search agents for resource discovery; the base architecture is different from our system.

For integrating ACO in our resource discovery algorithm, AntNet and Anthill are the most influential previous research works [27] [28] [29] [30]. In AntNet, a set of concurrent distributed agents collectively solves the adaptive routing problem. From each network node, mobile agents are asynchronously launched toward randomly selected destination nodes. This happens concurrently with the data traffic at regular intervals. Each agent searches for a minimum cost path joining its source and destination nodes. While moving, the agents collect information about the time, length, and the congestion status of the followed paths. When the agents arrive at the destination node, they go back to their source nodes through the same paths as they came but in the opposite direction. During the backward travel, the agents update the routing tables in the nodes along the paths with information that they have collected during their travels. Thus, they adaptively build routing tables [27] [28].

Michmayr has proposed SemAnt, an ACO base discovery algorithm [35], and more closely related to our work. SemAnt adopts the AntNet strategy of forward ants and backward ants for supporting the inherent dynamics of P2P networks where changes in the repositories occur frequently and peers can join and leave with arbitrary timings (churn). SemAnt explores a design where ants create pheromone trails that indicate the most promising paths for a given query with certain robustness in dynamically changing networks. Since Michmayr's purpose is to build robust resource discovery algorithms in dynamic environments, like

AntNet, search agents (ants) go back along the entire path to the node of origin. Although our experiments indicate creating entire pheromone trails produces too many messages and makes the trails too rigid, her numerical findings demonstrate that the algorithm converges fast, and its performance is acceptable with the great benefit of being stable and robust. Her approach is suitable for the inherent dynamics of P2P networks. She pointed out three features of her approach that contribute to build a self-repairing network that can adjust itself to network churn; they are 1) the evaporation feature is built-in by nature and automatically helps to remove outdated trails, 2) contents are not replicated and it is comparatively easy to cope with churn, and 3) the pheromone trails build a global map of the content distribution in the network so that it can adapt to changes in network topology [35]. We share the first two features, and her approach is instructive for our re-implementation of our methods in a real computer network.

More comprehensive information about the applicability of biological processes to distributed environments can be found in Babaoglu et al [36]. They discuss various methods based on the artificial ants and applications of the biological processes of proliferation for search in unstructured networks. As we have mentioned in the second section, Anthill is a notable framework for the design, implementation and evaluation of ant-based algorithms in P2P networks [29] [30].

6.8 Conclusion and Future Works

We have proposed a novel resource discovery method using mobile agents with ACO in a pure P2P system, and demonstrated its efficiency by comparing the quantities of message data transmitted in our P2P system with that of a simple flooding method and our method's previous implementation without ACO. In our proposed methods, static and mobile agents cooperatively perform their tasks to locate requested resources efficiently. In order to reduce communication traffics and at the same time to increase search flexibility, we use mobile agents. In order to improve performance, we also integrate ACO and clustering of network nodes. The pheromone values used in ACO are calculated by the number of resources and cluster words of each node.

We have implemented a P2P system integrating this resource discovery method on a distributed environment emulator in Overlay Weaver. Introducing ACO somewhat increases query message traffic, compared to our previous implementation without ACO, because ACO requires some maintenance of pheromone values of each link. Incorporation of ACO, however, provides more effective search than our previous implementation. Therefore, as the numerical experiments show, our new method with ACO produces fewer messages overall than our method's previous implementation without ACO. Even though the maintenance cost for pheromone values is added, all the maintenance activities are performed by NAs and not by any probe or reconnaissance agents, and therefore the additional overhead is tolerable. Even though the current implementation is not complete, the numerical experiments have strongly suggested the superiority of our new method with ACO.

We have made three design choices for our implementation. The first one is that we chose the threshold correlation value that determines a node is linked or not through preliminary experiments. It is an interesting task for future research to determine how to select the optimal correlation value to connect nodes in a P2P system, and it should be one of our future works. The second one is that we cause search agents who find the target go back three links. In a real computer network where the number of participants varies, the number of links that search agents go back may vary dynamically. This should be another research work. The third choice is the rule for the computation of the correlation value using the cluster words. When we employed node clustering, we defined a rule that a node evaluates two keywords when one includes the other, but does not when the two only partially overlap. The simulation system we have developed shows that all the nodes have similar correlation values when we allow both inclusion and overlapping to contribute to that value. We have to perform more experiments on a real computer network to support the rationale for the rule, and to explore other keyword systems.

We are aware of the incompleteness of our numerical experiments. In our simulation setting, resources are found faster by using a simple flooding method than by using our proposing method even though the simple flooding method produces an enormous number of messages. We are re-implementing our multiple-agent based resource discovery algorithm on a real computer network so that we can measure how much message congestion affects resource discovery time. We also plan to investigate the behaviors of search agents in an environment where they compete for network resources, how network bottlenecks may be increased, and the effect of the resulting queuing problems.

Upon re-implementing our multiple-agent based resource discovery algorithm on a real computer network, we must include some security features. We are prototyping a design in which each node has a *sand box* where mobile agents can visit and are under surveillance during their stay. Since our system consists of multiple agents, access control should be relatively easy to achieve. For example no mobile agents (search agents) can update the node management table. Every search agent who visits a node is only allowed to communicate with the static node management agent. The exchangeable information between them is strictly constrained. Any input/output from the mobile agents are not permitted. Even though it is possible to prevent illegal access from mobile agents through access control, it is difficult to prevent the propagation by mobile agents of untrue information, such as wrong IP addresses of target nodes. The only solution we can propose now is strict control over participation in the P2P system. The Turtle system accomplishes this issue through trust relationships among users [37]. We are designing a more open P2P network. Therefore we need to have some mechanism to detect malicious nodes. Constructing a P2P system that can detect malicious nodes and purge them is an interesting direction that we want to pursue. In that case, some mechanism to evaluate nodes' reputations precisely may be needed. Currently we put a constraint over mobile agents' life time so that unwelcome mobile agents cannot go around in the network too long.

Unlike many other ant-like agent based resource discovery methods, our method should be efficient, because our search agents can be dispatched to the destination directly by using DHT when enough location information is available. Nodes with DHT can be seen as super-peer and we need further investigation about the overall construction of a network with using them, as Yang and Garcia-Molina have done [38]. We may also need to have separate super-peers for efficient use of clustering.

Acknowledgment

We appreciate Kimiko Gosney who gave us useful comments.

References

1. Saroiu, S., Gummadi, K.P., Gribble, S.D.: Measuring and analyzing the characteristics of napster and gnutella hosts. Multimedia Systems 9, 170–184 (2003)
2. Harada, Y., Kambayashi, Y.: Designing a resource discovery method based on multi-agents in p2p systems. In: Proceedings of the IADIS International Conference WWW/Internet. vol. 2, pp. 196–200 (2006)
3. Kambayashi, Y., Harada, Y.: A resource discovery method based on multi-agents in p2p systems. In: Nguyen, N.T., Grzech, A., Howlett, R.J., Jain, L.C. (eds.) KES-AMSTA 2007. LNCS (LNAI), vol. 4496, pp. 364–374. Springer, Heidelberg (2007)
4. Adar, E., Huberman, B.A.: Free Riding on Gnutella, Technical Report, 10. Xerox PARC, Paro Alto (2000)
5. Saroiu, S., Gummadi, K.P., Gribble, S.D.: A measurement study of peer-to-peer file sharing systems. In: Proceedings of the Multimedia Computing and Networking (2002)
6. Maymounkov, P., Mazieres, D.: Kademlia: A peer-to-peer information system based on the xor metric. revised paper from the 1st international workshop on peer-to-peer systems. In: Druschel, P., Kaashoek, M.F., Rowstron, A. (eds.) IPTPS 2002. LNCS, vol. 2429, pp. 53–65. Springer, Heidelberg (2002)
7. Rowston, A., Druschel, P.: Pastry: scalable, distributed object location and routing for large-scale peer-to-peer systems. In: Proceedings of the 18th IFIP/ACM International Conference on Distributed Systems Platforms, pp. 329–350 (2001)
8. Stoica, I., Morris, R., Karger, D.: Chord: a scalable peer-to-peer lookup service for internet applications. In: Proceedings of the 2001 ACM SIGCOMM Conference, pp. 149–160 (2001)
9. Zhao, B.Y., Huang, L., Stribling, J., Rhea, S.C.: Tapestry: a resilient global-scale overlay for service deployment. IEEE Journal on Selected Areas in Communications 22(1), 44–53 (2004)
10. Castro, M., Costa, M., Rowstron, A.: Peer-to-Peer Overlays: Structured, Unstructured, or Both? Technical Report MSR-TR-2004-73. Microsoft Research, Redmond (2004)

11. Stutzbach, D., Rejaie, R., Sen, S.: Characterizing unstructured overlay topologies in modern p2p file-sharing systems. In: Proceedings of Internet Measurement Conference, pp. 49–62 (2005)
12. Jiang, H., Jin, S.: Exploiting dynamic querying like flooding techniques in unstructured peer-to-peer networks. In: Proceedings of IEEE Internet Conference on Network Protocol (2005)
13. Wooldridge, M.: An Introduction to Multiagent Systems. John Willey, New York (2002)
14. Cameron, R.D.: Using mobile agents for network resource discovery in peer-to-peer networks. SIGecom Exchanges 2(3), 1–9 (2001)
15. Kambayashi, Y., Takimoto, M.: Higher-order mobile agents for controlling intelligent robots. International Journal of Intelligent Information Technologies 1(2), 28–42 (2005)
16. Mizuno, M., Kurio, M., Takimoto, M., Kambayashi, Y.: Flexible and efficient use of robot resources using higher-order mobile agents. In: Proceedings of Joint Conference on Knowledge-Based Software Engineering, pp. 253–262 (2006)
17. Takimoto, M., Mizuno, M., Kurio, M., Kambayashi, Y.: Saving energy consumption of multi-robots using higher-order mobile agents. In: Nguyen, N.T., Grzech, A., Howlett, R.J., Jain, L.C. (eds.) KES-AMSTA 2007. LNCS (LNAI), vol. 4496, pp. 549–558. Springer, Heidelberg (2007)
18. Goss, S., Aron, S., Deneubourg, J.L., Pasteels, J.M.: Self-organized shortcuts in the argentine ant. Naturwissenschaften 76, 579–581 (1989)
19. Beckers, R., Deneubourg, J.L., Goss, S.: Trails and u-turns in the selection of the shortest path by the ant lasius niger. Journal of Theoretical Biology 159, 397–415 (1992)
20. Colorni, A., Dorigo, M., Maniezzo, V.: Distributed optimization by ant colonies. In: Proceedings of the European Conference on Artificial Life, pp. 134–142 (1991)
21. Dorigo, M., Gambardella, L.M.: Ant colony system: a cooperative learning approach to the traveling salesman. IEEE Transaction on Evolutionary Computation 1(1), 53–66 (1997)
22. Dorigo, M., Maniezzo, V., Colorni, A.: Positive Feedback as a Search Strategy, Technical Report 91-016. Dipartimento di Elettronica, Politecnico di Milano (1991)
23. Dorigo, M., Maniezzo, V., Colorni, A.: The ant system: optimization by a colony of cooperating agents. IEEE Transaction on System, Man, and Cybernetics-Part B 26(1), 29–41 (1996)
24. Costa, D., Hertz, A.: Ants can colour graphs. Journal of the Operational Research Society 48, 295–305 (1997)
25. Schoonderwoerd, R., Holland, O., Bruten, J.: Ant-like agents for load balancing in telecommunication networks. In: Proceedings of the First International Conference on Autonomous Agents, pp. 209–216 (1997)
26. Schoonderwoerd, R., Holland, O., Bruten, J.: Ant-based load balancing in telecommunication networks. Adoptive Behavior 5(2), 169–207 (1996)
27. Caro, G.D., Dorigo, M.: AntNet: A Mobile Agents Approach to Adaptive Routing, Technical Report 97-12. IRIDIA Universite Libre de Bruxelles (1997)
28. Caro, G.D., Dorigo, M.: Antnet: distributed stigmergetic control for communications networks. Journal of Artificial Intelligence Research 9, 317–365 (1998)
29. Montresor, A.: Anthill: A framework for the design and analysis of peer-to-peer systems. In: Proceedings of the 22nd International Conference on Distributed Computing Systems, pp. 15–22 (2002)

30. Babaoglu, O., Meling, H., Montresor, A.: Anthill: A Framework for the Development of Agent-Based Peer-to-Peer Systems, Technical Report UBLCS-2001-09 (revised). Department of Computer Science, University of Bologna, Bologna (2002)
31. Shudo, K., Tanaka, Y., Sekiguchi, S.: Overlay weaver: an overlay construction toolkit. In: Proceedings of Symposium on Advanced Computing Systems and Infrastructures, 183–191 (2006) (in Japanese)
32. Satoh, I.: A mobile agent-based framework for active networks. In: Proceedings of IEEE System, Man and Cybernetics Conference, pp. 71–76 (1999)
33. Dasgupta, P.: Improving peer-to-peer resource discovery using mobile agent based referrals. In: Moro, G., Sartori, C., Singh, M.P. (eds.) AP2PC 2003. LNCS (LNAI), vol. 2872, pp. 186–197. Springer, Heidelberg (2004)
34. Yang, K., Wu, C., Ho, J.: Antsearch: an ant search algorithm in unstructured peer-to-peer networks. IEICE Transaction on Fundamentals/Commun./Electron./Inf.&Syst. E85-A/B/C/D (1), 1–9 (2007)
35. Michlmayr, E.: Ant algorithms for search in unstructured peer-to-peer networks. In: Proccedings of the 22nd International Conference on Data Engineering Workshops, pp. 142–146 (2006)
36. Babaoglu, O., Jelasity, M., Canright, G., Urnes, T., Deutsch, A., Ganguly, N., Caro, G.D., Ducatelle, F., Gambardella, L.M., Montemanni, R.: Design pattern from biology for distributed computing. ACM Transaction on Autonomous and Adaptive Systems 1(1), 26–66 (2006)
37. Popescu, B., Crispo, B., Tanenbaum, A.S.: Safe and private data sharing with turtle: friends team-up and beat the system. In: Proccedings of the 12th Cambridge International Workshop on Security Protocols, pp. 213–220 (2004)
38. Yang, B., Garcia-Molina, H.: Designing a super-peer network. In: Proceedings of the 19th IEEE International Conference on Data Engineering, pp. 49–63 (2003)

7

Browsing Assistant for Changing Pages

Adam Jatowt[1], Yukiko Kawai[2], and Katsumi Tanaka[1]

[1] Graduate School of Informatics, Kyoto University
Yoshida-Honmachi, Sakyo-ku
606-8501 Kyoto, Japan
{adam, tanaka}@dl.kuis.kyoto-u.ac.jp
[2] Kyoto Sangyo University
Motoyama, Kamigamo, Kita-Ku
603-8555 Kyoto, Japan
kawai@cc.kyoto-su.ac.jp

Abstract. A number of Web agents have been recently proposed. One type of Web agents is an agent assisting users with browsing the Web. In this paper, we present such an agent that supports users during browsing to help them discover new content and better understand and orienteer themselves in the changes occurring in the Web. First, the agent enables finding novel content on visited Web pages by detecting and highlighting the information that is fresh for a user or that recently appeared. This involves the comparative analysis of the previously seen page views and the current state of the visited pages as well as the search in page histories that are reconstructed using data from Web archives. Second, using the historical versions of pages, the agent can create visual, concise summaries of pages' past content within specified time periods. These summaries can be used in order to provide a quick overview of main topics, for example, when the pages are accessed for the first time thus helping the user to better understand documents' main themes and characteristics. In the last part of the chapter we describe the results of an online survey conducted on a group of 1000 Web users. This study was made in order to provide more insight into the needs and expectations of users regarding the evolution of the Web as well as the various kinds of temporal information resulting from Web changes.

7.1 Introduction

The Web is a very dynamic environment where many changes occur continually. Several studies have confirmed this by measuring the frequency of Web changes over time [8,27]. In addition, other studies revealed that many Web users have favorite pages that they frequently re-visit [1,6,10,28]. Usually, such pages not only contain high quality content, but are also frequently changing, since otherwise they would soon have become uninteresting or even obsolete to users.

However, re-visiting pages can sometimes be costly or can be a waste of time. Users coming to a page in search for new content may have problems noticing it especially if the page has a large size and its changes are not easily visible. Often, fresh information may be hidden in the lower levels of a Web site's topology and thus become difficult to be found. For example, many top pages of popular Web sites (e.g. news sites) introduce only short sentences containing links which lead to the novel

N.T. Nguyen, L.C. Jain (Eds.): Intel. Agents in the Evol. of Web & Appl., SCI 167, pp. 137–160.
springerlink.com

content being published on separate pages. Noticing all such changes in the page may be troublesome and may take time, especially, for novice or unskilled users. Users who wish to obtain new content have to access pages one by one in order to check for new information. Such navigation is usually based on users' guess whether the linked pages are worth re-visiting, and may result in incurred costs and waste of time. The solution that we propose is a personalized, freshness-oriented browsing support due to which users are informed about the location and the amount of fresh content on visited pages or sites.

In addition, usually, there is lack of any temporal support related to content's age on pages during browsing. Content is often introduced to pages at different time points. However, for users who access them, it seems as if the content was created at the same time. Many times the users need to know how old certain content elements in pages are. For example, they may inquire since when a given person's name has been listed in a laboratory's home page or a financial statement has appeared in a company's home page. Such kind of temporal information may not be readily available. We propose enhancing browsers with the capability to allow users obtaining information about the age of Web content. This information is produced through efficient search in the repositories of archived data of Web pages. The agent employs a set of binary searches within the past copies of a page stored in Web archives. In consequence, it is possible for the system to approximate the actual age of the whole page and its content elements. The page is then annotated with age-related information so that users can receive a kind of temporal context for better understanding the current page content.

The last function provided is the one for generating summaries of page histories upon users' requests. Usually, users can see only the current content of pages and they have rather limited access to page histories. However, in order to better understand pages and their themes, and characteristics, it is necessary to analyze their past content. Since pages have often many past versions available in Web archives, then viewing them all for comparison and understanding is tiresome. Hence, there is a need for a summary-like presentation of page histories. The summaries that our system generates are presented, both, in a visual form as a series of thumbnails of past page versions and in a textual form as clouds of prevailing and active terms that occurred on pages over time. These kinds of summaries can be used to better characterize the pages. For example, page historical summary could be contrasted with the current content for determining if the latter is consistent with the main page topics.

The proposed agent integrates the three above-discussed temporal functionalities, a) freshness-centric browsing and navigation support for page re-visiting, b) age detection of page content using reconstructed page histories and c) summarization of page histories. It offers a combined temporal support for users during browsing pages which undergo changes over time. In general, we provide a new kind of contextual information that is determined by considering page historical content and user browsing histories. This temporal context is visualized together with visited pages for users to more easily find novel content and to better understand the pages or their current content by correctly locating it on a time scale, and, lastly, to obtain general knowledge on pages' topics and characteristics.

In the last part of this chapter we present the results of the questionnaire that was conducted on 1000 Web users in Japan in February 2008. The questionnaire was online and consisted of several multiple choice questions that respondents had to answer. This study aimed at understanding the needs of users for various kinds of temporal support that could be provided by agents assisting browsing and searching the Web. Through the results of this questionnaire, we hope to be able to offer more effective support and services for users with respect to continually changing Web content.

The remainder of this chapter is structured as follows. The next section provides the description of the related research. Section 7.3 presents our freshness-centric support for page re-visiting, while Section 7.4 discusses the concept and method behind the process of approximate age detection of content on pages. In Section 7.5 we describe the way to generate page historical summaries. In the next section we report the results of a survey study about the preferences of Web users related to utilizing data stored in Web archives as well as needs of users for different kinds of temporal support during browsing. The last section concludes the chapter and provides an outlook on our future research.

7.2 Related Research

In order to set a background of this work we overview the existing research in the areas of browsing agents, user browsing history, page re-visitation and document history navigation and visualization. Lastly, we broadly discuss resources that are used in this research and mention some key, related research initiatives.

7.2.1 Browsing Agents

Many proposals of agents assisting users during browsing the Web have been proposed until now. Letizia [20] is perhaps the best known such agent. It analyzes pages browsed by a user in order to find other related pages through breadth-first search within page's neighborhood. In this way, while the user watches the page, the agent may utilize free resources to analyze neighboring documents for other relevant and interesting content. The recommended documents are then shown in a window next to the main one.

Webwatcher [18] is an agent for guiding users along paths in a collection of documents based on the knowledge of user interests, the location of relevant resources and the typical way in which other users browsed the collection in the past.

Chen and Sycara have proposed WebMate [3] – an agent that learns user interests by tracking its activity in different domains. The modeling of user interests is done by computing sets of *tf*idf* (*term frequency * inverse document frequency*) vectors for different domains of interest.

Navigation-aided retrieval (NAR) [29] is a recent proposal of a system for merging searching and browsing for effective information discovery. After receiving query from users, the system, first, searches for pages that would constitute good starting points for exploration by browsing. Then, it annotates the links on these pages with information of their relevance to users' information needs.

Nadamoto and Tanaka [25] presented a comparative Web browser which automatically searches for similar, related pages to the one that a user watches. Thanks to it the user may obtain additional, complimentary content.

WebGuide [7] is an agent related closely to our proposal. It enables users to compare different versions of pages with respect to two dates. Our proposal is, however, different in many aspects. First, we seamlessly merge browsing and change detection together. Second, we improve navigation in the Web by guiding users to fresh content. Third, our system provides support for detecting age of visited pages and summarizing parts of their histories.

To the best of our knowledge, we are unaware of any work that would propose agents assisting in re-visiting and visiting Web pages by providing them with different kinds of temporal support. However we believe that such agents are necessary considering the high rate of users re-visiting Web resources and the high speed of Web changes.

7.2.2 User Browsing History and Page Revisiting

User browsing history has been frequently used for the purpose of personalization (e.g., [3,18,20]). The pages that a person has visited constitute an easy to use dataset and are an effective measure of the person's interests. Thus browsing history was often used in order to recommend new resources or tailor search performance to meet user needs.

In contrast, this paper proposes using the browsing history of users for the purpose of facilitating understanding of changes in the Web and for helping users to smoothly interact with Web content. Rather than searching for content similar to the previously viewed one that the user might be attracted to, we try to detect the fresh content on re-visited pages. Note that RSS or so-called *current awareness systems* [21] require users to specify the interesting pages or sometimes the type of changes beforehand and, in addition, they often generate high information overload by continuously sending new information. Since many users are accustomed to actively viewing the Web, we propose integrating the change detection and notification agents into standard Web browsers.

The studies of page re-visiting [1,6,10,28] reveal that about 50% to 80% of visits are actually page re-visits. In [28] a detailed analysis was done on how Web users re-visit pages and, especially, what sort of access methods they use. Although, short re-visits (hourly) constitute large fraction of re-visitation, users often re-visit important pages after longer time and their content was observed to change significantly [1,28]. In the most recent study aimed at analyzing re-visited pages [1], the authors found that re-visitation patterns differ due to personal interests, site structure and page content. Pages in the study were categorized into 12 different re-visitation patterns depending on the duration of inter-visit periods. The authors also observed that even quite similar pages may be characterized by different re-visitation patterns.

7.2.3 Document History Visualization

In this section we discuss previous efforts towards providing access points to past page content and, in general, studies aiming at analyzing and utilizing page histories.

Shipman and Hsieh [31] proposed the history navigation of hypertexts on an example of the system called *Visual Knowledge Builder* designed for organizing and interpreting information in hyperspaces. The objective of the authors was to allow readers observing and understanding the ways in which hypertexts were developed, the author's writing styles and the general context of documents.

Wayback Machine [19] and Web Archive Access[1] (WERA) are interfaces to the Internet Archive and Nordic Archive[2], respectively. They allow for accessing past snapshots of pages by displaying them in a directory-like page or by providing a clickable timeline for showing consecutive page snapshots. However, they do not allow for obtaining any aggregate knowledge about past content. Using the both applications a user has to access all page snapshots (or at least majority of them) to have any idea of overall topics or evolution of the page.

In our previous study [11,12] we proposed a past Web browser – a browsing application to Web archives similar to traditional browsers used for the live Web. An important feature was that the browser animated changes between snapshots drawing user's attention to the changed content by showing it as appearing or disappearing depending whether a change was an addition or deletion. The page was processed in this way line by line, snapshot by snapshot, providing the illusion of content transition much like in a slideshow or movie. The control units were similar to those used in VCR players (pause, stop, play and the control of the presentation's speed). Users could use two back and two forward buttons in order to navigate the spatio-temporal structure of the past Web.

Wexelblat and Maes [33] demonstrated a system called *Footprints* that utilizes historical data on user visits to documents for adding a novel social context to various browsed structures. The objective was to guide new users to useful and popular resources. These ideas formed later the basis for successful research area devoted to social navigation and social search.

McCown et al. [23] measured the availability of page copies inside the repositories of major search engines and the Internet Archive. Their research was motivated by the need to provide efficient methods for reproducing the latest versions of Web sites. The objective was to help Web authors with retrieving Web data in case of its sudden loss, for example, due to server crashes. Francisco-Revilla et al. [9] studied how users perceive changes in Web pages as the part of a project called *Walden Paths*, which investigated ways for managing collections of Web resources for educational purposes. The authors identified several key aspects that affect the perceived importance and usefulness of changes.

In addition, there were some efforts made for visualizing the evolution of large Web structures [5,32]. For example, Toyoda and Kitsuregawa [32] proposed displaying the evolution of page communities over time on the case of Japanese Web archive through the analysis of changes in link structure. The objective of these tools and the data used are, however, different from our research.

Generally, the Web archiving community has put relatively little attention to proposing applications that could successfully exploit the accumulated historical data,

---

[1] WERA: http://archive-access.sourceforge.net/projects/wera

[2] http://nwa.nb.no/

despite that the potential advantages resulting from utilizing Web archives have been noticed and emphasized [2,15,22,30].

7.2.4 Resources

Chi et al. [4] introduced an influential theory called "Infoscent" according to which Web surfers forage for information by browsing the Web and their actions are determined by current information needs, and by perceived utilities of links on visited pages. Although, no such theory has been proposed so far for the case of page re-visiting, we assume that the objective of users in re-visiting is a mixture of the need for re-finding old information and the need of finding new content on re-visited pages. Here, one of our goals is to support users with re-visiting pages to help them finding and re-finding information.

In this work we use external repositories of Web pages as well as we locally accumulate data that users generate during browsing. Along with the dramatic growth of the Web, various archiving institutions began large scale preservation projects for saving the Web for future. The Internet Archive[3] [19] is the best known public Web archive containing more than 2 Petabytes of data composed of periodical crawls delivered from Alexa[4] search engine. Other Web archives also exist, such as ones focusing on certain countries (e.g., the Australian Archive[5]) or thematic Web archives (e.g., the September 11 archive[6]). Besides, there are other repositories of past Web data such as local caches, site archives or search engine caches.

In practice, whether archiving document histories is done by specialized institutions (Web archives) or by users due to their activities on the Web, it is generally independent on documents' update patterns. In result, some historical content is lost and, usually, page histories can be reconstructed only to certain extent.

Web archiving continues to be a research area that poses many challenges. Masanes et al. [22] provides a thorough overview of many major issues related to archiving and preserving the Web. In addition, an annual event called the International Web Archiving Workshop[7] provides a forum for exchanging new ideas and experiences among practitioners and researchers in this area.

There have been some initiatives undertaken to predict the potential usage cases of Web archives. For example, the International Internet Preservation Consortium[8] (IIPC), which is a major organization involved in the task of preserving the Web for future, has identified some scenarios[9] in which data accumulated in web archives can be useful for both the average users and specialists. In one such a potential case, a local journalist wishes to study the content of a municipality homepage over a selected time period.

---

[3] Internet Archive: http://www.archive.org
[4] Alexa search engine: http://www.alexa.com
[5] Pandora, Australia's Web Archive: http://pandora.nla.gov.au
[6] September 11 Web Archive: http://september11.archive.org
[7] The International Web Archiving Workshop: http://www.iwaw.net
[8] The International Internet Preservation Consortium: http://netpreserve.org
[9] IIPC's Access Working Group. Use cases for Access to Internet Archives, 2006, http://netpreserve.org/publications/iipc-r-003.pdf

7.3 Temporal Support in Page Re-visiting

Several previous studies revealed that users often re-visit their favorite pages [1,6,10,28]. Due to the high frequency of Web change, the users may have troubles in spotting and understanding changes on visited pages as we have explained before. In addition, sometimes users may even be misled to visit unchanged pages when expecting new content. For example, a page might have a link labeled "new," causing a re-visiting user to think that some new content has been added since her or his last visit. The user might thus access this link only to find that the content is still old from her or his perspective. However, in contrast, some of the content on such a page may actually be new for the user, but, having already visited the page, the user may not visit it again since she or he may believe that the page content has not changed. In both cases, page viewing is ineffective, and the user can become frustrated. In the first case, the user might re-visit the page too often, thereby losing time and incurring costs, while in the second case, she or he might miss content updates. The "new" label on such a page is actually aimed at first time or infrequent visitors or it may be used in order to attract many visitors, despite the content being actually old. Automatically checking whether the content has been seen by the user and visualizing this information should thus result in making browsing more efficient. Here we show how to indicate fresh content on re-visited pages and how to support user navigation for novel content on the Web [14].

The idea behind our approach is quite simple. Each time a user accesses a given page its current view (snapshot) is stored in a local database. Next, when the user re-visits a page, its present version is compared with the one that was most recently visited. Any added content due to this comparison is marked to draw the user's attention by changing its background colour. We use here a difference computation algorithm proposed in [24].

Except for the content comparison of the currently viewed Web pages, there is a mechanism for estimating the degree of freshness of each page to which the current

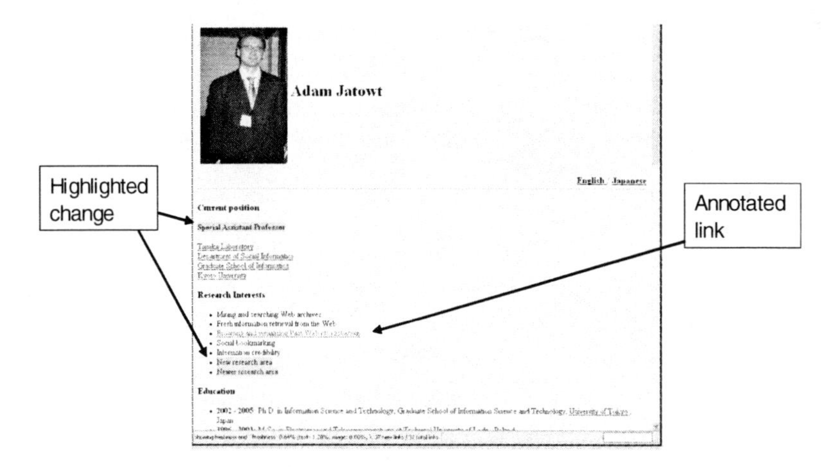

Fig. 7.1. Author's page annotated with information about user-dependent freshness

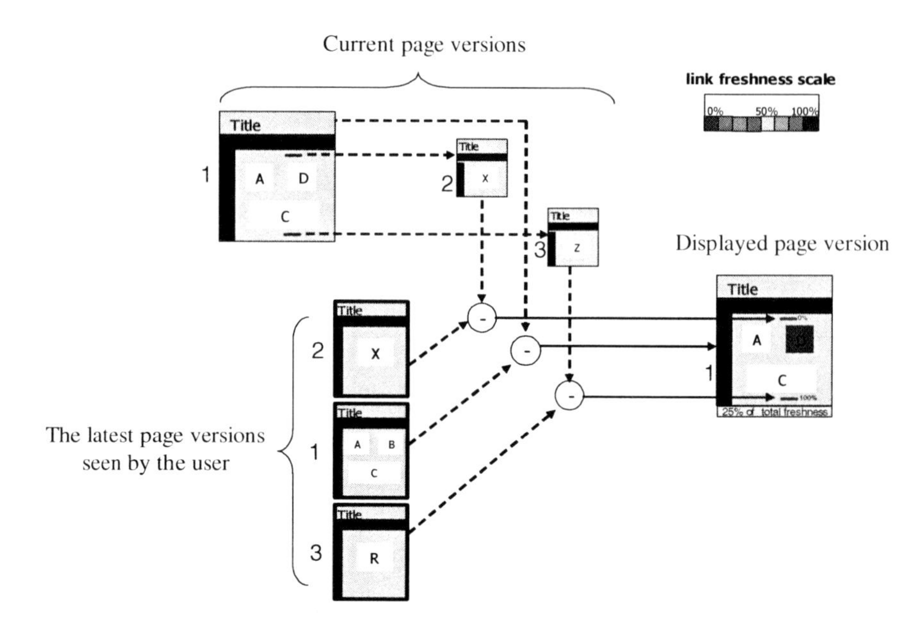

Fig. 7.2. Freshness-centric annotations of page content and links [14]

document links. If the system encounters any linked pages that have been visited by the user on the currently viewed Web document, then it automatically re-downloads the pages for comparison. Freshness degrees are then assigned to the pages and their links are annotated with these degrees. The agent attaches the freshness rates at the end of the link anchors and, at the same time, modifies the link color accordingly (Figure 7.1). The link color modification has been actually used on the Web for some time for the purpose of informing users about pages that have been already visited by them [26]. We modify here the standard link color mechanism in order to achieve finer granularity information. In a traditional link-coloring mechanism there are only two colors, red and blue. The red one indicates a visited link while the blue one denotes the page that has not been accessed by the user. Hence, the objective of this mechanism is supporting the Web navigation on the page level. However, if the contents of the visited page have been completely changed since the user's last visit, the corresponding link will still be displayed in the visited-link color (red). Thus, the user may not become aware of any changes in the content of the page already visited if she or he deems the page not worth re-visiting by just judging its freshness by the color of its link. Our link freshness visualization overcomes this problem by marking the extent to which the page contains new content thanks to using color scale ranging from blue to red; hence, it transforms the usual link color changing mechanism from a page level to the lower granularity level of page content.

The above algorithm is demonstrated in Figure 7.2. We summarize it below.

1. The current version of the accessed page is compared with the latest version viewed by the user that is stored in a database, provided the user has already visited the page before.

2. Any added content has background color changed and a total freshness degree of the page is displayed in the bottom bar.
3. Pages to which the accessed page has links are compared with their latest versions previously viewed by the user (provided there were any).
4. For each linked page that the user has viewed, its added content is identified, and the page's degree of freshness is computed as the ratio of novel content's size to the size of the whole page content[10]. The links to these pages are then annotated with their degrees of freshness[11] (optionally a date of the last visit is also displayed).
5. The pages viewed by the user are stored in a local database during browsing. If the pages have been already visited before, then, their past copies are overwritten with their current versions.

Below, we discuss the operation mechanism of the system in finer detail. The browser contains a "freshness mode" button, which, when clicked, triggers freshness computation and presentation. When the "freshness mode" button is on, the system takes the URL of the current page for comparison with the previously viewed version of the page that is stored in the database. The fresh content on the current page is highlighted, and the total freshness ratio of the page is shown in the bottom bar of the browser. At the same time, the system also fetches all the URL addresses of the links present on the current page and determines whether the linked pages have been previously accessed by the user. If the pages have been visited, then their current content is compared with that of the one in database. Depending on the amount of new content, a certain color is associated to the link based on the defined color scale, and the freshness ratio of the page is displayed to the right of the link's anchor text. Next, if one of the links is clicked, the browser loads the requested page with the changed content marked and displays its freshness ratio at the bottom of the page. If the user returns to the previous page from where the link was followed, the system displays this page with basically the same markings as before. The only change is the update of the color and freshness ratio of the visited link. The system does not re-compute the total freshness of this page as doing so would usually result in no fresh content being detected due to the short time period between consecutive accesses of this page. In addition, the user may not have finished viewing the fresh content and links and would probably have forgotten which ones they were. She or he likely needs more time to view all the fresh content on the page and that on the linked pages before the page freshness is re-computed. Thus, only when the user switches the "freshness mode" button off and on, the page freshness is recomputed again and the fresh content marked. Note that when the freshness mode button is off, the system works as a traditional browser. The only exception is that it preserves the content of visited page.

Using the above algorithm, users can spot which parts of pages are new to them and which linked pages contain large amounts of new content. The system thus facilitates the detection of fresh information in currently viewed pages and, at the same

[10] The size of the content is expressed as the number of terms.

[11] Note that the links for pages that have not been viewed by the user are still being shown in blue.

time, facilitates navigation of the Web for fresh information for users by directing them to new content.

The process can be fine tuned, for example, by setting a minimum time for considering the page content to have been viewed, similarly to the process in some mail programs, such as Microsoft Outlook. Additionally, page content viewed more than a certain period of time ago can be again considered as new or partially new by assigning time-dependent weights to the visited page versions. The freshness of the page content could be then indicated by using graded shades of colours denoting different levels of freshness. Finally, a scrolling-aware mechanism can be implemented to categorize page content into the one seen and unseen by the user.

Freshness rates can be also propagated and accumulated between pages through their connecting links and thus a freshness degree of larger page structures could be determined. In the current implementation we detect, however, only changes down to one level.

7.4 Estimating Age of Page Content

In this section we explain how the system estimates the age of content that users encounter on the Web [13]. Often the origin dates of content parts on pages determine their perception by readers. This is especially true for time-sensitive content such as the one of news articles. For example, dates of news articles inform about their freshness. Although often temporal references are present on pages, for example, in the form of timestamps indicating dates of content being added, in some cases there are no temporal clues on the age of the content. However, the information about the age of content may be required by users in many cases. For example, when readers encounter an important statement about company budget or its future plans on the company Web page they may wonder when this information was added to the page.

When a user requests the age of the content the system performs a search in the history of a page in order to find the approximate dates of content addition. Upon such a request, which is triggered by pressing "start/stop age detection" button, data is retrieved from archiving repositories[12] and the detection of content creation dates is done through the comparative search within the downloaded data. We defer the more detailed description of the search process to Section 7.3.1. When the search terminates, the approximate dates of content insertion are displayed in the form of annotations added to the bottom-right corners of content elements (Figure 7.3). The font size of the annotations is kept at the minimum level guaranteeing its readability, yet at the same time keeping the outlook of pages modified as little as possible.

Those content elements for which creation dates could not be found (due to the insufficient amount of data) are annotated with the expression "before *t_oldest_snapshot*" on the page, where *t_oldest_snapshot* denotes the timestamp of the oldest snapshot available.

The annotations are gradually visualized during the search process so that a user may stop the process at any time. The progress of the search is also displayed at the

[12] In our implementation the Internet Archive, Google, MSN caches and local cache were used as data sources.

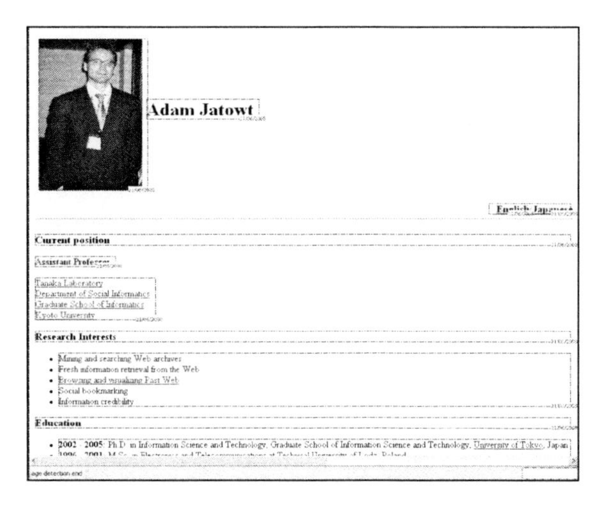

Fig. 7.3. The author's page annotated with the origin dates of content elements

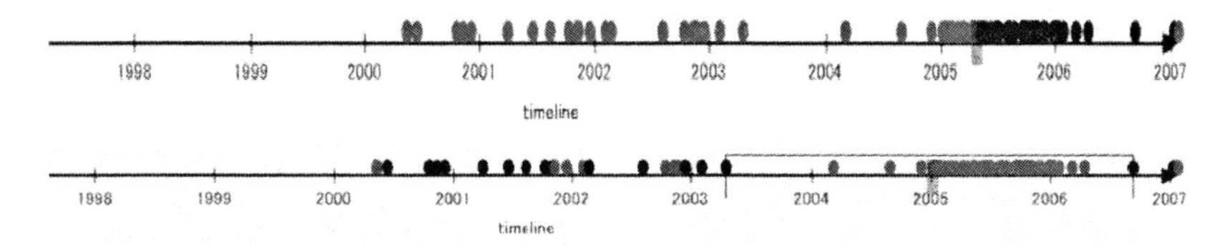

Fig. 7.4. Visualization of search progress for sequential search (top timeline) and for multi-binary search (bottom timeline) [13]

top of the page indicating on timeline those past snapshots that have been already downloaded compared to the total number of available snapshots (Figure 7.4). These implementations are made in order to decrease the time cost that the user may incur waiting for the results. Thus when the creation date of a given content of interest has been already displayed in the page or when there are too many snapshots left to be fetched, the user may stop the process by pressing again the "start/stop age detection" button. Pressing this button once more will make the browser return the original view of the page.

The browser displays also the creation date of the whole page (estimated as the timestamp of the oldest page snapshot) and the average creation date for the whole page content. The latter is computed by considering the relative sizes of elements in pages (Equation 7.1).

$$t_{ins}^{avr} = \sum_{i=1}^{m} \frac{n(e_i)}{N} * t_{ins}(e_i) \tag{7.1}$$

Here, $t^{avr}{}_{ins}$ is the average creation date of page content, $n(e_i)$ is the number of terms inside element e_i, $t_{ins}(e_i)$ is the creation date of the element e_i, N and m are the number of terms and the number of elements in the whole page, respectively. The information about the creation date of the page and the average creation date of its content should help users to more accurately assess the overall freshness of the page.

7.4.1 Search Process

Detecting content creation dates is done by a search for the oldest page snapshots containing the content elements that occur also in the present page version (Figure 7.5). As elements we consider the content of single HTML nodes[13]. The search is done in such a way that the system downloads selected snapshots and compares their content with the one that is in the present version of the page. We define the origin date of a certain content element as follows:

Definition 1. Let $(s_1,s_2,\ldots,s_n)$ be a sequence of past snapshots of a page where s_n is the present version and let $(t_1,t_2,\ldots,t_n)$ be their timestamps. The time point t_k is the creation date of an element within s_n if the element is contained in snapshot s_k while not being contained in s_{k-1} $(t_{k-1}< t_k)$.

In other words, the origin date of content's element is assumed to be the timestamp of the snapshot that is the oldest one to contain the element. The search for such a snapshot requires downloading past snapshots sequentially from a remote repository in order to compare them with the present page version[14]. The comparison is done using change detection algorithm [24] after the downloaded snapshots are stripped from any content that was added by archival repositories[15]. The whole age detection process terminates when all the content elements that occur in the present page version have their creation dates detected.

Figure 7.6 shows the example of the age detection algorithm where snapshots are downloaded sequentially starting from the most recent one. Page snapshots are depicted here as grey rectangles that contain elements represented as colored dots. For simplicity, the present page version is composed of one content element only, which is marked by a yellow dot. Thus, the search is aimed at finding the time point when the element has been added to the page. The figure shows the search cost in terms of search steps, that is, downloaded page snapshots.

Sequentially downloading past snapshots of a page can be, however, time consuming, especially, when the elements in the current page version are old and there are many snapshots. For improving the search efficiency a multi-binary search is proposed. It is a combination of several binary search processes. Each such process attempts at detecting the age of a single element. Since the searches are done on the same set of snapshots, thus the whole process is optimized so that the snapshots are downloaded only once and then used for discovery of origin dates of the remaining elements. First, the system looks for the oldest page snapshot which contains any element that also occurs in the current page version. Then, it searches for creation

---

[13] We focus on text here, however, other types of objects such as images can also have their age estimated.

[14] We assume here that Web archives provide basic metadata of stored snapshots such as their timestamps. This is the case of the Internet Archive.

[15] For example, cached pages in Google search engine have annotations added to their content containing the crawling dates among other information.

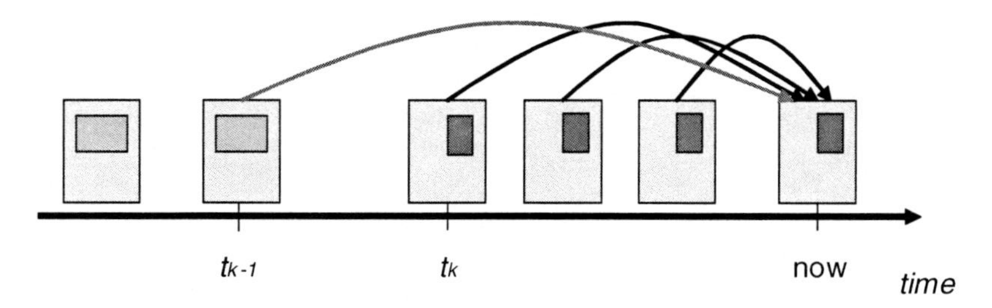

Fig. 7.5. Search process during the age detection (the arrows indicate comparison between past snapshots and the present version of the page; t_k is estimated to be an approximate creation date of an object represented by a smaller rectangle)

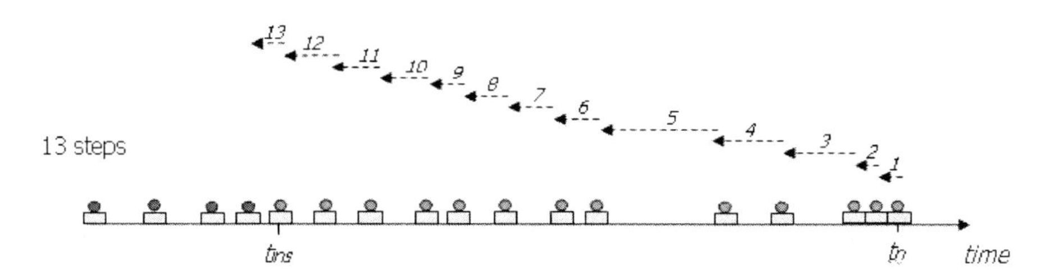

Fig. 7.6. Sequential search for one content element represented by a yellow dot (t_{ins} is the estimated element's creation date; t_n is the present time) [13]

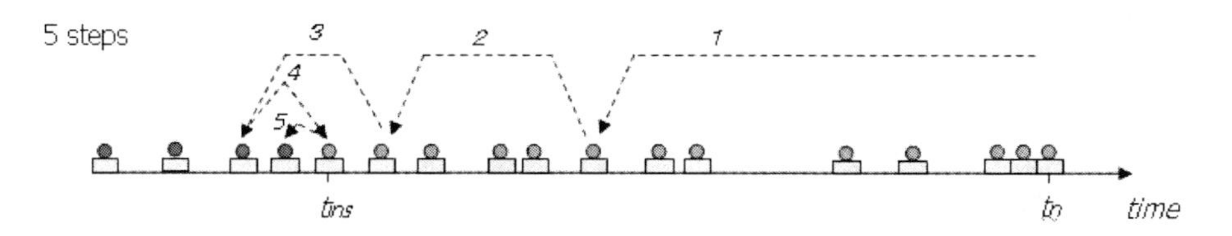

Fig. 7.7. Multi-binary search for one content element represented by a yellow dot (t_{ins} is the estimated element's creation date; t_n is the present time) [13]

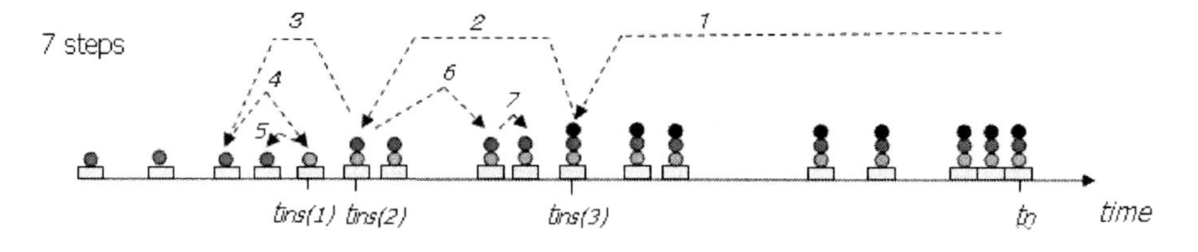

Fig. 7.8. Multi-binary search of three content elements represented by yellow, blue and black dots (t_{ins1}, t_{ins2} and t_{ins3} are estimated creation dates of the elements; t_n is the present time) [13]

dates of other elements utilizing the already downloaded snapshots. A more detailed overview of this search process can be found in [13].

The multi-binary search for the case of a page containing one object is displayed in Figure 7.7, while the case of several elements is displayed in Figure 7.8. By comparing Figures 7.7 and 7.8 with Figure 7.6 it can be seen that the multi-binary search is faster than the sequential one resulting in considerable cost savings, especially, for pages with many past snapshots and relatively old content age.

We need also to mention here the issue of trivial content modifications. Relatively insignificant changes (e.g., grammatical change, word order change) should not be considered as the creation of new content elements. Rather, the system should neglect such modifications and continue the search for detecting the actual insertion dates of the elements. In order to do so, a threshold is set up for each comparison step. If the similarity between two compared elements is higher than the specified threshold value, then the elements are considered to be the same. The threshold is defined here as the ratio of the number of common terms to the number of all terms occurring in both the compared elements.

7.4.2 Discussion

In the above search process the continuous occurrences of elements are assumed in pages. In other words, we neglect cases when a given object was first added and then deleted from a page to be later inserted again. Without this assumption, the multi-binary search method may generate wrong results. Some solution would be to download and analyze all available past snapshots or at least make a sequential download starting from the oldest available snapshot. These approaches can be however costly.

We also need to mention that there is usually some error involved with the age detection process as shown in Figure 7.9 (we thus use the word "approximate" for detected creation dates). This is because the actual time point of an element's insertion into the page lies between t_{k-1} and t_k, where t_k denotes the oldest page snapshot that contains the element. Hence, assigning t_k as the origin date of the element introduces certain error due to data scarcity, the maximum value of which is determined by the length of time period $[t_{k-1}, t_k]$. The maximum relative error is then descried as:

$$\sigma_{\max} = \frac{|t_{k-1}, t_k|}{|t_n, t_k|} \tag{7.2}$$

The amount of page snapshots and the lengths of distances between the consecutive snapshots determine the precision of the age detection process. In this paper we assume a uniform probability distribution of the element's creation event within $[t_{k-1}, t_k]$. However, this distribution could take other shapes in case of some additional information available such as the estimated change frequency of the page in the past.

Finally, we would like to point out that certain content could have been published on other pages and then migrated to the target page. Or, in another case, the content could have been expressed using a different surface form while having same meaning. Providing efficient solutions to these problems is, however, not trivial and is left for future research.

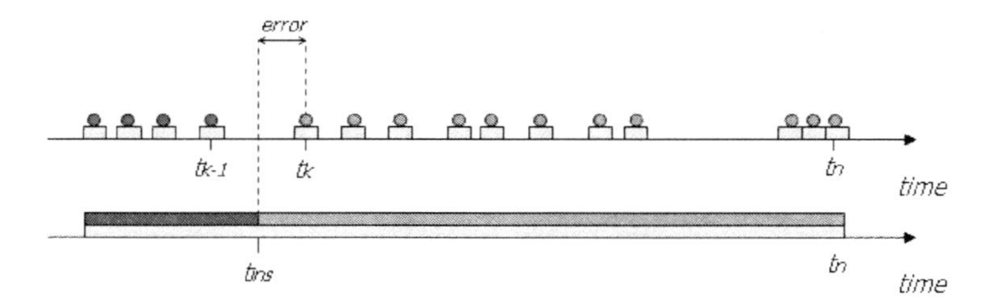

Fig. 7.9. Error in the process of creation date detection due to fragmentary data [13]

Table 7.1. Age detection done on May 21st, 2007 on sample pages (dates are shown in *dd/mm/yy* format; numbers in parentheses indicate running time in seconds) [13]

URL	# steps in sequential search	#steps in multi-binary search	Page creation date	Average creation date	# available snapshots	Creation date of the oldest element	Creation date of the youngest element
http://www.google.com/intl/en/about.html	67 (17s)	23 (15s)	10/05/00	02/12/05	74	15/06/01	22/05/07
http://www.dmoz.org	387 (124s)	28 (93s)	25/01/99	14/06/00	389	before 25/01/99	21/05/07
http://ir.iit.edu/cikm2004	13 (3s)	6 (3s)	12/02/04	05/04/04	13	before 12/02/04	09/10/04
http://www.worldchanging.com	2 (18s)	20 (120s)	06/12/03	01/03/07	367	16/05/07	22/05/07
http://www.yahoo.com	964 (489s)	28 (91s)	17/10/96	28/05/06	1015	03/10/99	21/05/07
http://www.businessweek.com	912 (859s)	65 (444s)	31/10/96	23/03/07	928	15/08/00	21/05/07
http://www.stanford.edu/	631 (231s)	20 (54s)	13/01/97	20/02/07	641	10/02/99	21/05/07
http://www.commonlaw.com	35 (6s)	7 (7s)	12/04/97	08/04/02	35	20/05/98	13/05/07
http://www.delaware.gov	236 (321s)	23 (257s)	18/04/01	27/03/06	302	24/03/04	21/05/07
http://www.state.nj.us	198 (336s)	40 (204s)	06/03/97	14/06/06	689	30/11/04	19/05/07

In Table 7.1 we show the results from a study done on sample Web pages on May 21st, 2007 using a computer equipped with 3GHz processor and 3GB memory. For each selected page we calculated the number of search steps (i.e. the number of downloaded past snapshots) for both search methods, the required time cost, the estimated creation dates of youngest and oldest elements, the estimated creation dates of pages and the estimated average age of content in pages. From the table, it can be seen that, on average, the multi-binary search required fewer snapshots for downloading and it was faster than the sequential search process.

7.5 Summarizing Page Histories

When users wish to see the historical content of pages they may refer to cached copies of pages stored in online Web archives. The Internet Archive provides a directory-like page where past page snapshots can be accessed through the list of links. This means of access is, however, rather impractical for users who would like to have a quick and general overview of content that a certain page had within a given, selected time period. When directly using the data provided by the Internet Archive, users would need to access selected page copies one by one and compare their content to receive some information about the page evolution. Since this is not only time consuming but also imposing quite high cognitive load to users, we propose the creation of historical summaries of pages that, in a visual way, provide users with an overview of page historical content and, in general, page evolution during selected time periods [17].

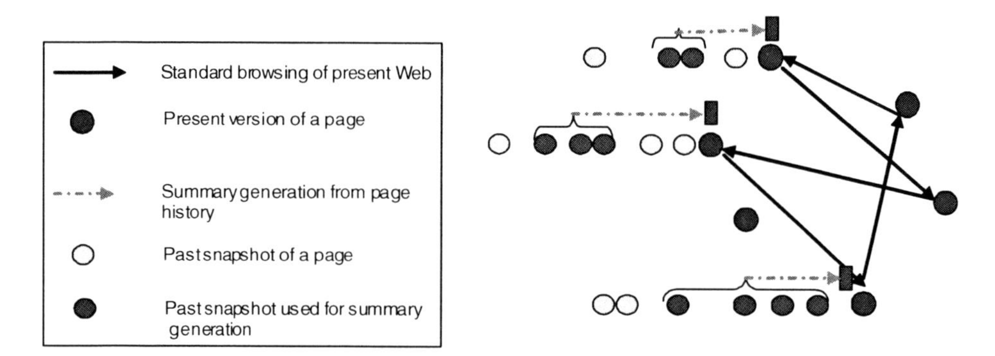

Fig. 7.10. Browsing model in which users can obtain historical summaries of visited pages [16]

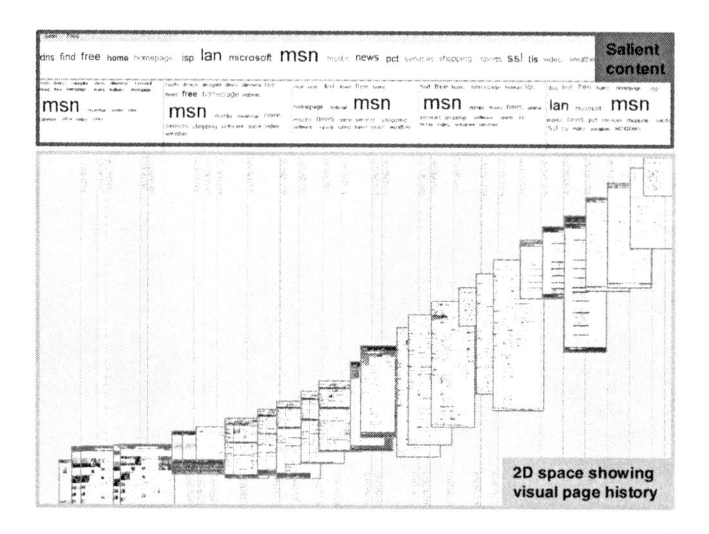

Fig. 7.11. Example of summarized page history (MSN homepage[16])

[16] MSN: http://www.msn.com

When browsing the current Web, users could occasionally request the generation of such historical summaries for visited pages (Figure 7.10).

In Figure 7.11, we show the visual summary generated for an example page. For a user-selected time period a certain number of past page snapshots are withdrawn from a Web archive. They are then mapped on 2D space that represents time and accumulated change degree. This allows for roughly portraying various aspects of page evolution and indicating time periods when large content changes occurred. Second, the system summarizes historical content of pages by detecting prevailing and active terms over time. This kind of a temporal summary is visualized as clouds of salient terms that characterize historical page content.

7.5.1 Methodology

In this section we describe the method for providing overviews of historical content of pages. The system uses term clouds, which are similar to tag clouds, in order to visualize past content of pages. Tag clouds have become a common method for representing popular content in many Web 2.0 applications where tag fonts indicate the levels of content popularity. In the case of a single document, term clouds can be used to provide a quick overview of document's content. We extend the usage of term clouds as a means for generating aggregate information on past content of pages.

We distinguish here two types of temporal term clouds, prevalence- and activity-based term clouds. The former is used to represent the most common content appearing in a page over time. The latter indicates the content that frequently appeared inside the changing (active) parts of the page. In addition, we propose a series of short-term clouds that enable viewing the distribution of salient terms inside shorter time segments.

7.5.1.1 Prevailing Content

Prevalence term cloud is constructed based on the prevalence scores of terms. These inform us how commonly terms occurred over time. The prevalence score of a term is estimated as the weighted average of the term's frequency function over time.

$$S^{pr}(a;T) = \frac{1}{T}\sum_{i=0}^{N-1}(t_{i+1} - t_i) * TF_a([t_i, t_{i+1}]) \tag{7.3}$$

$TF_a([t_i,t_{i+1}])$ is the estimated frequency of a given term a in the time period $[t_i,t_{i+1}]$ where t_i and t_{i+1} are the timestamps of the pair of consecutive page snapshots. In a simple approach, we assume $TF_a([t_i,t_{i+1}])$ to be equal to $1/2*(TF_a(t_i) + TF_a(t_{i+1}))$ where $TF_a(t_i)$ and $TF_a(t_{i+1})$ are the frequencies of the term a in the page at t_i and t_{i+1}, respectively. T is a user-selected time period of page history and N is the number of downloaded page snapshots that have timestamps within T. By default N is equal to 30 but it can be set by users.

According to Equation 7.3, the longer are the time frames during which a given term had a relatively high frequency in the page, the higher is its prevalence score. Thus, if the term occurred frequently in the page over lengthy time frames, it is considered to be prevailing in its history.

7.5.1.2 Active Content

Prevalence scores do not distinguish between static and changing content in pages. Rather, they only reflect the general level of occurrence of terms, without the distinction whether the terms were static or rather frequently added or deleted over time. Thus, if a given term (e.g., a term "copyright") appeared always in the static content parts, it may not be of much interest to users. On the other hand, terms that were often added and deleted may indicate important content. Hence, representing "active" content in pages can be also interesting. We calculate the activity scores of terms in order to construct the activity term cloud.

The activity score of a term a within the user-selected period T is computed as the probability that the term occurred in added or deleted content parts in the past. This likelihood is estimated as the combination of the probability of a change occurrence in T and the probability of the term a appearing in this change.

$$S^{ac}(a;T) = \frac{M}{T} * TF_a^{ch}(T) \tag{7.4}$$

M is the number of snapshots with any content changes and $TF^{ch}_a(T)$ is the average frequency of term a calculated over the changed parts (added and deleted content) inside all snapshots.

7.5.1.3 Unit Term Clouds

In addition, we calculate term clouds for a series of unit time segments within T in order to enable a finer granularity analysis of past content.

The calculation of term scores in this case is derived from a well-known *tf*idf* weighting scheme which is adapted to time series data. First, the page history is divided into a series of equal time segments. Scores are then computed for all terms appearing in each time segment. This is analogous to *tf*idf* calculation with the difference that a single document is represented here by a unit segment of page history. The sequence of such segments corresponds to the collection of documents. In general, we score terms in each unit segment according to the following rule. Salient terms in a given time segment T_w are terms that have high scores inside T_w and, at the same time, have low scores inside other time periods (Equation 7.5).

$$S^X_{unit}(a;T_w) = X(a;T_w) * \log\left(\sum_{j=1}^{R}\left[\frac{X(a;T_w)}{X(a;T_j)+1}\right]+1\right) \tag{7.5}$$

Depending on user's choice, $X(a;T_w)$ denotes either prevalence or activity term scores. R is a user-specified number of unit time segments.

7.5.1.4 Visualization

Selected page snapshots after being downloaded from the Web archive are converted to thumbnail images and overlaid on 2D space. The horizontal axis represents time distance and the vertical one indicates the cumulative degree of added change (Figure 7.12). Added change degree is calculated as the number of terms inside content added to past page snapshots. The distance of snapshot s_{i+1} from snapshot s_i is the combined distance in temporal and change-degree dimensions. Note that the

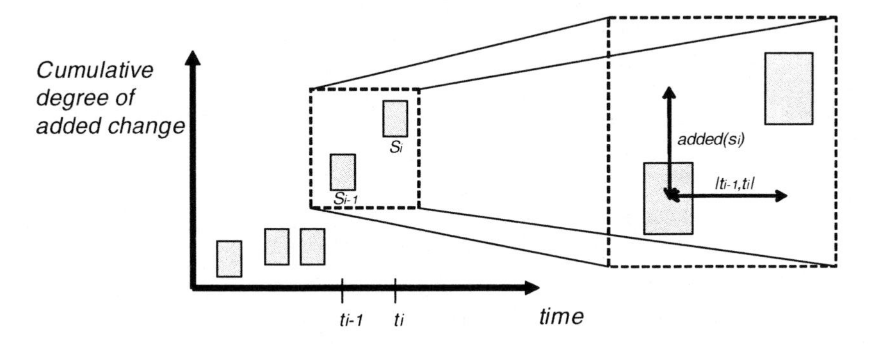

Fig. 7.12. Visualization of page snapshots on 2D space

vertical axis shows the cumulative change of the page since the beginning of the analysis period.

With this kind of visualization, a user can roughly grasp the outlook of a page at different time points in the past. Also, since the relative sizes of pages are retained, thus users can compare page sizes at different time points. More interesting, however, is the capability to portray the frequency and the amount of changes occurring over time. For example, a relatively long distance in the vertical dimension may be quite surprising when accompanied by a short horizontal distance between snapshots. This could imply a sudden burst of new content. By looking at a change distribution over time users can spot the time periods with large additions of new content. In contrast, the lack of any content changes in longer time frames translates to the higher probability of content obsoleteness or even page abandonment.

In order to characterize the past content for the whole analyzed period the system displays top 20 terms calculated according to their prevalence or activity scores. The font sizes of the terms reflect their scores. In addition, the sequence of unit term clouds is visualized directly over a 2D space. Each such unit term cloud contains top 20 terms calculated according to Equation 7.5. Unit term clouds indicate the top terms for shorter time frames and can be contrasted with the main term cloud computed over T. Users can change the number of unit term clouds in real time by adjusting parameter R[17].

7.6 User Survey

In this section we report the results of the survey that we had done regarding temporal aspects of the Web and user needs for access to document histories [16]. The study aimed at understanding the needs of users and their willingness to have the kind of support in browsers that we propose. The questionnaire was made between in February 2008 on a group of 1000 Internet users in Japan. Subjects were divided into equal categories of 250 respondents depending on their age: 20-29, 30-39, 40-49 and 50-59 years old. In each category, half respondents were males and half females. For

[17] By default R is made equal to $N/4$.

completing the survey the respondents received certain financial gratification. The survey was done in Japanese language and we show here translated results.

In Figure 7.13, we show the results on an analysis aimed to find out how many users use Web archives in comparison to other types of Web content. The percentage of users that use any Web archive, such as the Internet Archive, at least once in a month is actually quite low. Only 19 respondents confirmed that they do so (1.9%). To some extent, the reason for this may be the lack of publicly open Web archive in Japan. In general, many users seem not to be aware of the existence of Web archives.

In Figure 7.14, we show the answers to the question on temporal context that users would like to obtain when encountering information on the Web. There were 7 potential answers provided and the subjects had to choose 3 most important to them. From the figure, we can see that the recency of the information is the most important temporal factor that users pay attention to. In total, 66.8% of the respondents selected it as their first choice. The age of content on a page was another frequently chosen aspect (12.8% users selected it as their first choice, while 26.5% and 23.5% selected it as their second and third choice, respectively). The popularity of the information on the Web and its evolution were less important to users.

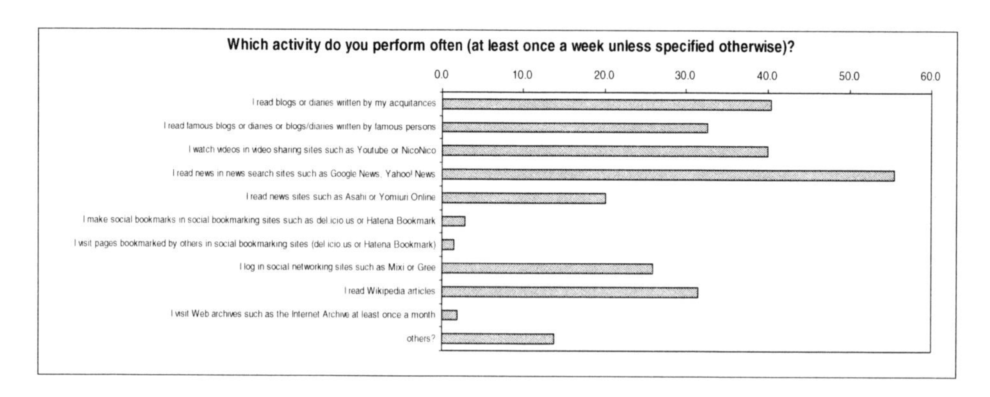

Fig. 7.13. Question about frequent user activities on the Web [16]

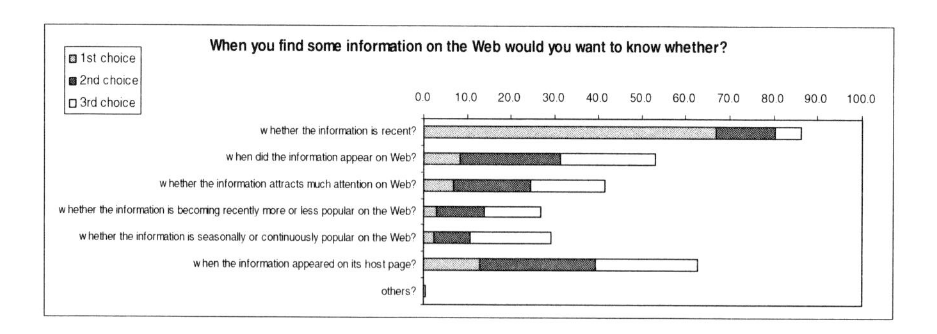

Fig. 7.14. Question related to the types of temporal support for information found on the Web [16]

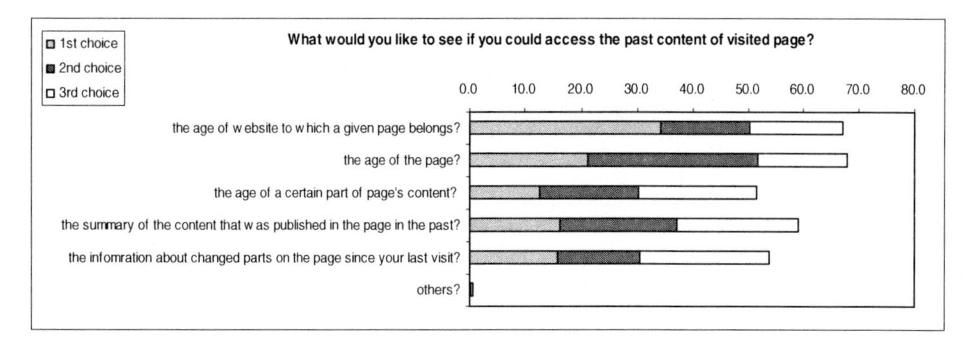

Fig. 7.15. Question related to the temporal support for visiting pages (specific needs) [16]

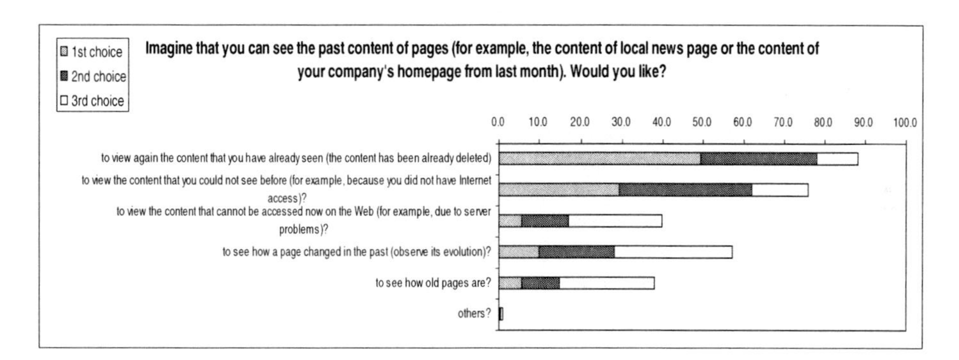

Fig. 7.16. Question related to the temporal support for visiting pages (specific situations) [16]

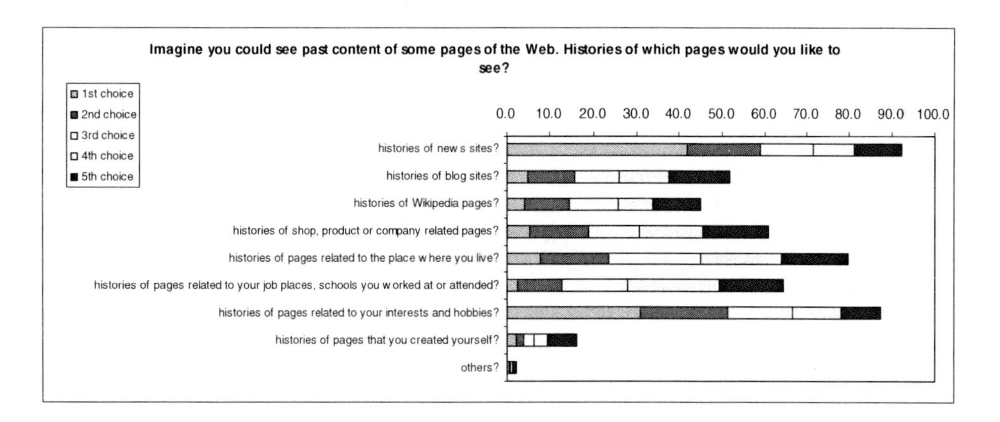

Fig. 7.17. Question related to the types of pages for which temporal support is needed [16]

Our next study focused on page content. We asked users what sort of information they would like to have if they could freely access page histories (Figure 7.15). Users had to choose 3 answers here among 7 answers in total. We asked whether the users would like to know the age of page, the age of its site, the age of certain content elements on the page, the summary of past content of the page and the information on the content parts that changed on the page since its last visit by users. Generally, all the answers were chosen quite frequently. The most popular choice was the information about the age of the site and the age of the page (34.2% users selected the former and 21.1% selected the latter as their first choices).

In Figure 7.16, we show the answers to another question about the accessibility of past content of pages. The users could select 3 top answers among 6 in total. Three answers had a personal aspect: a) to view the content that users have already seen (but it is no more accessible on page), b) to view the content that the users could not see (for example, due to limited Web access) and c) to view the content that is inaccessible now on its hosting Web server (for example, due to server problems). The other three answers were: d) to see how old the page is, e) to see how the page changed in the past and f) others. This time, the answers that do not have any personal aspect were selected less frequently when contrasted with the answers featuring personal aspects. The top two answers were to re-visit the already disappeared content and to see the content that could not be accessed before (49.4% users selected the former and 29.2% selected the latter as their first choices).

Our last question concerned the type of pages the histories of which users would like to view (Figure 7.17). In this question the respondents were asked to select 5 answers out of 9 in total. Histories of news sites were selected most often as the first answer (42% of users). This is not surprising as news articles contain time-dependent content. Users also would like to view the histories of pages related to their interests and hobbies (30.7% of the responders has chosen it as the first answer). Histories of pages related to living places or places and schools where users worked at or attended were definitely less interesting (7.8% and 2.5%, respectively). These results, to some extent, imply that users are interested in histories of pages with time-sensitive content and of pages with the content that is related to their interests.

7.7 Conclusions

The Web is constantly and rapidly changing. It is thus necessary to help users to keep track, spot and understand the changes. In this chapter we have proposed an integrated support framework for providing a continuous assistance to users browsing the Web regarding its temporal context. The system has three key features. First, it adapts views of re-visited pages to help the users notice new content and navigate the Web more efficiently. Second, upon request, it searches within the reconstructed page histories in order to detect approximate origin dates of content elements on visited pages. In this way, the users can learn about the age of page content and the information that is contained in it. Lastly, the system produces visual, concise summaries of page histories over arbitrary time periods. Such summaries inform the users about page evolution, typical page topics and general characteristics in the past.

We believe that users should have more control and power when browsing documents on the Web and they should be allowed free interaction with past content on visited pages. Previous browsing enhancements generally neglected the temporal aspect of content on pages, despite that the Web changes very rapidly, thus, imposing much cognitive burden on users. We think that the proposed system not only facilitates browsing and information discovering on the Web but also empowers users to judge the trustworthiness of the encountered information.

Acknowledgements

This research was supported by the MEXT Grant-in-Aid for Scientific Research in Priority Areas entitled: Content Fusion and Seamless Search for Information Explosion (#18049041, Representative: Katsumi Tanaka), the Kyoto University Global COE Program: Informatics Education and Research Center for Knowledge-Circulating Society (Representative: Katsumi Tanaka) and by the MEXT Grant-in-Aid for Young Scientists B (#18700111, Representative: Adam Jatowt; #18700110, Representative: Yukiko Kawai).

References

[1] Adar, E., Teevan, J., Dumais, S.T.: Large scale analysis of web revisitation patterns. In: Proceedings of CHI 2008, pp. 1197–1206 (2008)

[2] Arms, W.Y., Aya, S., Dmitriev, P., Kot, B.J., Mitchell, R., Walle, L.: Building a research library for the history of the web. In: Proceedings of the Joint Conference on Digital Libraries, pp. 95–102 (2006)

[3] Chen, L., Sycara, K.P.: WebMate: A Personal Agent for Browsing and Searching. Agents, 132–139 (1998)

[4] Chi, E.H., Pirolli, P., Chen, K., Pitkow, J.E.: Using information scent to model user information needs and actions and the Web. In: Proceedings of CHI 2001, pp. 490–497 (2001)

[5] Chi, E.H., Pitkow, J., Mackinlay, J., Pirolli, P., Gossweiler, R., Card, S.K.: Visualizing the evolution of Web ecologies. In: Proceedings of Conference on Human Factors in Computing Systems, pp. 400–407 (1998)

[6] Cockburn, A., McKenzie, B.: What Do Web Users Do? An Empirical Analysis of Web Use. International Journal of Human-Computer Studies 54(6), 903–922 (2001)

[7] Douglis, F., Ball, T., Chen, Y.-F., Koutsofios, E.: WebGUIDE: Querying and Navigating Changes in Web Repositories. In: Proceedings of the 5th International World-Wide Web Conference on Computer Networks and ISDN Systems, pp. 1335–1344 (1996)

[8] Fetterly, D., Manasse, M., Najork, M., Wiener, J.L.: A large-scale study of the evolution of Web pages. In: Proceedings of the 12th International World Wide Web Conference, Budapest, Hungary, pp. 669–678 (2003)

[9] Francisco-Revilla, L., Shipman, F.M., Furuta, R., Karadkar, U., Arora, A.: Perception of content, structure, and presentation changes in Web-based hypertext. In: Proceedings of the 12th ACM Conference on Hypertext and Hypermedia, pp. 205–214 (2001)

[10] Herder, E., Weinreich, H., Obendorf, H., Mayer, M.: Much to Know about History. In: Proceedings of the Adaptive Hypermedia and Adaptive Web-based Systems Conference, pp. 283–287 (2006)

[11] Jatowt, A., Kawai, Y., Nakamura, S., Kidawara, Y., Tanaka, K.: A browser for browsing the past web. In: Proceedings of the International World Wide Web Conference, pp. 877–878 (2006)

[12] Jatowt, A., Kawai, Y., Nakamura, S., Kidawara, Y., Tanaka, K.: Journey to the past: proposal for a past Web browser. In: Proceedings of the 17th ACM Conference on Hypertext and Hypermedia, pp. 134–144 (2006)

[13] Jatowt, A., Kawai, Y., Tanaka, K.: Detecting Age of Page Content. In: Proceedings of the 8th International Workshop on Web Information and Data Management, pp. 137–144 (2007)

[14] Jatowt, A., Kawai, Y., Tanaka, K.: Personalized detection of fresh content and temporal annotation for improved page re-visiting. In: Proceedings of the 17th Conference on Database and Expert Systems Applications, pp. 832–841 (2006)
[15] Jatowt, A., Tanaka, K.: Towards mining past content of web pages. New Review of Hypermedia and Multimedia 13(1), 77–86 (2007)
[16] Jatowt, A., Kawai, Y., Ohshima, H., Tanaka, K.: What Can History Tell Us? Towards Different Models of Interaction with Document Histories. In: Proceedings of the 19th ACM Conference on Hypertext and Hypermedia, pp. 5–14 (2008)
[17] Jatowt, A., Kawai, Y., Tanaka, K.: Visualizing Historical Content of Web Pages. In: Proceedings of the International World Wide Web Conference, pp. 1221–1222 (2008)
[18] Joachims, T., Freitag, D., Mitchell, T.M.: Web Watcher: A Tour Guide for the World Wide Web. IJCAI (1), 770–777 (1997)
[19] Kimpton, M., Ubois, J.: Year-by-year: from an archive of the Internet to an archive on the Internet. In: Masanes, J. (ed.) Web archiving, pp. 201–212. Springer, Heidelberg (2006)
[20] Lieberman, H.: Letizia: An Agent That Assists Web Browsing. Proceedings of IJCAI (1), 924–929 (1995)
[21] Liu, L., Pu, C., Tang, W.: Continual Queries for Internet Scale Event-Driven Information Delivery. IEEE Knowledge and Data Engineering 11(4), 610–628 (1999); Special Issue on Web Technology
[22] Masanes, J. (ed.): Web archiving. Springer, Heidelberg (2006)
[23] McCown, F., Diawara, N., Nelson, M.L.: Factors affecting website reconstruction from the web infrastructure. In: Proceedings of the Joint Conference on Digital Libraries, pp. 39–48 (2007)
[24] Myers, E.W.: An O(ND) difference algorithm and its variations. Algorithmica 1, 251–266 (1986)
[25] Nadamoto, A., Tanaka, K.: A comparative web browser (CWB) for browsing and comparing web pages. In: Proceedings of WWW 2003, pp. 727–735 (2003)
[26] Nielsen, J.: Change the color of visited Links. Jakob Nielsen's Alertbox (retrieved April 1, 2008) (2004), http://www.useit.com/alertbox/20040503.html
[27] Ntoulas, A., Cho, J., Olston, C.: What's new on the Web? The evolution of the Web from a search engine perspective. In: Proceedings of the 13th International World Wide Web Conference, pp. 1–12 (2004)
[28] Obendorf, H., Weinreich, H., Herder, E., Mayer, M.: Web page revisitation revisited: implications of a long-term click-stream study of browser usage. In: Proceedings of CHI 2007, pp. 597–606 (2007)
[29] Pandit, S., Olston, C.: Navigation-aided retrieval. In: Proceedings of WWW 2007, pp. 391–400 (2007)
[30] Rauber, A., Aschenbrenner, A., Witvoet, O.: Austrian online archive processing: analyzing archives of the world wide web. In: Proceedings of the 6th European Conference on Digital Libraries, pp. 16–31 (2002)
[31] Shipman, F.M., Hsieh, H.: Navigable history: a reader's view of writer's time: Time-based hypermedia. New review of hypermedia and multimedia 6, 147–167 (2000)
[32] Toyoda, M., Kitsuregawa, M.: Extracting evolution of Web communities from a series of Web archives. In: Proceedings the 14th Conference on Hypertext and Hypermedia, Nottingham, UK, pp. 28–37 (2003)
[33] Wexelblat, A., Maes, P.: Footprints: History-Rich Tools for Information Foraging. In: Proceedings of Conference on Human Factors in Computing Systems, pp. 270–277 (1999)

8

Considering Resource Management in Agent-Based Virtual Organization*

Grzegorz Frąckowiak[1], Maria Ganzha[1], Maciej Gawinecki[1], Marcin Paprzycki[1], Michał Szymczak[1], Myon Woong Park[2], and Yo-Sub Han[2]

[1] Systems Research Institute, Polish Academy of Sciences, Warsaw, Poland
{maria.ganzha,marcin.paprzycki}@ibspan.waw.pl
[2] Korea Institute of Science and Technology, Seoul, Korea
{myon,emmous}@kist.re.kr

Abstract. In this chapter we discuss a system designed to support workers in a virtual organization. The proposed approach is based on utilization of software agents and ontologies. In the system all *Users* are represented by their *Personal Agents* that help them in fulfilling their specific roles. At the same time all entities that the organization is comprised off (human and non-human) are represented as instances of resources in an ontology of the organization. Furthermore, each resource is associated with one or more profiles and these profiles are adapted to represent changes in resources (e.g. new experience/knowledge gained by a human resource, or approval of a duty trip application). The aim of this chapter is to describe basic functions of our system with special attention paid to software agents, their roles and interactions as well as utilization of ontologies in support of worker needs.

8.1 Introduction

Let us consider an organization in which teams of researchers are engaged in R&D projects and share a common virtual work-space (regardless if they are geographically distributed or not). Obviously, team work requires cooperation between members and support of collaborative research has to go beyond, even most sophisticated forms of, document versioning and flow of resources in the hierarchical structure of the organization. What needs to be taken into account is: (1) representation of domain specific knowledge (e.g. geological sciences); to provide context for management of resources pertinent to running projects (e.g. establishing a specific "location" of a resource within the domain knowledge allows for resource indexing, clustering; it also allows to establish *who* within the organization should receive a notification that a new resource—such as a book—has been acquired); (2) representation of structure of interactions and flow of resources in the project (and, more generally, within the organization); to route resources, based on project needs and responsibilities of team members (e.g. who should receive a report that a given task is completed, or to whom

* Work was partially sponsored by the KIST-SRI PAS "Agent Technology for Adaptive Information Provisioning" grant.

N.T. Nguyen, L.C. Jain (Eds.): Intel. Agents in the Evol. of Web & Appl., SCI 167, pp. 161–190.
springerlink.com

an application for a business trip should be routed); (3) representation of user profiles (situated within the domain knowledge and the structure of the project); to specify *interests*, *needs* and *skills* of individual workers (e.g. to establish who needs to be proactively trained in view of an upcoming project); (4) adaptability of the system; to deal with the fact that as the time passes the scope of the project may expand, contract or shift; functional interrelationships between team members (or within the whole organization) can change; their interests, needs and skills evolve; and, team members may be added, removed or replaced.

It is relatively easy to see that these four points can be generalized beyond the initial collaborative research scenario. Let us assume that for the second point we utilize a notion of a virtual organization (*VO*) [16, 17, 18, 19, 20, 21], which allows us to define roles, interdependencies and interactions of participants. Here, it is important to note that while most conceptualizations of a *VO* stress the importance of its workers being spatially distributed, in the proposed approach "virtualization" involves realization of an actual organization as an e-organization. Therefore, it does not matter if the organization itself is actually geographically distributed or not. In such an organization its members need access to resources to complete their individual tasks and to facilitate completion of projects. In our approach, any entity within the organization, human and non-human, is considered to be a resource. Obviously, access of resources to resources should be, among others, adaptive (change with the task) and personalized (each team member—human resource—requires access to different resources; furthermore access is likely to be restricted by the organization policy/structure). The aim of our work is to develop a software infrastructure for such a virtual organization. The basic assumption underlying our approach is that emergent technologies such as software agents [42] and ontologies [36] should be utilized as a foundation around which the proposed system should be conceptualized. Let us stress that we do realize that these assumptions are not uncontroversial. However, our aim is to develop a system on their basis, and *in this way* to add to the discussion of viability of this approach (instead of getting involved in theoretical discussions). This being the case we assume that: (i) the organizational structure, consisting of "roles" played by various entities within the organization and interactions between them, should be represented by software agents and their interactions (i.e. the complete structure of an actual organization is mapped into the structure of an agent-based virtual organization), and (ii) domain knowledge, organization structure, resource profiles and resource matching should be based on ontologies and reasoning machinery associated with them. For instance, in a company that installs and services satellite TV antennas, the domain specific knowledge consists of a complete body of knowledge concerning such antennas. The structure of the company involves, among others, antenna installing teams, their equipment, the way that work orders are delivered to them, and the reporting upon task completion. Software agents represent each worker and support them in completing assigned tasks (e.g. managing a team of installers). Finally, human resource profiles describe skills of each member of service team, while the adaptability involves situation when a new antenna is to be introduced to the market and installation crews have to be trained in its features.

The aim of this chapter is to summarize main results obtained thus far within the project and is based on [37, 8, 9, 38]. To this effect we proceed as follows. In the next section we present a general description of the the system. Then, we discuss the issues concerning interactions between software agents of human workers. Following the discussion concerning agents in the system, we concentrate our attention on ontologies and ontological demarcation of resources. We start with the generic ontology of the virtual organization, and follow with description of its extensions to the areas facilitated by applications proposed by an Institute of Science and Technology. Finally, we discuss processes involved in ontological matchmaking proposed in the system.

8.2 System Overview—Introducing Project into the System

Before discussing the main features of the system let us first stress that in the proposed approach each worker in the organization is represented by her/his *Personal Agent* (*PA*). This agent plays two roles: (a) it is the interface between the *User* and the system, and (b) it supports its owner in all *roles* that (s)he is to play within the organization. Let us now present birds-eye view of the system, by discussing processes involved in introducing and running a project. To focus our discussion, in Figure 8.1 we present the use case diagram of the system. Note that the following discussion is written in terms of *entities with specific roles*, and such units can consist of one (or more) humans, agents, or "teams" consisting of humans and agents. We will return to the issue of interactions between humans and agents later in the chapter.

When a service/project is requested from an organization (which can be anything from a one-person business to a 50,000+ employees corporation) a *Project Manager* (*PM*) is associated with it. The *PM* is a *role* that is associated, for instance, with a person who in the *VO* is represented by the *Personal Agent*, which will support that person in fulfilling the role of the *PM*. *PM*s first task is to make sure that the request is thoroughly analyzed and on the basis of such analysis to make a decision if the job should be accepted. This task is delegated to the *Analysis Manager* (*AM*). At the same time a *Task Monitor Agent* (*TMA*) is created to oversee the task performed by the *AM* (for more details about role of the *TMA*, see below). It should be noted that the structure of the *AM* can be either very complicated and consist of a number of humans and agents (e.g. in the case of a corporation that is evaluating a multi-million euro construction project) or very simple (e.g. in case of a small business assessing acceptance of a brake pads replacement job). Finally, it is even possible that the *PM* can play the role of the *AM* (e.g. in the case of a very small business or self-employment). Regardless of the specific situation, the most important deliverable prepared by the *AM* is a set of reports that support the decision to accept or reject the

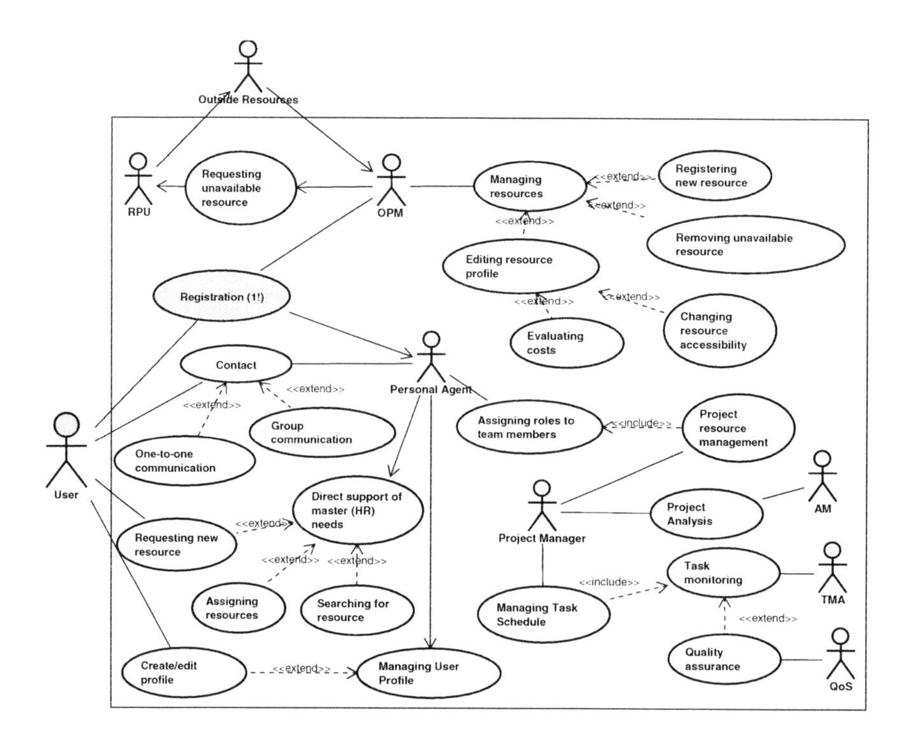

Fig. 8.1. Use case of the system

requested project. The report(s) prepared by the *AM* is(are) backed up, among others, by the cost, resource and income analysis.

Since we assume that data processed in the system is based on appropriate (domain and organization) ontologies, one of crucial tasks of the *AM* is to "translate" the common language requirements originating from the user (project proposer) into a set of requirements specified utilizing ontologies employed in a given organization. To fulfill this need, at the beginning of its work, the *AM* instantiates a new resource called the *Project Request* (which has its own profile). This resource is used when the *AM* performs initial analysis of the proposed task and creates the first version of the *System Requirements Specification* (*SRS*) which, again, is a resource with its own profile. During its work the *AM*, among others, analyzes resources available in the organization (their profiles, availability and accessibility). For instance, in the case of a cable TV installation job, this step is going to be rather simple and involves steps like: (1) checking whether the customer who requests installation lives in the service-covered area, and (2) if there are resources available to install the cable TV at her address, within a specific time-frame. Note that such simple analysis could easily be performed by a software agent with a build-in expert system. On the other hand in the scenario of the project involving development of an intranet and a knowledge portal for a company, analysis would involve more elaborate actions, such as: checking technological requirements for the project, existing similar solutions, organization or customers experience etc. It can be conjectured that in this case the *AM* would most likely involve human resources as well as software agents helping them completing the tasks. Let us mention, that in the case of a large corporation the *AM* may not have permission to know about all resources available in the organization. In this case, the *AM* will specify resources that the project needs and which, according to its best judgment, are unavailable. It will be then the role of the *PM* (and possibly its supervisors in the organization) to assess if the resources are actually available, or if they need to be found outside of the organization, and what is the effect of such search on the viability of project acceptance. Note that this process may involve interaction with the *Organization Provisioning Manager*, which is aware of all resources available in the organization (see below).

If the *AM* recommends that the project is rejected, and the *PM* concurs (which may, or may not be the case; it is a well-known fact that there exist projects which are accepted even if they should not have been accepted, for instance for "political" reasons), the requester is informed about the decision, the *PM* is disassociated from the role, and this ends the process. Let us now assume that the *SRS* and other supporting documentation prepared by the *AM* suggests that the project should be accepted. As a result the *PM* prepares an initial *Project Schedule* and on its basis works to establish if the *Resource Reservation* can be completed (note that the fact that John, the Java coder, works for the organization does not mean that John is available starting from May 4th). To achieve this goal, the *PM* has to analyze available resources (its own and provided by the customer). It may involve, for instance, checking availability of programmers who have the required competence in PostgreSQL, object oriented programming and recent web technologies, as well as availability of resources such as: servers, (e-)learning materials for software to be used in the project, licenses and requirements for both test and final deployment environments etc. Again, the *PM* can analyze only these resources which it has access to (is allowed to know about, as established by the ontology of the organization). If resources that the *PM* knows about are not sufficient, the *PM* requests the *Organization Provisioning Manager* (*OPM*) to facilitate the missing resources (e.g. C# programmer(s), or a DB2 e-learning course). Note that such resources may be available in the organization, but the *PM* may not have access to this knowledge. *OPM*'s role is to provide resources for other resources which request them, as well as to deal with resources that are being delivered to the organization (e.g. books/papers/reports send to its digital library). Here, we assume that, to fulfill its role, the *OPM* has access to information about all resources available in the organization. Since the *OPM* can be queried by authorized (where the authorization is also ontologically specified) resources that play various roles in the organization, it has to analyze available resources using various patterns of reasoning and possibly some expert systems. Note also that, again, the *OPM* can be either an agent, a human represented by its *PA*, or a composite structure consisting of multiple agents and humans (e.g. it can have in its disposal a resource that indexes and routes incoming documents / books / journals, a search engine, a library material acquisitor, etc.). Again, if the resource (a) is found, and (b) can be reserved (for a specified time), it can then be assigned to the requesting *PM*. Otherwise, the *OPM* triggers action of a *Resource Procurement Unit* (*RPU*), which is responsible for finding an appropriate resource. Assuming, for instance, that C# programmers and DB2 e-learning materials were not found within the organization the *OPM* may generate a (ontologically demarcated) request to the *RPU* to acquire specific resources. The *RPU* in turn will communicate it to the "world outside of the organization." For simplicity we omit situations which clearly have to involve human intervention. Let us assume that company needs construction workers to start a project in Lublin, Poland. If it does not employ large enough number of such workers it, most likely, will be the role of human managers to assess if they can be hired for the time of the project. Therefore, at this stage, we assume that the *RPU* can immediately provide information about availability and cost (estimate) of requested resources. As a result of these processes two outcomes are possible. First, it is established, that the initial *Project Schedule* cannot be supported with necessary resources (which may result in project schedule (re)negotiation(s) with the client, or project rejection). Second, the *Project Schedule* can be completed (with possible minor modifications) in such a way that the project can be accepted and a contract signed.

Let us now discuss processes that take place after the final version of the *Project Schedule* is created and contract signed. First, the *Project Schedule* is used by the *PM* to assign tasks to appropriately reserved (human or non-human) *Resources*. Note that each *Resource* can be either a "single resource," or a collection of resources treated as a single unit. For instance, team that is responsible for the back end of the portal may consist of 4 coders and a manager, while the team dealing with user interface could consist of 2 coders and an artist, etc. In the hierarchical structure of the organization, at one level, both teams will be treated as a single resource, with their own tasks, and a *Task Monitor Agent* (*TMA*) associated with it. At the same time, inside these composite resources an appropriate organizational structure (based on the same ontology of the organization) will be realized, and individual (sub)tasks and their *TMA*s instantiated.

The *PM* monitors status of all tasks (including their start and completion) by assigning to each task a *TMA* and by communicating with them. Each *TMA* monitors a specific task until its completion (then it is killed by the *PM*; currently, we assume that creation of a new *TMA* is easier to achieve than adapting a given *TMA* to manage a different task). While working on the task, *Resources* might be interrupted by unexpected circumstances which either can be dealt with "locally" (e.g. by finding tips on how to deal with a "heap memory exceeded" error in Java, or how to build a DB2 cluster) or ones that will probably influence other parts of the project (e.g. customer requested that a different data structure is to be interfaced with, or some additional unavailable resources turn out to be needed, or a particular employee has to immediately take a family leave of absence, etc.). These circumstances are expected to involve *PM*'s reaction and should be tagged appropriately by the *Task Monitor Agent.* Let us stress that not every interruption requires an immediate *PM*'s intervention. Across the system we assume that resources can interact with each other (which resources can communicate directly is specified by the organization and represented in the organizational ontology), among others, to solve basic problems occurring during task execution. For instance, to find a manual for software used in the project given resource can contact other members of its group. Finally, each resource might generate multiple interrupts, but as long as these do not require the *PM* to react (e.g. the schedule of the project is not affected) they are going to be tackled locally.

Obviously, at a certain moment each (sub)task comes to an end. Upon completion of a task, the task-specific *Quality of Service* (*QoS*) module analyzes the work. A *Os* module might consists of a team of humans, or be instantiated as an expert system. Here, consider testing functionality of the company portal, or a test of a completed unit of a Python code, or checking quality of the TV signal after the TV is installed. Each of these quality tests requires different testing and different resources to complete the quality assessment. Unless the quality of the work is not satisfactory and further improvements are needed, the *PM* is informed about completion of the (sub)task. If the result of the task does not satisfy the requirements, there is a necessity to repeat some part of, or even the whole task. This can take more time and resources than it was specified in the *Project Schedule*. However, only conflicts with the schedule should result in the *PM* being "alarmed." Note that a "major interrupt" that results in changes in the *Project Schedule* may need to be propagated within the structure of the team that works on the project.

Obviously, completion of a (sub)task may trigger execution of another (sub)-task specified in the workflow of a given project. Upon completion of all (sub)-tasks specified in the *Project Schedule*, the project is completed. This means that the *Human Resource* that played the role of the *Project Manager*, will no longer play this role (for *that* project) and the functionality of its *Personal Agent* has to be appropriately adjusted. Similarly, functionalities and profiles of *all* agents involved in the project have to be adjusted (e.g. experience-related information).

8.3 Agents in the System

Thus far we have described processes that take place within the Virtual Organization, considered from the point of view of "roles" existing in the system and their interactions. In this context let us recall, that one of our assumptions is that each *Worker* will be represented by a *Personal Agent*, while a number of auxiliary agents may be instantiated as well. In this way, the proposed approach is grounded not only in general agent notions (see, for instance, [26]), but also in role-oriented agent system development methodologies (e.g. Gaia [43], or Prometheus [33]). Here, the problem space is initially defined in terms of (1) roles that are to be fulfilled, and (2) interactions between entities playing these roles. In the second step each identified role is functionalized by a single agent, or is further divided into a number of cooperating (sub)agents (see, also, [25]).

However, we are well aware of the fact that not all roles can be fulfilled by software agents alone. Therefore, let us consider roles that have been distinguished thus far: *PM*, *AM*, *RPU*, *OPM* *TMA*, and *QoS*. As noted above, in some cases these roles may be fulfilled by software agent(s), some of them are likely to be played by one or more humans (supported by their *Personal Agents*), while some are likely to be completed by teams consisting of software agents and humans. Note that while specific arrangements may depend on the particular organization (and its domain of operation), processes described above remain unchanged. In this context we have identified a few situations that are expected to trigger a necessary reaction of a human actor (however, this list is not exhaustive):

1. project requirements analysis
2. accepting a particular person to become a manager of a project
3. changes in customer requirements
4. the *OPM* not being capable of finding required resource(s) within the organization
5. negotiating and accepting the *Project Schedule*
6. accepting the *Resource Reservation* document
7. final task acceptance

Even though human intervention is likely to be required, it has to be stressed that our interest is in performing as many tasks as possible utilizing software agents alone and thus completing them in an autonomous fashion or to provide support for humans in fulfilling the above specified roles. In this context, the role-based approach allows us to specify sets of functions associated with each role and then select which of them can be fulfilled by software agents and which have to involve human participation. For instance, consider the *TMA* that makes sure that a Cable TV Box was successfully installed before 16:13 at a specific address and if this is not the case, raises an alarm, and a *Personal Agent* that helps the human *PM* managing a team of coders. The process is as follows: the autonomous agent, when created to fulfill a given role is provided with required modules to accomplish it, e.g. the *TMA* obtains information about the deadline it is to observe and what to do if it is, or is not met. The situation is somewhat more complicated in the case of the *PA*. First, let us recall that every worker is represented in the system by her/his own *PA*. Furthermore, upon joining the system (and thus the organization) the *PA* registers with the *OPM*—the resource manager—and becomes one of available resources. In Figure 8.1 we have depicted the *PA* and conceptualized it as an interface between the human and the remaining parts of the system; as well as a "helper" that supports user in fulfilling her role. Since role can change, the *PA* has to be able to support user in anyone of them. However, in Figure 8.1 we were able to identify core functions of the *PA*, which are used regardless of a specific role. To provide support for the user who is assigned a specific role, modules facilitating functions associated with that particular role are then loaded into the *PA*, extending its functionality (for more details see [10]). Note that in the, somewhat more complicated case, when a team of analysts (*AM*) estimates feasibility of a project, each team member will be represented by its *PA*. Therefore, a role-specific set of interactions between

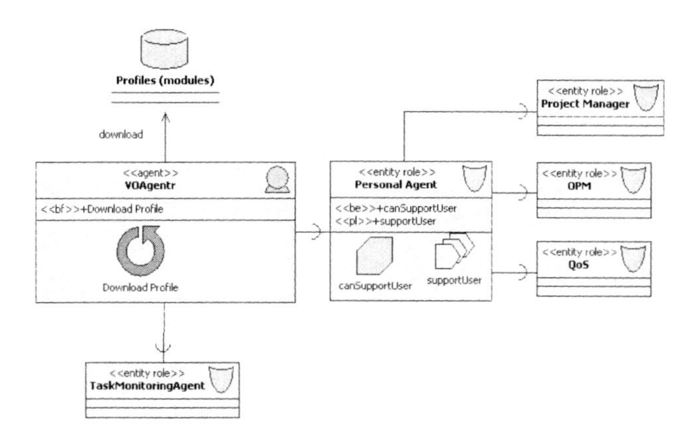

Fig. 8.2. AML Mental Diagram for the VOAgent

these *PA*'s and humans they represent will constitute fulfillment of the role *AM*. We utilize an AML [2] diagram in figure 8.2 to illustrate this general schema. In this figure we can see the generic agent the *VOAgent* which utilizes appropriate modules stored in the module / profiles library to become a *Personal Agent* first, and then play any role in support of its *User*.

Observe that in this way, we have only "one basic type" of an agent in the system: the *Personal Agent* (supplemented by possibly some auxiliary agents). Furthermore, the *PA* can support user in playing *any* role identified in the organization. This in turn matches very nicely with the real world organization, where we also have only one main entity: human being that can play various roles identified in the organization.

8.4 Ontologies in the System

Thus far we have focused our attention on specific roles and their interactions involving various entities (human and non-human resources) in the system, and associating these entities with software agents alone and with agent-human teams. The other important developmental decision that was made was to utilize ontologies to represent (1) the domain of interest, (2) the structure of the organization, and (3) resource profiles. Let us now focus our attention on ontologies that are to be used in the system and start with a brief analysis of related work.

8.4.1 Toronto Virtual Enterprise (TOVE)

TOVE project run at the Enterprise Integration Laboratory of the University of Toronto. Its main goal was to establish generic, reusable enterprise data model that was to have the following characteristics [3, 1]:

- to provide a shared terminology for the enterprise that each entity within it can jointly understand and use,

- to define the meaning of each term in a precise and unambiguous manner,
- to implement the semantics in a set of axioms, to enable *TOVE* to automatically deduce the answer to "common sense" questions about the enterprise,
- to define a set of symbols for depicting a term or concept constructed thereof in a graphical context.

According to documents found within the project WWW site, ontology developed by the project included terms such as: resource, requirement, time, state or activity; and was created in Prolog. We thought about relying on the *TOVE* project and utilizing data model constructed there. Especially, since *TOVE* was based on extensive research and considered work of an enterprise from the design and operations perspectives [27]. Unfortunately, inability to find actual ontologies (except of conference papers), a long list of important features to be added found at the project web site, and the fact that the last update of that site was made on February 18, 2002, let us to believe that the TOVE project has died sometime in 2002. Therefore, we have decided to utilize only the theoretical part of *TOVE*.

8.4.2 OntoWeb

The *OntoWeb Network* is an initiative aiming at building a bridge between academics and the industry in order to promoting the Semantic Web [44]. The *OntoWeb Portal* of the *OntoWeb Network* project allows to insert and retrieve information about academic and industry employees, projects and documents [46]. Within the project the *OntoWeb* ontology was developed and made available at [45]. Unfortunately the *OntoWeb* ontology has also important drawbacks:

- The *OntoWeb* ontology is created in RDF Schema, which does not have rich enough semantics. Our experience with the RDF Schema shows that it is undeniably well suited for building conceptualizations [39]. However, in the case of a more complex software system, richer semantics and guaranteed computational completeness are desired. In particular, semantics of the RDFS which lacks quantifiers is hardly suitable for defining a data model (which involves defining cardinalities of entity relations) of the system. Therefore, reusing the *OntoWeb* ontology as the system core ontology would result in restricting types of reasoning available in the system.
- The *OntoWeb* ontology does not support resource profiles and information access restrictions, while they are necessary for the proposed system [9, 11].

Summarizing, we dropped the idea of reusing the *OntoWeb* ontology due to the limited expressivity of the RDF Schema and lack of necessary concepts. Instead, we followed guidelines and results obtained within both *TOVE* and *OntoWeb* projects and developed an ontology matching our project's needs. Let us therefore look into ontologies that have been developed within our system.

8.4.3 Generic Top-Level Ontology of the Organization

Before we start let us note that delivering a comprehensive ontology for modeling an organization is beyond the *current* scope of our project. Our main aim is to deliver a framework for adaptive resource management (information provisioning in particular). Hence, the proposed ontology may not include all the necessary features to design model of any organization. However, we believe that ontology requirements considered at this stage have been specified to support currently-necessary functions of the system. Furthermore, they are flexible enough to support its future extension in order to support comprehensive organization modeling. Keeping this in mind, let us look into main ontologies of the proposed system.

We have decided to use OWL-DL as the ontology demarcation language, as it guarantees computational completeness and rich semantics—utilizing the Description Logic [4]. As mentioned above, one of main ideas of our approach is to model *every* entity within the organization (including humans) as a *resource*. Furthermore, each resource will have a *profile* and, depending on its type and role, may appear in a context of multiple profiles. For instance, knowledge about a person may be described with any of the following (and not limited to these) profiles: professional experience, education, personal, accommodation preference or dining preference. In Figure 8.3 we depict the generic resource and the generic profile concepts.

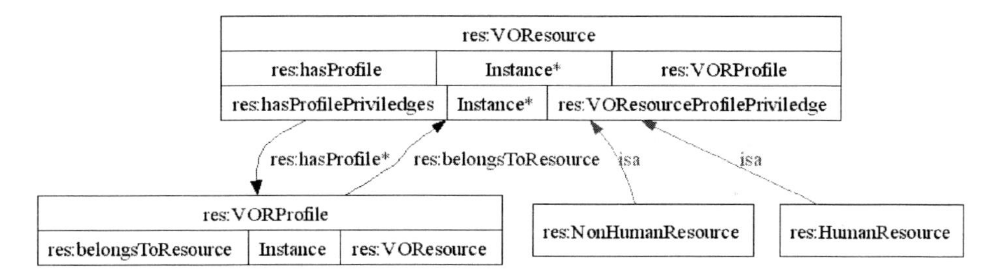

Fig. 8.3. Generic resource and generic profile concepts

A *resource profile* provides detailed information about any resource (human or non-human). It is composed of a resource specific data and "opinions" about other ontology concepts or ontologically demarcated objects [22]. Classes *VOResource* and *VOProfile* are designed to be extended by any organization specific resources and their profiles (assuring that the concept is robust and flexible). Deriving these core concepts in domain ontologies allows to define more organization-specific resource, such as: an *employee* of the cable TV installation company or academic institution; a *book* in a library; *requirements specification* (*SRS*) document in an IT company; or a *Duty Trip Report* in an organization that requires its employees to deliver such reports.

Note that some resource profiles may consist of private or classified information (e.g. personal data) therefore it is necessary to build an infrastructure which can restrict access to the information. This is also important since accessibility to certain documents depends on employees "position" within an organization (e.g. annual evaluation of a worker should be visible only to that worker and her supervisors, but not her co-workers). A *VO Resource Profile Privilege* is a class which describes restrictions established for a profile. It binds a profile with a restriction type which is applied to all resources from a particular *Organization Unit (OU)*—whenever information is requested by, or matched with, resources. The binding of the *OU* and a particular *Profile Privilege Type* is realized by the *Profile Privilege* class.

The *Profile Privilege Type* is an enumerable type specifying supported access privileges: *Read*, *Write* and *Admin*. Names of the first two are self-explanatory, while the third (*Admin*) type represents an administrative privilege which allows to modify access restrictions of the profile. Here, for instance, the *HR Department* is expected to have *Write* privileges for worker profiles, while the *PA* is going to have *Read* privileges for information provided by the *OPM* (see Figure 8.1). The design of the *Profile Privilege* is depicted in Figure 8.4.

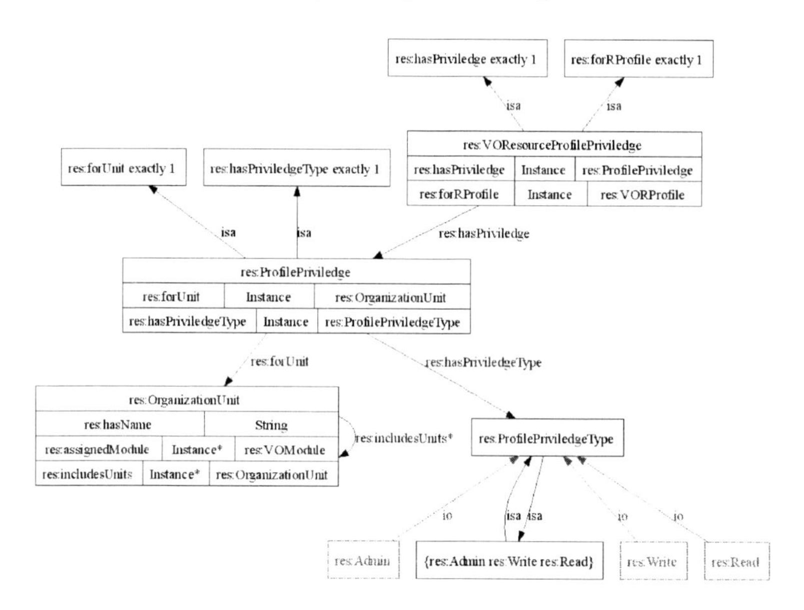

Fig. 8.4. Profile Privilege design

8.4.4 Demonstrator Applications

To extend our discussion of ontologies to be used in the system, let us introduce two applications depicted in Figure 8.5. In the *Grant Assistant System* (*GAS*), the *OPM* of a university (or a research institute) receives grant announcements and its role is to deliver them to these and only these *PA*s that represent *Users* that may be interested in them. In other words, the *OPM* (see figure 8.1) has to decide who (which *PA*(s)) should receive a given announcement, based on ontologically demarcated profiles describing faculty in the university (researchers in the institute) and profiles of grant announcements. Here, we assume that the announcement is a resource that has already an assigned profile based on the internal domain ontology (note that specifying entity inside, or outside, of the system that performs profile demarcation is of no importance here).

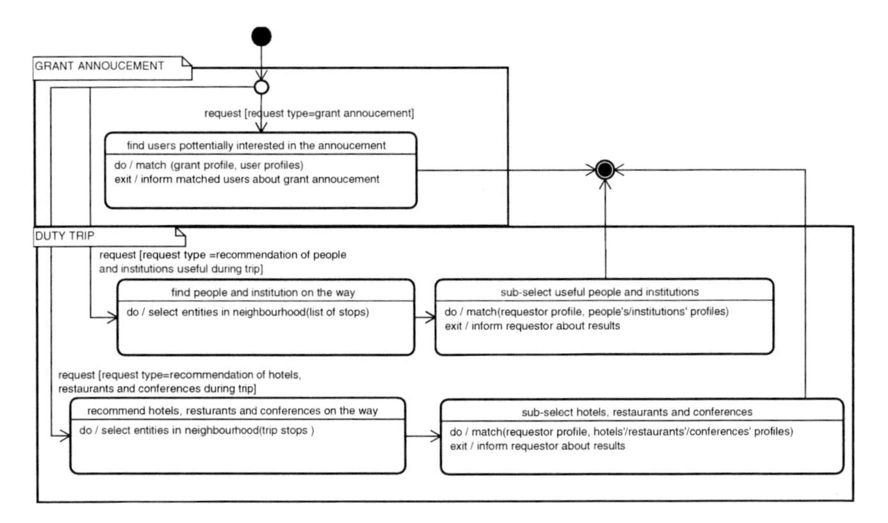

Fig. 8.5. Matching scenarios in the proposed system

The *Duty Trip Support* (*DTS*) scenario (specific to a Research Institute in East Asia, but easily generalizable) is more involved. Here, workers use the intranet to apply for a *Duty Trip* and to submit trip report. Our aim is to utilize results obtained in our project to provide them with additional functionalities. First, note that for the Institute in question, cost of air travel (to most destinations outside of East Asia) is much higher—in a relative sense—than costs of a stay extended by a few days. Thus, an employee traveling to a given city (e.g. in Europe or America), may visit also near-by-located institutions (e.g. universities or companies), or persons that her institute has contacts with. Second, a recommender where to stay and eat could be of value (e.g. consider Indonesian researchers confronted with typical Irish food). In addition to personalized information delivery, the system is expected to help researchers in all phases of duty trip participation; from the preparation of the initial application until filing the final report. Note that the *Trip Assistant* is actually a role played by the *OPM*, which provides the requested personalized input to the *PA* (see function *Searching for resource* in Figure 8.1). In Figure 8.6 we present the activity diagram of the *Duty Trip Support*. In this diagram we can see two moments when the *PA* communicates with the *Trip Assistant* (*OPM*), first when the application for the trip is prepared (and institutions/people to visit are sought), second, when actual details of the trip (e.g. hotels) are to be selected.

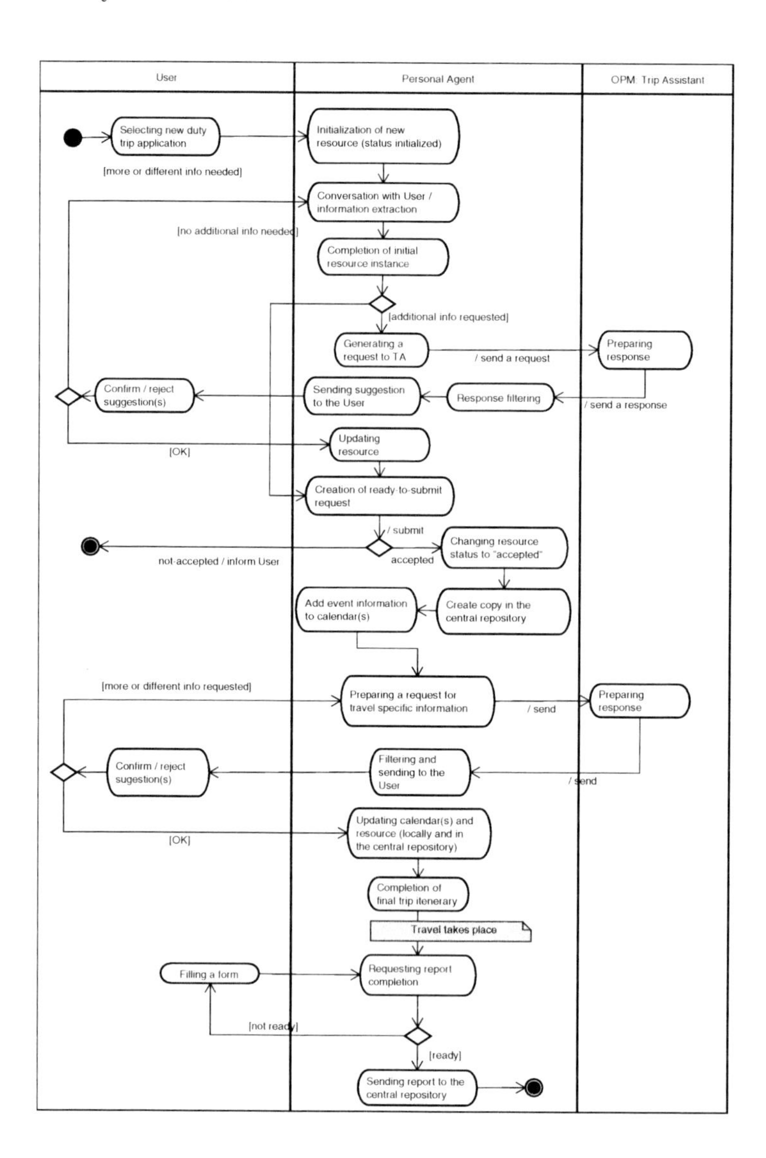

Fig. 8.6. Activity Diagram of the *Duty Trip Support* subsystem

8.4.5 Ontologies for the *Institute of Science and Technology* (*IST*)

To illustrate how the proposed ontology can be utilized in a specific organizational setting, let us discuss briefly its application to selected features of an ontological model of an Institute of Science and Technology (the *IST* ontology). In the architecture of our system, the *Domain Ontology* is an extension of the *Generic Ontology* outlined in Figure 8.3. Here, human resources are modeled in a way that is specific to the *East Asian Institute of Science and Technology*,

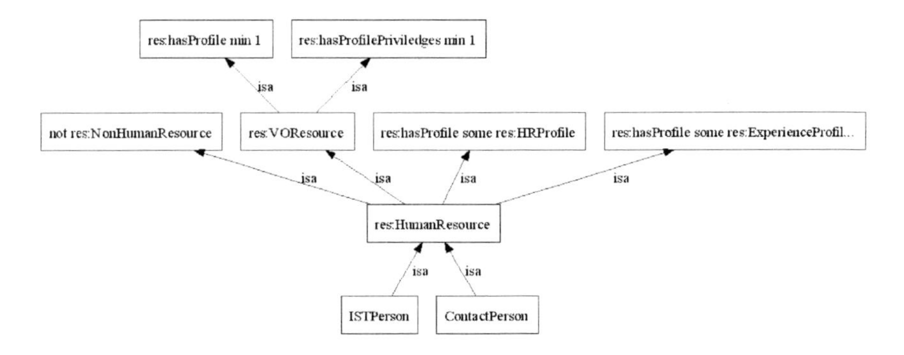

Fig. 8.7. Ontological description of the person on the Virtual Organization

though similarities with general human resource descriptions can be seen. Let us start from the *ISTPerson*, which is a class describing all employees of the Institute. Figure 8.7 illustrates the *ISTPerson* concept.

While human resources have (multiple) general profiles, according the ontology, currently the following profiles can be assigned to employees of the Institute:

- *FIST Experience Profile.*
- *FIST Person Profile.*
- *Organization Profile.*
- *Dining Preference Profile.*
- *Accommodation Preference Profile.*

Here, the *FIST Experience Profile* allows to describe both educational and professional experience of the employee (and is depicted in figure 8.9). Professional experience is represented as a "project history" in which a given worker participated, while working in the organization. The educational experience lists academic degrees of the employee.

Additionally, specification of (multiple) research field(s) further describes employees competences (research fields used here are based on the South Asian RFCD [28]). Note that, as described in the last section, it is also possible to assign level of competence for each research field [12].

The *Personal Profile*, presented in figure 8.8, is a set of data typically stored by the *HR Department*. It represents personal data of an employee.

The *Organization Profile* specifies, for instance, a division in which the employee works; it can be also used to establish who is the supervisor of an employee. Finally, *Accommodation Preference* and *Dining Preference* profiles represent ones attitude toward restaurants and hotels visited thus far. These concepts establish a link between the *Travel Support System* ontology ([14]) and the *IST* ontology. They utilize the *Hotel* and the *Restaurant* classes which are defined in the *TSS* ontology. While the *ISTPerson* is an example of a human resource in the domain of the *IST*, the *ISTDutyTrip* (*DTR*) is an example of a non-human resource (see figure, 8.10). This class represents a duty trip description from the *DTS* scenario and, as a resource child class, all its instances may

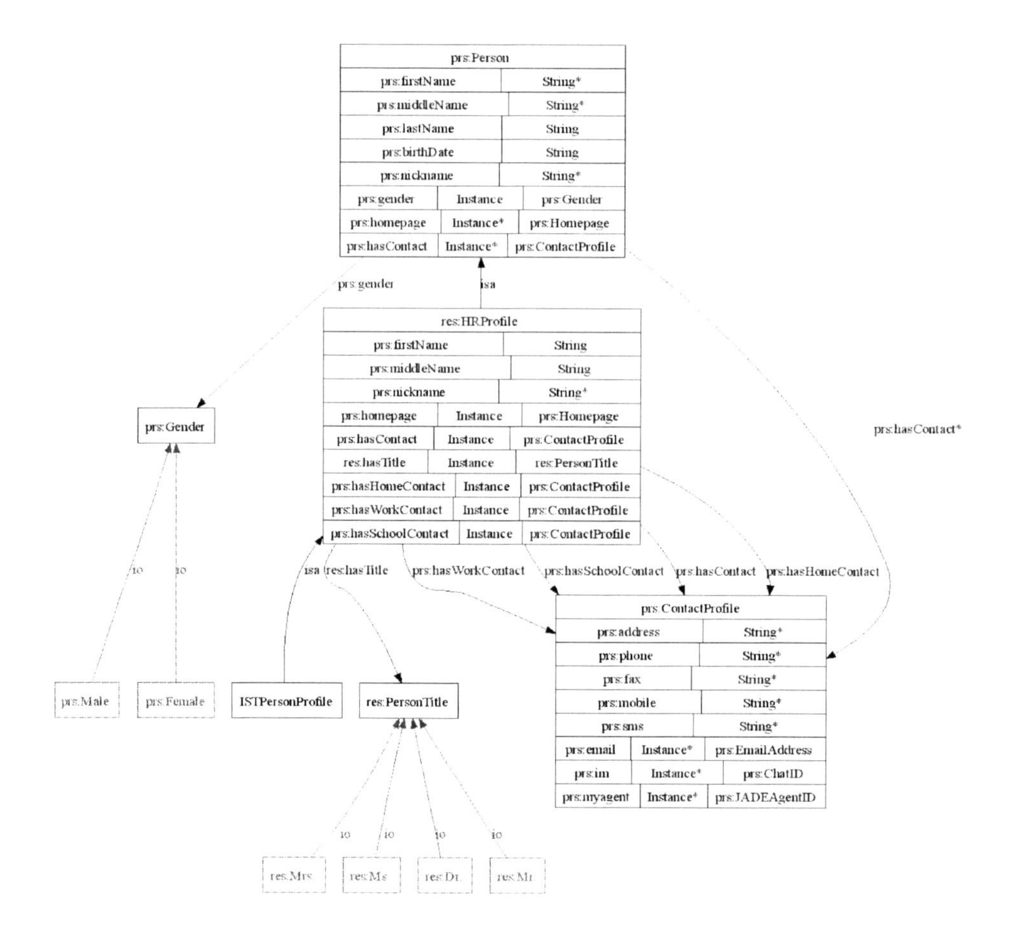

Fig. 8.8. Ontological Description of the Personal Profile

have various profile instances assigned. However, the *DTR* concept is restricted to have all assigned profile instances of no other class than the *ISTDutyTripProfile*, which is designed to describe a potential duty trip in possibly big detail (for more details on the *Duty Trip* and this profile type, see below). The discussed profile may carry also information about accommodation and dining preferences which refer to the *Hotel* and the *Restaurant* class instances. It is easy to notice that, similarly to the case of *Accommodation Preference* and *Dining Preference* profiles, this is another link between the *IST* ontology and the "travel objects" of the *TSS* ontology. Additionally, the OWL class range of the *ISTDutyTripProfile destination* property, which is defined in the *IST* ontology, refers to the *PlaceOnEarth* class which derives the *SpatialThing* class of the *TSS* ontology.

Concerning the *TSS* ontology, let us note that we are currently using only its minimalistic OWL-DL version. In the near future, for description of "travel objects," we intend to utilize the full version of the *TSS* ontology. However, since

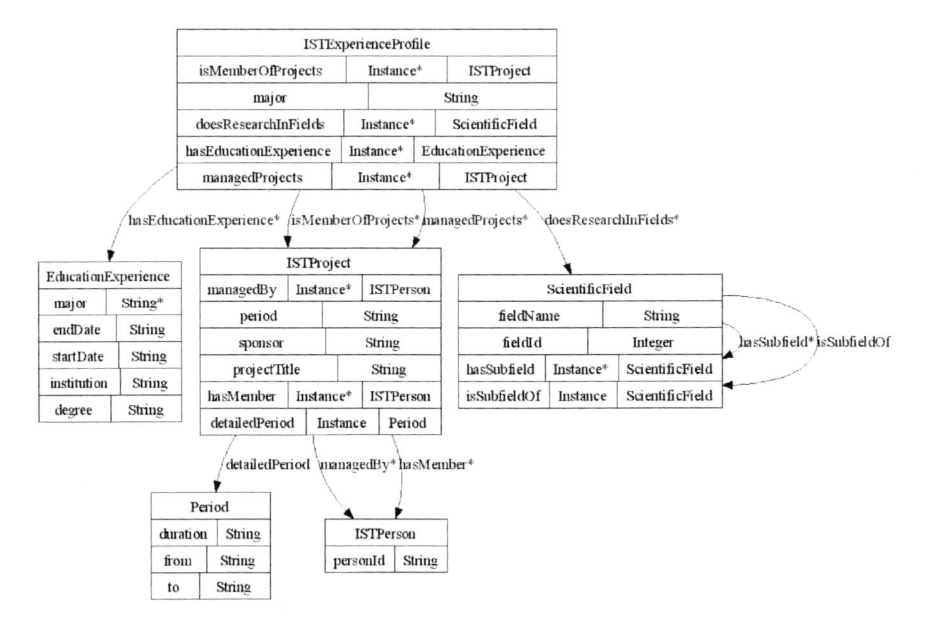

Fig. 8.9. Ontological Description of Employees Competences

it was created in the RDF Schema [5], this will require an adequate migration to OWL-DL; for the sake of compatibility with the *Duty Trip Support* and the *Grant Announcement* subsystems. In figure 8.11 we present the simplified version of the *TSS* ontology used in the system today.

Another example of a non-human resource subclass is the *ISTAnnouncement* which represents grant opportunities in the *GSA* scenario. This class has a property *refScientificField* which refers to an instance of the *ScientificField* class. Please note that instances of the same class are also referenced by instances of the *ExperienceProfile*. Hence, a relation between instances of the *ISTAnnouncement* and the *ISTPerson* who has her *ExperienceProfile* defined may be established. This issue will appear again in detail when we discuss ontological matching applied in the system.

8.4.6 Using Proposed Ontology to Demarcate Sample Resources

Let us now present a collection of samples of demarcating specific resources in the *Virtual Organization*, using concepts introduced thus far. Note that due to the limited space, we can only point to a few aspects and we hope that the reader will be able to follow the example and find more features. The initial context is provided by a *Duty Trip* to a conference in Oulu, Finland, where Mr. Jackie Chan (who comes from Hong-Kong) will stay in a Radisson SAS Hotel (and visit also Mikka Korteleinen in Rovaniemi). We start by illustrating (1) how the geolocation will be demarcated (following the travel ontology proposed in [14]), and (2) the direct connection between the travel ontology ([14]) and the organization

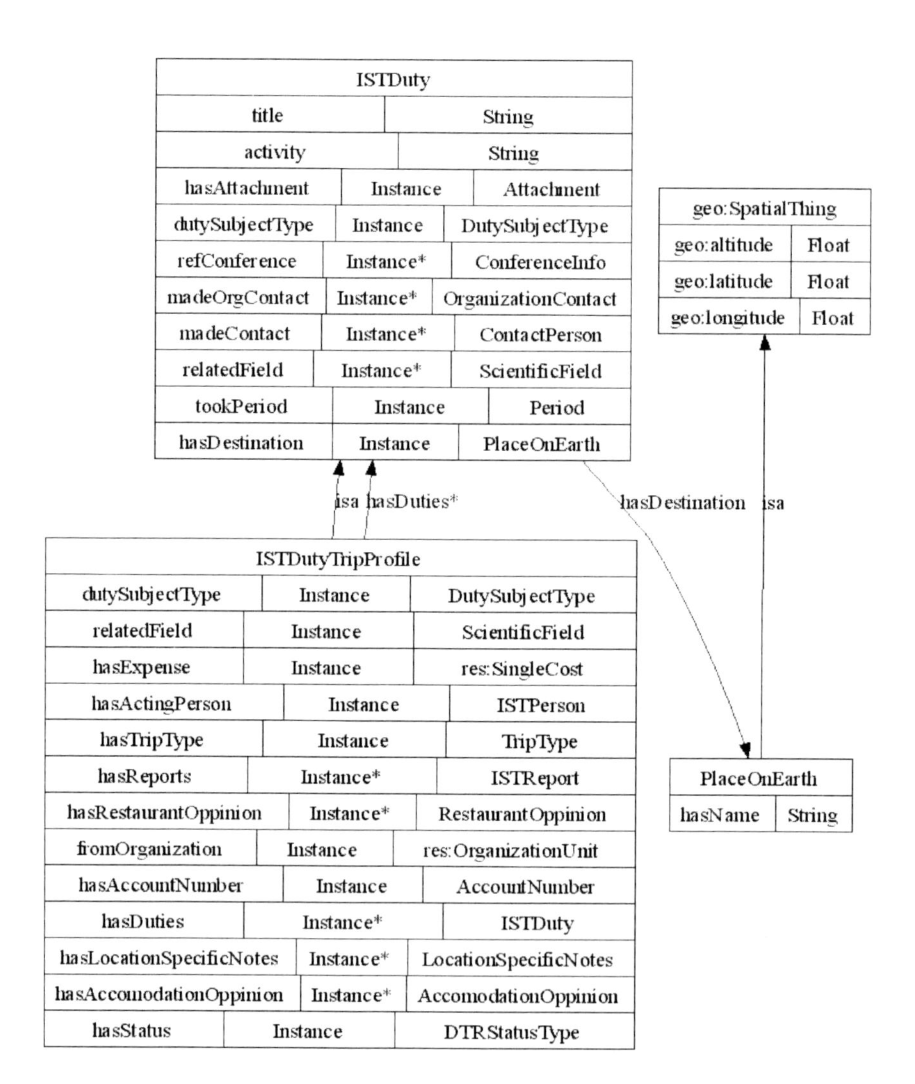

Fig. 8.10. Duty Trip Report Profile

ontology as the cities *geo:OuluCity*, *geo:HongKongCity* and *geo:RovaniemiCity* are instances of travel ontology element: *SpatialThing* and organization ontology class *City*.

```
geo:FinlandCountry a onto:Country;
    onto:name "Finland"^^xsd:string.
geo:ChinaCountry a onto:Country;
    onto:name "China"^^xsd:string.
geo:OuluArea a onto:Area;
    onto:name "Oulu"^^xsd:string;
    onto:isInCountry :FinlandCountry;
    onto:adjacentArea :LappiArea.
geo:LappiArea a onto:Area;
    onto:name "Lappi"^^xsd:string;
```

```
    onto:isInCountry :FinlandCountry;
    onto:adjacentArea :OuluArea.
geo:HongKongArea a onto:Area;
    onto:name "Hong_Kong"^^xsd:string;
    onto:isInCountry :ChinaCountry.
geo:OuluCity a onto:City;
    onto:name "Oulu"^^xsd:string;
    onto:long "25,467"^^xsd:float;
    onto:lat "65,017"^^xsd:float;
    onto:isInArea :OuluArea.
geo:RovaniemiCity a onto:City;
    onto:name "Rovaniemi"^^xsd:string;
    onto:long "25,8"^^xsd:float;
    onto:lat "66,567"^^xsd:float;
    onto:isInArea :LappiArea.
geo:AberdeenCity a onto:City;
    onto:name "Aberdeen"^^xsd:string;
    onto:long "114,15"^^xsd:float;
    onto:lat "22,25"^^xsd:float;
    onto:isInArea :HongKongArea.
```

Here we have defined the *FinlandCountry* and the *ChinaCountry*. The presented snippet includes also instances of *OuluCity*, *RovaniemiCity* and *AberdeenCity* cities, and their region related data: *OuluArea*, *RovaniemiArea* and *HongKongArea*. The first two instances represent countries in which these cities and regions are located: China and Finland. Country, Region and City are three levels of administrative land division that we initially intend to support in the system. The issue of populating database with real life geospatial and administrative information will be discussed in the future as the most suitable methods for this purpose are still being researched. Let us note that we do not claim the above proposed representation of geospatial information is the most efficient solution to the problem, but we assume that it is sufficient enough for the purpose of our information provisioning system and the *Duty Trip Support* application.

The listing that follows shows a simple instance of the Radisson SAS Hotel in Oulu, demarcated according to the simplified *TSS* schema. Note that the hotel feature *locatedAt* references instance of a *City* and *SpatialThing* classes—the *OuluCity*. It is the direct connection of the *TSS* ontology and *VO Ontology* which was discussed above (see also, 8.11).

```
hot:OuluRadisonSAS a tss:Hotel;
  onto:locatedAt geo:OuluCity.
```

ContactPerson#1 represents a human resource that is not employed at the Institute but is recognized because it has been introduced in the past to the system by one of the Institute's employees. According to the example beneath, Mikka Korteleinen is a researcher who specializes in Paleontology and is located in Rovaniemi, Finland.

```
:ContactPerson\#1 a onto:ContactPerson;
    onto:hasProfile :ContactPersonProfile\#1.
:ContactPersonProfile\#1 a onto:ContactPersonProfile;
    person:fullname ‘‘Mikka Korteleinen’’^^xsd:string;
    person:gender person:Male;
    person:birthday "1967-11-21T00:00:00"^^xsd:dateTime;
    onto:doesResearch science:Paleontology-13108;
    onto:locatedAt geo:RovaniemiCity;
    onto:belongsTo :ContactPerson\#1.
```

In the next snippet we introduce instances of the *ISTPerson* and *OrganizationUnit* classes. These instances represent Mr. Jackie Chan and Ms. Mi Lin who are employees of the Institute. The organization units to which Mr. Chan and Ms. Lin belong to reflect their positions in the organizational structure of the Institute.

```
:HROU a onto:OrganizationUnit;
  onto:name ``Human Resource Management Organization Unit''^^xsd:string.
:GOU a onto:OrganizationUnit;
  onto:name ``General Organization Unit-suitable
        for all employees''^^xsd:string.
:Employee\#1 a onto:ISTPerson;
        onto:id "1234567890"^^xsd:string;
        onto:hasProfile (:Employee\#1PProfile, :Employee\#1EProfile);
        onto:hasProfilePriviledges :ResProfPriv\#2.
        onto:belongsToOUs (:GOU).
:Employee\#2 a onto:ISTPerson;
             onto:id "011111111"^^xsd:string;
             onto:hasProfile (:Employee\#2PProfile);
             onto:belongsToOUs (:GOU, :HROU).
```

Detailed personal information of each of these employees is described in separate instances of the *ISTPersonalProfile* class. Such profiles could look as follows:

```
:Employee\#1PProfile a onto:ISTPersonalProfile;
  onto:belongsTo :Employee\#1;
  person:fullname "Jackie_Chan"^^xsd:string;
  person:gender person:Male;
  person:birthday "1982-01-01T00:00:00"^^xsd:dateTime.
:Employee\#2PProfile a onto:ISTPersonalProfile;
  onto:belongsTo :Employee\#1;
  person:fullname "Mi_Lin"^^xsd:string;
  person:gender person:Female;
  person:birthday "1981-02-01T00:00:00"^^xsd:dateTime.
:Employee\#1EProfile a onto:ISTExperienceProfile;
  onto:belongsTo :Employee\#1;
  onto:doesResearchInFields
    scienceNamespace:Volcanology-13105,
    scienceNamespace:Paleontology-13108,
    scienceNamespace:Geochronology-13204;
  onto:knowsFields
    [a onto:Knowledge;
    onto:knowledgeObject scienceNamespace:Volcanology-13105;
    onto:knowledgeLevel "0.75"^^xsd:float],
    [a onto:Knowledge;
    onto:knowledgeObject scienceNamespace:Paleontology-13108;
    onto:knowledgeLevel "0.40"^^xsd:float],
    [a onto:Knowledge;
    onto:knowledgeObject scienceNamespace:Geochronology-13204;
    onto:knowledgeLevel "0.90"^^xsd:float];
  onto:managesProjects (:Project1).
:Project1 a onto:ISTProject;
  onto:managedBy :Employee\#1;
  onto:period
    [a onto:Period;
    onto:from "2008-06-01T00:00:00"^^xsd:dateTime;
    onto:to "2009-05-31T00:00:00"^^xsd:dateTime];
 onto:fieldsRef scienceNamespace:Volcanology-13105;
 onto:projectTitle``Very Important Volcanology
    Scientific Project''^^xsd:string.
```

Note that from the snippet above we can establish that a person identified as *Employee#1* specializes in *Volcanology* and his *level of knowledge* is identified as 0.75 (for more info about assigning levels of skills, or more generally "temperature" to a feature, see [12, 14, 22]), *Paleontology* (level of knowledge identified as 0.4), and *Geochronology* (level of knowledge 0.9). Additionally, this person is

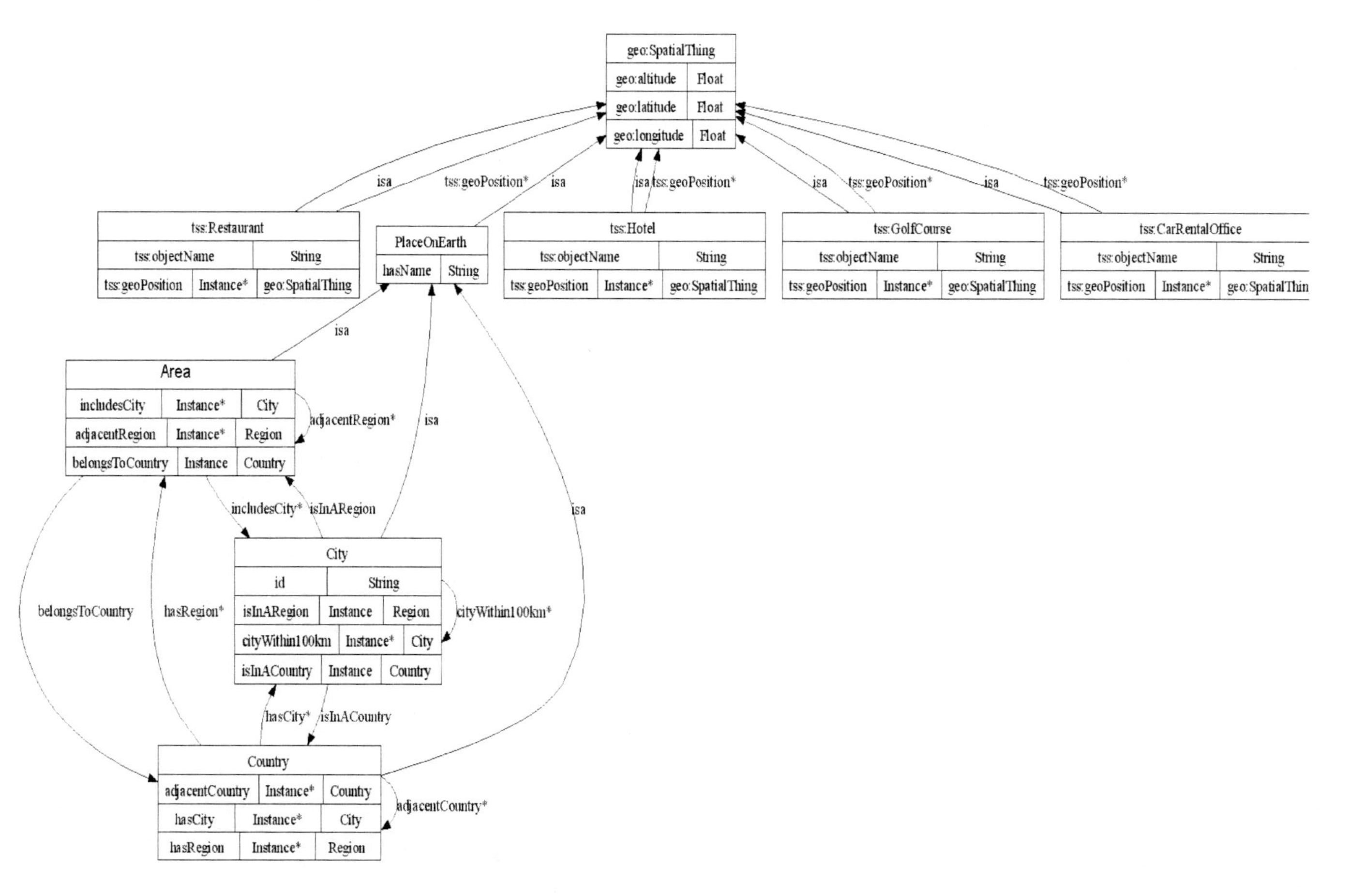

Fig. 8.11. Minimalistic version of the *TSS* ontology used in the system

scheduled to manage a project entitled: "Very Important Volcanology Scientific Project", which starts on June 1st, 2008 and ends on May 31st, 2009.

Obviously, scientific interests of a given employee (*onto:knowsFields* in the above example) can be replaced by professional skills describing worker in any discipline. For instance, they could as well be used to specify that a given programmer has knowledge of Smalltalk (level 0.7), Fortran (level 0.5), dBase (level 0.65), etc. In this way the proposed approach is both robust and flexible.

Having defined human resources, in what follows, we define an exemplary *Duty Trip Report* as a non-human resource.

```
:DTR\#1 a onto:ISTDutyTripReport;
  onto:hasProfile (:DTRProfile\#1);
  onto:hasProfilePriviledges :ResProfPriv\#1.
:DTRProfile\#1 a onto:ISTDutyTripReportProfile;
  onto:destination geo:OuluCity;
  onto:traveler :Employee\#1;
  onto:status dtStatusNamespace:Application;
  [a onto:Period;
  onto:from ''2008-06-07T00:00:00''^^xsd:dateTime;
  onto:to ''2008-06-19T00:00:00''^^xsd:dateTime.].
  onto:stayedAt hot:OuluRadisonSAS
  onto:expense [a onto:SingleCost;
   ''4000''^^xsd:float;
   onto:expenseCurrency ''USD''^^xsd:string.]
  onto:duty :AdditionalDuty\#1;
  onto:purpose ''Conference''^^xsd:string;
  onto:belongsTo :DTR\#1.
:AdditionalDuty\#1 a onto:ISTDuty;
   onto:destination geo:RovaniemiCity;
   onto:madeContact :ContactPerson\#1.
```

Here the *DTRProfile#1* is a profile of resource represented by the *DTR#1*. In our example the latter is a *Duty Trip Report* resource. The employee who this profile directly refers to, is represented by the *:Employee#1*. Hence, we can tell that a person represented in the system as *:Employee#1* applied for a duty trip (*DTR#1*). The current status of that Duty Trip Report is *Application* and the trips destination is Oulu, Finland. The researcher intends to stay there for twelve days (from June 7th till June 19th, 2008). Again, properties of the *ISTDutyTripProfile* refer to the system schemas (organization and domain ontologies) and fulfill the data model requirements set by the Duty Trip Support System which we develop (see also [38]). Please note that, in order not to overly complicate the example, the snippet above does not cover all properties of the *ISTDutyTripProfile* class which deprecated the *DTProfile* class defined in [38].

```
:ProfPriv\#1 a onto:ProfilePriviledge;
          onto:forUnit :HROU;
          onto:hasPriviledgeType priv:Admin.
:ProfPriv\#2 a onto:ProfilePriviledge;
          onto:forUnit :GOU;
          onto:hasPriviledgeType priv:Read.
:ResProfPriv\#1 a onto:VOResourceProfilePriviledge;
          onto:forRProfile :PersonalProfile\\#1;
          onto:hasPriviledge (<:ProfPriv\\#1>).
:ResProfPriv\#2 a onto:VOResourceProfilePriviledge;
          onto:forRProfile :DTRProfile\\#1;
          onto:hasPriviledge (<:ProfPriv\\#1> <:ProfPriv\\#2>).
```

In the snippet above we define a set of resource profile privileges which were presented in figure 8.4. Observe that the defined privileges allow members of the

HR unit (in case of our example: Ms. Mi Lin) to *administer* selected profiles (e.g. *:PersonalProfile#1*), while members of the *General Organization Unit* are only allowed to *read* it (by default all access is forbidden). On the other hand, the *:DTRProfile#1* can be read by all employees of the *GOU*. In this way we assure control of access rights within the organization.

Finally, in order to discuss matchmaking that is going to take place in the system, let us introduce one more example of a non-human resource—an exemplary grant announcement. The main topic of this grant is: *Geochemistry*.

```
:SampleGrant a onto:ISTAnnouncement;
  onto:hasDescription ''Description of the exemplary
   grant announcement. It should be really interesting.''^^xsd:string;
  onto:refScientificFields
          (<scienceNamespace:Geochemistry-13200>).
```

Note that the *SampleGrant* does not have its own profile and all its attributes are defined as its bare properties because there is not need to restrict access to information about *Grant Announcements* in the system with the use of *VOResourceProfilePriviledge* instances.

8.5 Matchmaking in the System

In the *Duty Trip Support* scenario, the *OPM* undertakes the role of a *Travel Assistant*, while in the *Grant Assistant* scenario it plays the role of a *Document Dispatcher* (in a real organization the first role may be played by a human supported by its *PA*, while the latter by a sfotware agent alone). Let us now describe how the desired results (finding personalized information or delivery of the document to the correct set of workers) are to be obtained. Before we proceed let us stress that we assume that all data *within* a given organization is demarcated utilizing a common (for that organization) ontology. Therefore, in what follows we do not have to deal with matching differing and potentially incompatible (external) ontologies. All that we are interested in is: how to establish "distances between resources" within a single ontology and how to use this information in the above described scenarios.

Let us now consider, introduced above, sample profile of a human resource—the *Employee#1* human resource and his profile; and demarcated non-human resources—the Duty Trip Report (*:DTRProfile#1*) and the *SampleGrant*. These profiles will allow us to introduce and briefly discuss matchmakings that are to take place in the system.

8.5.1 Calculating Distances between Resources

From the scenarios described above and summarized in figure 8.5, we can easily see the need for resource matching (finding distances between two or more resources). To focus our attention, let us present a few examples of types of resource matching operations that have to be implemented in our system (this list is not intended to be exhaustive, but rather to point to some classes of needed resource matching and/or distance calculations):

1. computing distance between two geographical locations; to be able to establish if a given location is close-enough to the place where the employee is to travel (so that she can attempt at visiting another institution and/or colleague),
2. matching a non-human resource (e.g. a grant, hotel, restaurant, conference) with a human-resource; to find if a person who is planning a trip could be interested in a given nearby located conference, or if an employee is potentially interested in a grant announcement,
3. matching two human resources to find out who are the researchers that a person planning a trip may be interested in visiting.

Upon further reflection it is easy to notice that the way the distance between resources should be calculated depends on types of objects which are arguments of calculations. For example, the distance between value of *onto:destination* property of the *DTRProfile#1* and the value of the *onto:locatedAt* of the *ContactPersonProfile#1* instance will be calculated in a different way than the distance between values of *onto:refScientificField* property of the *SampleGrant* and the *ExperienceProfile#1* instances. The following object types that appear in our work can be distinguished, based on different approach to calculate their distance (calculations specified here involve the above presented ontology snippets):

1. Objects which represent geographical locations—distance between the *onto:destination* property range of the *onto:ISTDutyTripProfile* class and the *onto:locatedAt* of the *onto:ContactPersonProfile* class.
2. Numeric objects—distance between *onto:long* property values.
3. Date objects—distance between *onto:from* and the *onto:from* (or the *onto:to*) property values.
4. Enumerable objects—distance between *onto:refScientificField* property values/range of the *onto:ISTAnnouncement* and the range of *onto:doesResearchInFields* property of the *onto:ISTExperienceProfile* class.

Let us now discuss possible approaches to distance calculations/resource matching for the four distinguished classes of properties.

Location based calculations. *City*, *Country* and *Area* are classes designed to represent geographical locations which may be visited by the *User*. These classes have properties which allow to build a tree structure of countries, areas and cities. For instance, *FinlandCountry*, *LappiArea*, *OuluArea*, *ChinaCountry*, *HongKongArea*, *RovaniemiCity*, *OuluCity* and *AberdeenCity* were samples of geo-locations introduced above. They represent a part of an administrative division of Finland and China. First level in our structure is a country, the second level is an area and finally city is the third one. Available properties allow to query for neighbor (adjacent) instances of the same class. This approach requires access to administrative divisions of the world data, otherwise it may be of little value in terms of facilitating a location based advice. Apart from the administrative division tree, these classes allow to describe actual geo-coordinates of objects. Location based advising can be performed by calculating object's distances using the general formula (*long* - longitude, *lat* - latitude, *alt* - altitude):

$$\sqrt{(long_0 - long_1)^2 + (lat_0 - lat_1)^2 + (alt_0 - alt_1)^2}$$

Note that in most business travel scenarios the altitude (alt_0 and alt_1) is of little relevance and can be omitted. Obviously, similar calculations can be performed not only for conferences and/or institutions, but also for all other geo-objects (e.g. restaurants and hotels) as their coordinates are described in the same way as cities (hotel, restaurant and city are subclasses of the *Spatial Thing* class in our travel ontology; see [14, 39]). Therefore, the *DTS* system will be able to provide at least the following geo-info-based advice:

1. Location notes and tips (textual information about a location which was added to Duty Trip Reports - class in the ontology: *Location Specific Notes*),
2. Organizations and people that can be visited (objects of *Organization Contact* and *Contact Person* classes, these objects are created by the employees during the Duty Trip Report's creation),
3. Information about nearby conferences of possible interest (based on location of the trip and the conference as well as on the personal interest and conference topics),
4. Hotels and restaurants (based on the *Hotel* and *Restaurant* TSS ontology classes),
5. Car rental and golf courses (ontology extensions based on the OTA specification [31], also included in the *TSS* ontology).

8.5.2 Numeric and Date Object Calculations

Computing distance between numeric and date object is rather obvious. The distance will be represented by the result of difference operation on these objects. In the first case, the result will be a number, in the latter case the result will be a time period (e.g. of a stay in a given place). Note that currently most major programming languages provide date calculation support hence we believe this issue should not be discussed in more detail (assuming there are no problems with date representation and deserialization).

8.5.3 Enumerable Object Calculations

In case of an ontology, enumerable values can be more complex than *enums* known from popular programming languages. In an ontology, class instances can also be enumerable values. In that case complex structures can be constructed, representing relations between objects. For instance, presented above *ScientificField* class falls under the OWL *oneOf* restriction, however each instance of that class has property values which refer to other instances of that class. This results in a graph-like structure of enumerable values.

To calculate distance between two object of enumerable type, let us note first that if the structure of *enum* values is flat (plain list with no relations between objects) it can be assumed that the distance is 0 if the values match, otherwise it equals to 1. An example of such simple enumerable is the *Gender* property, which is utilized in the human resource profile. Here we have two values: *Male* and *Female* and if they match the distance is 0, and 1 otherwise.

Let us now present a method for calculating distance between class instances which involve transitive, non-symmetric properties. Here, a path in a directed graph is calculated for all relations. Let us assume that R is such a transitive, not symmetric relation (*property* in the OWL notation). Then the distance between two vertices of a graph of relation R: v_0 and v_k ($dist_R(v_0, v_k)$) is calculated according to the following algorithm:

1. If there exists $path_R(v_0, v_k)$ in the graph of relation R, then the shortest one can be found and

$$dist_R(v_0, v_k) = length(shorthestPath_R(v_o, v_k));$$

otherwise go to 2nd step.

2. Let $X = \{x : path_R(x, v_0) \text{ and } path_R(x, v_k) \text{ exist}\}$. Find such $y \in X$, that $length(path_R(y, v_0))$ is minimal among all vertices belonging to X (i.e. this is the shortest path):

$$dist_R(v_0, v_k) = 10^{length(path_R(y,v_0))} + length(shorthestPath_R(y, v_k))$$

Note, that this is a simplified case of a method introduced in [34]. The basic difference between them is as follows. The method proposed in [34] assumes existence of multiple relations linking any class instance (node) from a single node and merging edges which represent relations of the same direction between the same nodes; thus, the distance is computed including all properties (relations) of classes (concepts). The algorithm presented above, on the other hand, is restricted to one selected property (relation) of a class (concept) and an inverse of the selected property. This pair represent generalization and specialization relations between concepts. The algorithm presented here can be substituted for the algorithm of [34] by adjusting appropriate weights to concepts relations. Specifically, used here weights of 1 for specialization and 10 for generalization.

Let us now describe calculation of distance between research interests of a human resource and grant announcement topics, while utilizing examples introduced above. According to the proposed algorithm the following distance values can be found (here we calculate all-against-all distance values):

$dist_{SF} = path_{isSubfieldOf}$
$dist_{SF}$(Volcanology −13105, GeologicalScience −13100)=10
$dist_{SF}$(Volcanology −13105, Geochemistry −13200)=101
$dist_{SF}$(Volcanology −13105, Geochronology −13204)=102
$dist_{SF}$(Volcanology −13105, Paleontology −13108)=11
$dist_{SF}$(Paleontology −13108, GeologicalScience −13100)=10
$dist_{SF}$(Paleontology −13108, Geochemistry −13200)=101

```
distSF(Paleontology -13108, Geochronology-13204)=102
distSF(Paleontology -13108, Volcanology -13105)=11
distSF(Geochronology-13204, Geochemistry-13200)=10
distSF(Geochronology-13204, GeologicalScience -13100)=101
distSF(Geochronology-13204, Volcanology -13105)=102
distSF(Geochronology-13204, Paleontology -13108)=102
distSF(Geochemistry-13200, GeologicalScience -13100)=11
distSF(Geochemistry-13200, Volcanology -13105)=12
distSF(Geochemistry-13200, Paleontology -13108)=12
distSF(Geochemistry-13200, Geochronology-13204)=1
distSF(GeologicalScience -13100, Geochemistry-13200)=11
distSF(GeologicalScience -13100, Volcanology -13105)=1
distSF(GeologicalScience -13100, Paleontology -13108)=1
distSF(GeologicalScience -13100, Geochronology-13204)=12
```

These values allow us to utilize a number of strategies to establish "closeness" of two resources. The simplest one would be, if for any two properties the distance is below a certain threshold, then a exemplary grant announcement should be recommended as potentially interesting. In case of Mr. Chan who is interested in *Volcanology*, *Paleontology* and *Geochronology* we may measure the distance between his interests and the exemplary grant announcement which main topic is *Geochemistry*. The results, which are part of the all-against-all calculation presented in the listing above, are as follows:

```
distSF(Geochemistry-13200, Volcanology -13105)=12
distSF(Geochemistry-13200, Paleontology -13108)=12
distSF(Geochemistry-13200, Geochronology-13204)=1
```

Here, since in one of the areas the distance is equal to 1, this grant announcement should be delivered to Mr. Chan. Note that distance could be also scaled by the level of knowledge of the specialist in the field. Furthermore, a number of more involved considerations are also possible. In this context let us note that values of the $dist_R(v_0, v_k)$ function allow us to specify how far are the graph nodes located from each other in terms of a transitive, not symmetric relation R. In the case of research specialization modeling relation we can assume that the maximum length of $path_R(v_0, v_k)$ is 9. In our ontology an example of such relation is the *isSubfieldOf* property of the *ScientificField* class, where the maximum length of $path_{SF}(v_0, v_k)$ is 2. Additionally, infinite distance is not considered. With such assumptions we are able to distinguish following groups of conclusions which can be drawn from the function values:

1. If $dist_R(v_0, v_k) = 0$, then $v_0{=}v_k$
2. If $dist_R(v_0, v_k) = n$ and $0 < n < 10$, then $R(v_0, v_k) = true$ and v_k is n-deep specialization of v_0
3. If $dist_R(v_0, v_k) = n$ and $n = 10^k, k > 0$, then $R(v_0, v_k) = false$ and v_0 is k-deep specialization of v_k
4. If $dist_R(v_0, v_k) = n$ and $10^k < n < 10^{(k+1)}, k > 0$, then $R(v_0, v_k) = false$ and v_0 is $n - 10^k$-deep specialization of k-deep specialization of v_k

For instance, if

$$dist_{SF}(Volcanology - 13105, Paleontology - 13108) = 11,$$

we may say that there is a node of which both *Volcanology* and *Paleontology* are direct specializations. In terms of the exemplary grant announcement and Mr Chans experience profile we may conclude that the grant announcement may be of potential interest to him. In particularly the distance between *Geochemistry* and *Geochronology* equals to 1, meaning that *Geochronology* is the direct offspring of the *Geochemistry* field of science.

These observations allow us to develop a number of reasoning scenarios that utilize not only information about numerical distance, but also following form it knowledge about the structure of relations. Developing a reasoning engine utilizing this information is next step in our work.

8.6 Concluding Remarks

The aim of this chapter was to summarize our work concerning development of system responsible for resource management in an agent-based virtual organization. First, we have outlined the proposed system in the context of a task introduced to the organization. We followed with a discussion of interrelations between entities belonging to the real-world organization and software agents that belong to the virtual organization. Next, we have focused on ontologies that are to be used in the system. We have introduced the general ontology of a virtual organization and followed it with discussion of the way that these ontologies can be extended to deal with a specific case of a research institute. We have concluded the chapter with an overview of matchmaking procedures that are needed in the proposed system.

In the paper we have outlined a number of future research directions. At this stage the most important ones are: (1) completing subsystem for geospatial data processing, (2) implementing interface between the web-based data input and the agent-based system, (3) completing implementation of ontological matchmaking (as pertinent to the two application areas), (4) completing implementation and testing of two sample applications (*Duty Trip Support* and *Grant Announcement Support*). We will report on our progress in subsequent publications.

References

1. http://www.eil.utoronto.ca/enterprise-modelling/index.html
2. Cervenka, R., Trencansky, I.: AML: The Agent Modelling Language, Birkhäuser (2007)
3. Fox, M.S., Gruninger, M.: Enterprise Modelling. In: AI Magazine, pp. 109–121. AAAI Press, Fall (1998)
4. http://www.w3.org/TR/2004/REC-owl-guide-20040210/#OwlVarieties
5. http://www.w3.org/TR/rdf-schema/

6. Biesalski, E., Abecker, A.: Human Resource Management with Ontologies. In: Althoff, K.-D., Dengel, A.R., Bergmann, R., Nick, M., Roth-Berghofer, T.R. (eds.) WM 2005. LNCS (LNAI), vol. 3782, pp. 499–507. Springer, Heidelberg (2005)
7. Bizer, C., Heese, R., Mochol, M., Oldakowski, R., Tolksdorf, R., Eckstein, R.: The Impact of Semantic Web Technologies on Job Recruitment Processes. In: Proc. International Conference Wirtschaftsinformatik (WI 2005), Bamberg, Germany (2005)
8. Ganzha, M., Paprzycki, M., Popescu, E., Bădică, C., Gawinecki, M.: Agent-Based Adaptive Learning Provisioning in a Virtual Organization. In: Advances in Intelligent Web Mastering. Proc. AWIC 2007, Fontainebleu, France. Advances in Soft Computing, vol. 43, pp. 25–40. Springer, Heidelberg (2007)
9. Ganzha, M., Paprzycki, M., Gawinecki, M., Szymczak, M., Frackowiak, G., Bădică, C., Popescu, E., Park, M.-W.: Adaptive Information Provisioning in an Agent-Based Virtual Organization—Preliminary Considerations. In: Proceedings of the SYNASC 2007 Conference, pp. 235–241. IEEE CS Press, Los Alamitos (2007)
10. Ganzha, M., Gawinecki, M., Szymczak, M., Frackowiak, G., Paprzycki, M.: Generic Framework for Agent Adaptability and Utilization in a Virtual Organization—Preliminary Considerations. In: Proceedings of the 2008 WEBIST Conference (to appear) (May 2008)
11. Szymczak, M., Frackowiak, G., Ganzha, M., Gawinecki, M., Paprzycki, M., Park, M.: Resource Management in an Agent-based Virtual Organization — Introducing a Task Into the System. In: Proceedings of the MaSeB Workshop, pp. 458–462. IEEE CS Press, Los Alamitos (2007)
12. Gawinecki, M.: Modelling User on the Basis of Interactions with a WWW Based System, Master Thesis, Adam Mickiewicz University (2005)
13. Gawinecki, M., Gordon, M., Paprzycki, M., Vetulani, Z.: Representing Users in a Travel Support System. In: Kwasnicka, H., et al. (eds.) Proceedings of the ISDA 2005 Conference, pp. 393–398. IEEE Press, Los Alamitos (2005)
14. Gawinecki, M., Gordon, M., Nguyen, N.T., Paprzycki, M., Szymczak, M.: RDF Demarcated Resources in an Agent Based Travel Support System. In: Golinski, M., et al. (eds.) Informatics and Effectiveness of Systems, pp. 303–310. PTI Press, Katowice (2005)
15. Gawinecki, M., Gordon, M., Nguyen, N.T., Paprzycki, M., Zygmunt Vetulani, Z.: Ontologically Demarcated Resources in an Agent Based Travel Support System. In: Katarzyniak, R.K. (ed.) Ontologies and Soft Methods in Knowledge Management, Advanced Knowledge International, Adelaide, Australia, pp. 219–240 (2005)
16. Warner, M., Witzel, M.: Zarzadzanie organizacja wirtualna, Oficyna Ekonomiczna (2005)
17. Barnatt, C.: Office Space, Cyberspace and Virtual Organization. J. of General Management 20(4), 78–92 (1995)
18. Bleeker, S.E.: The Virtual Organization. In: Hickman, G.R. (ed.) Leading Organizations. Sage, Thousand Oaks (1998)
19. Grenier, R., Metes, G.: Going Virtual: Moving Your Organization into the 21st Century. Prentice-Hall, Englewood Cliffs (1995)
20. Goldman, S.L., Nagel, R.N., Preiss, K.: Agile Competitors and Virtual Organizations, Van Nostrand Reinhold (1995)
21. Dunbar, R.: Virtual Organizing. In: Warner, M. (ed.) International Encyclopedia of Business and Management, pp. 6709–6717 (2001)
22. Greer, J., McCalla, G.: Student modeling: the key to individualized knowledge based instruction. NATO ASI Series F, 125 (1993)
23. HR-XML Consortium, `http://www.hr-xml.org/`
24. Jena Semantic Web Framework, `http://jena.sourceforge.net/`

25. Jennings, N.R.: An agent-based approach for building complex software systems. In: Commun. ACM, vol. 44(4), pp. 35–41. ACM Press, New York (2001)
26. Â ă Agent Technology: Foundations, Applications, and Markets. In: Jennings, N.R., Wooldridge, M.J. (eds.). Â ă Springer, 2002
27. Kim, H.M., Fox, M.S., Gruninger, M.: An ontology for quality management—enabling quality problem identification and tracing. BT Technology Journal 17(4), 131–140 (1999)
28. Korea Science and Engineering Foundation, http://www.kosef.re.kr/english_new/index.html
29. Mochol, M., Wache, H., Nixon, L.: Improving the Accuracy of Job Search with Semantic Techniques. In: Abramowicz, W. (ed.) BIS 2007. LNCS, vol. 4439, pp. 301–313. Springer, Heidelberg (2007)
30. Montaner, M., López, B., de la Rosa, J.L.: A taxonomy of recommender agents on the Internet. Artif. Intell. Rev. 19(4), 285–330 (2003)
31. Open Travel Alliance, http://www.opentravel.org/
32. PostGIS : HOME, http://postgis.refractions.net/
33. http://www.cs.rmit.edu.au/agents/#prometheus
34. Rhee, S. K., Lee, J., Park, M. W.: Ontology-based Semantic Relevance Measure. In: Kim S., et al. (eds.) Proceedings of the The First International Workshop on Semantic Web and Web 2.0 in Architectural, Product and Engineering Design, 6 pages (2007), http://ftp.informatik.rwth-aachen.de/Publications/CEUR-WS/Vol-294/
35. Schmidt, A., Kunzmann, C.: Towards a Human Resource Development Ontology for Combining Competence Management and Technology-Enhanced Workplace Learning. In: Meersman, R., Tari, Z., Herrero, P. (eds.) OTM 2006 Workshops. LNCS, vol. 4278, pp. 1078–1087. Springer, Heidelberg (2006)
36. Semantic Web, http://www.w3.org/2001/sw/
37. Szymczak, M., Frackowiak, G., Ganzha, M., Gawinecki, M., Paprzycki, M., Park, M.-W.: Resource Management in an Agent-based Virtual Organization—Introducing a Task Into the System. In: Proc. of the WI/IAT 2007 Workshops, IEEE CS Press, Los Alamitos (in press, 2007)
38. Szymczak, M., Frackowiak, G., Ganzha, M., Gawinecki, M., Paprzycki, M., Park, M.-W.: Adaptive Information Provisioning in an Agent-Based Virtual Organization—Ontologies in the System (submitted for publication)
39. Szymczak, M., Gawinecki, M., Vukmirovic, M., Paprzycki, M.: Ontological Reusability in State-of-the-art Semantic Languages. In: Olszak, C., et al. (eds.) Knowledge Management Systems, pp. 129–142. PTI Press, Katowice (2006)
40. Training, http://en.wikipedia.org/wiki/Training
41. Tzelepis, S., Stephanides, G.: A conceptual model for developing a personalized adaptive elearning system in a business environment. In: Current Developments in Technology-Assisted Education, vol. III, pp. 2003–2006. Formatex Publishing House (2006)
42. Wooldridge, M.: An Introduction to MultiAgent Systems. John Wiley & Sons, Chichester (2002)
43. Wooldridge, M., Jennings, N.R., Kinny, D.: The Gaia Methodology for Agent-Oriented Analysis and Design. Autonomous Agents and Multi-Agent Systems 3(3), 285–312 (2000)
44. http://ontoweb.org/About/Deliverables/ppOntoweb.pdf
45. http://ontoweb.aifb.uni-karlsruhe.de/Ontology/index.html
46. Spyns, P., Oberle, D., Volz, R., Zheng, J., Jarrar, M., Sure, Y., Studer, R., Meersman, R.: OntoWeb—a Semantic Web Community Portal. In: Proc. Fourth International Conference on Practical Aspects of Knowledge Management, Vienna, Austria (2002)

9

BDI Agents: Flexibility, Personalization, and Adaptation for Web-Based Support Systems

Maria Fasli[1] and Botond Virginas[2]

[1] University of Essex, Department of Computing and Electronic Systems
Wivenhoe Park, Colchester CO4 3SQ, UK
mfasli@essex.ac.uk

[2] Intelligent Systems Research Centre, British Telecom
MLB1 pp12, Adastral Park, Ipswich, IP5 3RE, UK
botond.virginas@bt.com

Abstract. Users increasingly value personalized, flexible and interactive forms of support. In this chapter we consider the problem of providing decision support to mobile knowledge workers through a mixed-initiative multi-agent system. We describe an abstract architecture for developing mixed-initiative MASs and identify the main components and underlying ontologies that are required to support automated reasoning, problem solving and adaptation. We propose the BDI model of agency for developing agents to whom goals can be delegated and who can work collaboratively with the users and each other. Users and agents interact in a mixed-initiative mode to establish a common ground and the shared goals to be achieved. Plans as the means to bring about the user's goals, are utilized to enable the refinement and fine-tuning of business processes to better conform to the work style of the individual knowledge worker. Furthermore, we consider possible ways of dealing with exceptional circumstances as they arise in the process of executing plans and how an agent can essentially learn to deal with such situations by interacting with other agents as well as the user who is considered an expert.

Keywords: BDI Agents, Planning, Digital Service Space, Business Processes, Personalization, Adaptation, Flexibility.

9.1 Introduction

In our ever increasing in complexity information society, users require personalized, flexible and interactive forms of support. Agent technology is promising in this respect as agents – smart autonomous pieces of software that can interact with the user – are tasked with the achievement of the user's goals. But, despite the great advances in related sub-fields of Artificial Intelligence, fully automated reasoning in complex, dynamic and time-constrained environments is beyond the capabilities of current methods and techniques and nowhere near the flexibility of human reasoning. To address this limitation, mixed-initiative systems propose to take advantage of the strengths and complementarities of human and automated reasoning and create systems in which software agents interact seamlessly with the users to solve complex problems more efficiently [45].

N.T. Nguyen, L.C. Jain (Eds.): Intel. Agents in the Evol. of Web & Appl., SCI 167, pp. 191–221.
springerlink.com

In this chapter, we consider the problem of providing decision support to mobile knowledge workers through a mixed-initiative multi-agent system. In particular, we will explore the use of a specific genre of agents whose cognitive state is represented in terms of beliefs, desires and intentions (BDI). The BDI model of agency is a suitable candidate for developing agents to whom goals can be delegated and who can work collaboratively with users as they have the ability to represent the user's information state and objectives and execute plans to bring them about. Our aim is to enhance the user experience by providing flexible, personalized and interactive support. One of the issues that we are interested in is how agents can learn from the interaction with the user and through other means, especially in situations that the agent does not know what to do.

The chapter is organized as follows. Next we discuss the role of agents as a key-enabling technology for enhancing the user experience. The following section introduces the problem domain of providing tailor-made support to mobile knowledge workers through their digital service space. The subsequent section discusses theoretical and practical issues of the BDI model of agency. Section 9.5, presents an abstract architecture for mixed-initiative multi-agent systems, and following on from that we discuss the details of a mixed-initiative BDI multi-agent system for providing support to mobile knowledge workers that we designed and developed as a proof of concept. The penultimate section discusses related approaches that have enhanced the BDI model with learning capabilities and we close with the conclusions and a brief discussion on current and future work.

9.2 Agents: A Key Enabling Technology

The widespread deployment of computing systems has inevitably led to changes in how people regard and use them. When such changes are major and have a significant impact on people's lives and way of thinking they constitute a *paradigm shift*. It is widely acknowledged that computers have gone through two major paradigm shifts: from powerful calculation and logictics machines to personal tools [49]. The third phase of computing, ubiquitous computing, which is slowly dawning upon us, is characterized by the ever presence of digital and other computational technologies. The signs of this new paradigm shift abound around us today. Users are increasingly surrounded by numerous tiny processing units while computing is almost everywhere and has become part of nearly every human activity. The huge growth of the Internet since the middle 1990s has meant increased connectivity and the flourishing of distributed systems. The Internet has redefined the role of computers as windows to the world: they provide us with a new means to communication and interaction as well as access to information, resources and services in a scale never before possible or imaginable [42].

Inevitably, the users' needs have changed: the standard user-computer interaction model, which is based on the direct manipulation metaphor, no longer suffices. Instead of *acting upon* computer systems, we are increasingly *delegating to* and *interacting with* them in more complex, conversational ways. The nature of computation has changed from mere calculation to delegation and continuous interaction.

Consequently, new theoretical models and software engineering paradigms are required to support the design and development of such systems.

This trend to view computation more as delegation of our goals and continuous interaction with computer systems assumes building systems that can act on one's behalf. As the tasks being delegated increase in complexity, the developed systems need to become smarter, anticipate and learn from their users and actively seek to further their goals. Hence, systems are required to act almost autonomously without the constant and direct intervention from the users, observe their environment and respond to any changes that occur and which affect the achievement of their goals. Such smart computer systems may interact with or seek the assistance or the services of other systems to accomplish their goals.

The natural metaphor for such inevitably intricate and complex systems that aid us in our everyday lives and can react, learn and adapt to the environment and our needs and have the ability to interact with each other is that of an *agent*. Put simply, an agent is a computer system that acts on behalf of its user and attempts to achieve the user's goals and objectives by acting autonomously. An agent is embedded within an environment and it continuously interacts with it as well as other entities, including agents and humans. What distinguishes agents from other pieces of software is that computation is not simply calculation, but delegation and interaction; users do not act upon agents as they do with other software programs, but they delegate tasks to them and interact with them in a conversational rather than in a command mode. Intrinsically, agents enable the transition from simple static algorithmic-based computation to dynamic interactive delegation-based service-oriented computation.

Increasingly, users require services that are tailored to their own individual needs and preferences. The role that agents are called upon to take is that of our digital assistants that facilitate the achievement of our tasks and goals with the minimum effort. As such, they are a key-enabling technology for enhancing the user experience as they are able to mask the underlying complexity of our use of computer systems.

9.3 Problem Description

Agents acting as digital assistants have found applications in the tourist industry [10] and in more traditional office environments [35]. The domain of application that we are concerned with is that of providing configurable and self-adaptable digital services spaces (DSS) in a business context. A digital service space (DSS) is a cohesive and user-specific bundle of information, communication and collaboration services ranging from commodities (news feeds, maps, etc.) and communications (VoIP, presence, push-to-talk, etc.) to complex or domain-specific components (search engines, task lists, mailing lists, business reports, etc.). DSS configuration is the process whereby service consumers/knowledge workers (end-users and organizational entities) and producers (designers and providers) exert control over the presentation and content of DSS, i.e., over the situational logic driving the inclusion, composition and orchestration of services in the DSS.

Personalization is the aspect of configuration concerned with the handling of requirements expressed by end-users. DSS self-adaptation is the process whereby a DSS automatically and continuously re-configures itself.

Mobile knowledge workers in fieldforce and salesforce domains perform different types of jobs, each involving a number of tasks. These tasks may have a strict order of execution (i.e. the underlying business process is fixed) or the order may not be of the essence (i.e. the action plan is loose). An organization would want to provide its knowledge workers with a support system that would assist them in their daily operations. But such a support system should not be overly prescriptive and ideally it should allow knowledge workers to retain some control over the execution of their jobs. In other words, the system should provide a flexible mode of assistance, one that is not rigid and stifles the individual, but provides them with some flexibility in making decisions. Staff empowerment, a central concept in the management of organizations and people, suggests that by recognizing an individual employee's needs and preferences and accommodating them within a well-defined organizational framework, job satisfaction increases [50]. Increased job satisfaction leads to increased productivity and even a sense of ownership of the organization. A support system in this domain needs to be flexible for another important reason. Although jobs may usually involve a number of steps/tasks that need to be performed, the environment in which a knowledge worker operates is dynamic and unavoidably exceptions may arise. Even though responses for some common exceptions can be pre-programmed in the system, inevitably, there are going to be situations that the designer/developer of the system would not have anticipated.

Our objective here is to empower the mobile knowledge worker by providing flexibility in decision-making as he is performing his jobs. His DSS, typically accessed through a mobile device, will be the main channel and decision support capability. Agent technology can be used to support the mobile knowledge worker, by providing him with an agent who can guide him through the various jobs that need to be completed. Each job may be accomplished in a number of different ways, and therefore the designer needs to encode different "recipes", or plans[1], that would enable the knowledge worker to perform jobs. But the designer of such an agent may not be in a position to describe all possible ways of performing a job or all possible exceptions that can be encountered by a knowledge worker (even if a detailed requirements elicitation process has been undertaken). Even if this was indeed possible at a particular moment in time, when the system would be put in use, new exceptions would probably arise in the field which would not have been encountered before. Constantly redesigning the agent or manually creating appropriate responses to deal with exceptions is not an attractive solution.

This is something that ideally the system needs to learn to do gradually. Hence, the agent designer can provide a set of basic "recipes" and then the agent can build a gradually more complex and richer library of recipes that can deal

[1] We use "plans" here loosely to describe a series of actions or tasks that lead to the completition of a job.

with different situations as these arise and which better conform to the work style of the particular knowledge worker. At the same time, the organization would like to maintain some level of control over the activities of its knowledge workers, therefore it is important that boundaries are established as to what they are permitted to do in terms of tasks.

In summary, we require an agent-based system that would support the mobile knowledge worker and satisfy the following:

- Control: The purpose of the system is ultimately to serve the users' needs. Although the agent should be autonomous, the user should have a high degree of control over the system's behavior. The system should guide and support the user, but accept input and instruction from the user when necessary. Hence, the user should be allowed to alter the system's behavior, though within well-defined boundaries within the organizational framework.
- Personalization: Providing personalized services is a key requirement. Agent technology should be used to enhance the user experience by providing flexibility and conforming to individual work styles as these may be expressed in the way that individual knowledge workers perform their jobs. Personalization in this context also includes the appearance of the agent interface.
- Teachability/Adaptation: The user is not simply someone who requires help and support. He is an expert and can intervene and inject new knowledge and capabilities (operational knowledge on performing a job) into the system. Thus, the system should be able to learn from the user's interventions.
- Transparency: For any supportive technology to be successful, it has to be transparent. The user needs to understand what the agent is doing, and therefore providing succinct, but useful information through a simple but well-designed interface is very important.

Support systems that endeavour to integrate and harness automated and human reasoning in order to take advantage of their complementary strengths, have been termed mixed-initiative systems [16, 21, 45]. Mixed-initiative systems enable human and software agents to interact seamlessly and to solve problems more effectively and efficiently. They draw from a number of areas in Artificial Intelligence, such as agents and multi-agent systems, knowledge representation, problem solving and planning, learning and human-computer interaction. One fundamental issue in such systems is that of communication between users and agents. Users and agents need to work collaboratively and hence an effective means of communication is important to establish the common ground as well as the shared goal to work towards.

The problem of providing decision support to a mobile knowledge worker lends itself to mixed-initiative agent-based systems as these can provide both support and guidance but also enhance the user experience by providing more flexibility. In particular, we will explore the use of a specific genre of agents that have explicit mental representations for beliefs, desires and intentions. The BDI model of agency has been used in mixed initiative systems such as TRIPS [16] and PExA [35] and it seems to be an ideal candidate for developing agents to whom goals can be delegated and who can work collaboratively with users. BDI agents can achieve their goals and design objectives through executing plans. Concomitantly, knowledge workers that perform jobs can be regarded as executing workflow for business processes. A business process can be viewed as a high level plan, i.e. a series of tasks (actions, subplans) that have to be executed to bring about a state of affairs where the user's goal will have been successfully completed. However, whereas workflows tend to be rigid, plans are more flexible, in that different plans and subplans may be executed as the environment changes and different conditions are applicable. The problem then of how to handle exceptions during the execution of a business process, is transformed into the problem of handling exceptions arising while an agent executes a plan.

9.4 BDI Agents

The Belief Desire Intention (BDI) model of agency emanated from the work of Bratman [4] in practical reasoning and views agents as intentional systems [12]. An agent according to the intentional stance possesses information about its environment and is able to reason with and about this information and make decisions and take actions in order to achieve its design objectives. Hence, an agent performs two forms of reasoning: theoretical reasoning and practical reasoning. The former is the process of deriving knowledge or reaching conclusions using one's beliefs or knowledge, whereas the later has to do with the process of deciding what to do. Practical reasoning involves two aspects: deciding what we want to achieve also known as *deliberation*, and deciding how we are going to achieve it, also known as *means-ends reasoning* or planning.

An agent's cognitive state according to the BDI paradigm is characterized by three attitudes: beliefs represent the agent's information about the world which may not necessarily be accurate; desires are the agent's potential options or states of affairs that the agent would like to bring about; and intentions express states of affairs that the agent is committed to bringing about through acting.

Intentions are key in practical reasoning. They describe states of affairs that the agent is actually committed to bringing about and as a result they are action-inducing: an intention constitutes reason for action and it is a conscious wish to carry out an act. Inevitably, a philosophical theory of actions must include an account of what it is for an agent to do something intentionally [4, 20].

Although there is much debate in the philosophical literature over whether a reduction of intentions to combinations of beliefs and desires can be vindicated, in his philosophical investigation Bratman [4] argues convincingly that intentions play a major role in an agent's practical reasoning by being conduct-controlling and not simply potential influencers of conduct as desires can be. Intentions provide strong reasons for action, reasons that are over and above ordinary desire and belief reasons together. According to Bratman, intentions play three characteristic functional roles in an agent's behavior:

(i) an agent needs to determine ways to achieve its intentions;

(ii) an agent's intentions lead it to adopt or restrain from adopting further intentions;

(iii) an agent is interested in succeeding in its intentions and as a result it needs to keep track of its (successful or unsuccessful) attempts to do so.

Before we move on, we need to pose and clarify a couple of issues that can be the source of confusion when we talk about BDI-style agents. Unfortunately the BDI concepts are not interpreted in a uniform way. The first clarification point concerns desires and the closely related concept of goals. In particular, there is confusion between the two, as some authors use the same term to refer to slightly different concepts and in some languages, architectures and frameworks the concept of desire may be implicit, not present at all, or replaced by that of goals. Desires describe states of affairs that an agent would ideally like to bring about, and in this sense the set of the agent's desires may not be consistent. For instance, one may have the desire to be an academic and be rich at the same time – the two are not usually compatible. Goals represent states of affairs that the agent would like to bring about, but these are consistent. Hence, goals may be regarded as a consistent subset of desires or refined desires.

The second point requiring clarification concerns goals themselves. Goals (desires/objectives) are central to agent theories and programming languages as they form the basis of the agent's proactive behavior. Goals can take two forms: *declarative* when they describe states of affairs that the agent wants to bring about and *procedural* when they describe a series of steps (or a plan) which when executed achieves the goal. The procedural aspect of goals is important for implementing agents. In AgentSpeak [3], goals are *event-goals*, i.e. every event generates a goal that needs to be handled. If there is an appropriate plan in the agent's plan library to handle the event-goal, this is placed in the intention structure and the agent attempts to execute the plan to completion. These are also known as *achievement goals* as they attempt to achieve a state of affairs. Achievement goals drive the agent to perform actions in order to reach the required state of affairs. *Maintenance goals* [14] define states of affairs that need to remain true, i.e. the agent only needs to take action if they ever become false.

A number of formal logics have been developed to formalize the properties of BDI-style agents. Cohen and Levesque [8] recognize the importance of intentions in the agent's practical reasoning, but instead of accepting intentions as a primary attitude they propose a reduction of intentions to what they call "persistent goals" which involve a sense of commitment and beliefs. Based on this idea they developed a logic in which they attempt to capture what they call rational balance: how the agent's beliefs, goals and intentions should be related to its actions. Rao and Georgeff [40, 41] formalized Bratman's ideas of practical reasoning using modal and Computational Tree logics in what they are now known in the literature as BDI logics. These can also be viewed as specification languages for BDI agents. In the context of BDI logics, various interesting problems have been studied such as for instance the connections between the three attitudes and the types of agents that are characterized by considering different relations as well as the asymmetry thesis and the side-effect problems [41, 15].

9.4.1 From Theory to Implementation

Although formal languages have their place in understanding BDI-style agents, in practical terms, executable languages that one could use to write and implement such agents are also required. Similarly to the object-oriented paradigm which is supported by a number of object-oriented programming languages, one would expect that there would be appropriate agent-oriented programming languages supporting the agent-oriented paradigm. Such languages would allow us to specify the agent's attributes as well as behavior at a high level of abstraction. In particular, BDI agent-oriented languages would need to make use of explicit constructs for beliefs, desires and intentions. A BDI agent architecture would then determine the agents' internal workings: how these components are represented, updated and processed to generate actions to be executed [43].

The holy grail of BDI research and development is to develop a high level language based on the mentalistic notions of belief, desires and intentions for which a one-to-one correspondence could be shown between the model theory, proof theory and abstract interpreter [39]. Rao endeavoured to bridge this gap by proposing the AgentSpeak(L) BDI language [39] which formalizes the operational semantics of BDI systems like PRS and dMARS. The Procedural Reasoning System (PRS) [19] and its successor the distributed Multi-Agent Reasoning System (dMARS) [13] were among the first implemented systems that were built based on practical reasoning and the BDI paradigm. These systems included internal structures that emulate beliefs, desires and intentions and decision making components that enable the agent to achieve its objectives through the execution of appropriate plans that are instantiated from a plan library. The language developed by Rao is based on a restricted first order language with events and actions. Agents have a set of beliefs and a set of context-sensitive plans allowing for the hierarchical decomposition of goals. The agent responds to changes in the environment that generate events which trigger goals which can be handled through appropriate plans. Jason [27] is a Java-based interpreter for an extended version of AgentSpeak. Jason implements the operational semantics of AgentSpeak and it is the first fully-fledged interpreter which also includes inter-agent communication based on speech acts (KQML messages).

The CAN (Conceptual Agent Notation) [51] language and its successors CANPLAN [43] and CANPLAN2 [44] offer a conceptual notation for agents with procedural and declarative goals. Another approach which provides a formal, and semantically grounded model of agent behavior is 3APL [11].

Other approaches which have been designed to support the development of practical multi-agent systems include Jack Intelligent Agents [25], JAM [26] and Jadex [38] among others. A framework which attempts to combine a formal characterization of the agent's behavior with programming constructs for the development of practical multi-agent systems is SPARK [33]. For a recent survey on BDI agent-oriented languages the reader is referred to [30].

Despite the significant number of languages and frameworks that have been developed, there is no widely adopted and accepted standard – for the time being a high level BDI agent-oriented language remains an elusive goal.

9.4.2 Means-End Reasoning in BDI Agents

For agents to demonstrate "smart" behavior they need to be able to: (i) reason about their environment and use this information to decide on actions to execute to achieve their goals; and (ii) react to the changes that occur in the environment and may affect the achievement of these goals. Building reactive agents that constantly react to their environment is not difficult, and the same applies to building proactive agents. The difficulty lies in building agents that achieve a balance between proactive and reactive behavior.

BDI agents are in essence cognitive agents who can reason about the world and use this information to choose among their desires and commit to intentions to achieve them. Once an agent commits to achieving a goal, this becomes an intention and the agent needs to perform means-end-reasoning. Means-end-reasoning is better known as planning. A plan is a sequence of steps (actions, subgoals, or tasks) or a "recipe" that when an agent performs can potentially lead to the required state with respect to some goal.

Unfortunately the efficiency of planning algorithms has been the most major obstacle in their deployment as part of an agent's means-end reasoning mechanism. Planning from first principles in fairly complex domains is inefficient as planning is known to be undecidable [6] and planning problems have PSPACE complexity [5]. To resolve this issue, a number of BDI-based implemented systems and languages rely on a pre-compiled library of plans which is built by the designer/developer of the system. However, although the use of such a plan library is computationally efficient, this inevitable restricts the flexibility of the agent and limits its capabilities. Unless there is a pre-specified plan in its library, an agent cannot deal with a situation that the designer/developer has not planned for or when an exception arises.

Approaches in which a BDI system or architecture is enhanced with planning have been explored. For example, in [32] a BDI agent based on the X-BDI framework is enhanced with a propositional planning algorithm based on Graphplan. In [31] a STRIPS planning component is incorporated into AgentSpeak BDI agents. BDI states are translated into STRIPS problems and then an action invokes the planner that attempts to come up with an appropriate plan. If a plan is found, it is converted back to AgentSpeak code and is executed by the agent.

In [43] the authors explore the similarities between Hierarchical Task Networks (HTN) planning and the BDI paradigm in order to augment the CAN [51] BDI programming language with planning. The enriched language is called CANPLAN. However, the designers of a CANPLAN BDI agent still have to specify the rules for HTN planning, in much the same way as designers of a traditional BDI agent would. CANPLAN has been further extended to CANPLAN2 [44] which addresses a number of issues with regard to the semantics of goals that previous versions of the language failed to capture. In particular three types of goals are captured: event goals (typical BDI-style goals as supported by AgentSpeak), declarative goals and planning goals. The authors develop a new semantics for the language which also better captures commitment strategies.

An alternative when agents have no relevant plans and lack a traditional planner is to enquire other agents and obtain plans from them [1]. As an agent may have multiple foci of attention, it will in essence be pursuing a number of goals simultaneously. Selecting new plans in the context of existing plans is explored from a theoretical point of view in [23]. A number of approaches to exploit potential synergies among goals and detect possible conflicts and negative interference have been explored in [47] and [48] respectively. In [22] the idea of equipping an agent with appropriate mechanisms so that it is able to respond to and plan for emergency goals that need to take priority and interrupt the execution of other goals is explored. The planning mechanism proposed is based on the Dynagent HTN planner.

9.5 A Mixed-Initiative Adaptive BDI System

The problem domain that we are concerned with is that of providing a configurable and self-adaptable digital services space (DSS) in a business context. Mobile knowledge workers in fieldforce and salesforce domains perform different types of jobs, each involving a number of tasks. To provide a more concrete motivation for our work consider the following use case. A telecommunications provider such as British Telecom (BT) employs knowledge workers such as for instance engineers who undertake the maintenance of the telecommunications network and also deal with customer issued requests. Each engineer can perform a number of operations/jobs which fall under his job description. Each job requires the engineer to perform a number of steps (tasks) that would enable him to complete the job. For instance, consider the job of installing a new line[2]. This job requires a number of tasks to be completed: first the engineer needs to arrive at the customer's address, next he has to collect information from the customer, then he has to perform the installation etc. These tasks may or may not have a strict order of execution and different engineers may have different ways of performing the same job. Assume that an engineer visits a customer and his task is to install a phone line. His personal agent publishes through the DSS a bundle of info-com services required to perform the job. However, the customer informs the engineer that he does not require the installation of a new telephone line but rather would like his old line to be re-enabled. The engineer will be provided with the ability to communicate the exception back to his agent. Based on past situations where this exception has arisen and from which the agent learnt recovery actions, the latter will adapt the engineer's plan. For instance, it will check whether the engineer is skilled to perform the new task, it will request real-time approval and will analyze historical examples. The results will materialize through a reconfiguration of the engineer's DSS, e.g., dropping service applications that relate to new line installations, exposing an application that provides access to technical data about line reenactment, updating the IM/VoIP contact list with the names of experts about the subject or peers who have come across the problem before, and so on.

---

[2] Please note that the examples of jobs and associated tasks are fictional.

Our objective here is to empower the knowledge worker by providing flexibility in decision-making as he is performing his jobs. His DSS, typically accessed through a mobile device, will be the main channel and decision support capability. In this context, we will explore how BDI technology can provide the underlying support in the form of personal agents to achieve flexibility in process execution. The intelligence of the system is encompassed into an agent's plans, but also other processes that enable agents to learn from each other as well as the user. The problem of how to handle exceptions during process execution will translate into the problem of handling exceptions when an agent executes its plans. Learning will also be an important aspect and we propose to explore solutions whereby agents learn by observing users and making associations. We would like to design a system that starts with an initial set of plans which can perform some jobs/tasks but as time goes by and the knowledge worker goes about performing the various specific jobs, the system learns from this interaction and learns new plans which it can consider using when a similar situation arises.

9.5.1 An Abstract Architecture for Mixed-Initiative MAS

Although a number of mixed-initiative systems have been developed, there is no standard architectural framework that we can draw on. In [45] a number of recommendations are made based on currently implemented systems:

- Separation of communication from control: Control in terms of which entity takes the initiative in the interaction and exhibits proactive behavior, needs to be separated from the communication protocol that facilitates the exchange of information and knowledge between the agents, i.e. communication and control should be handled by different parts of the system.
- Representation and reasoning. The system should be able to reason over its competencies and capabilities (i.e. tasks and courses of action) so that it can adapt and improve.
- Asynchronous execution. The agents should perform asynchronously.
- Component reuse. Code reuse should ne maximized.
- Uniform interface. A uniform interface should be provided across the system components.

Taking into account the above recommendations, we propose a high level multi-agent architecture for developing mixed-initiative systems as illustrated in Figure 9.1. The architecture consists of the following agents:

Personal Agent (PA): Users interact with their personal agents that guide and support them and facilitate the achievement of their goals. There may be several PAs in the system.

Problem Solving Agent (PSA): A PSA is an agent who has expertise in solving a particular problem and can provide the solutions to other agents or access to these methods which can be regarded as procedures or plans. There may be several of these PSAs in the system, and each PSA may be specializing

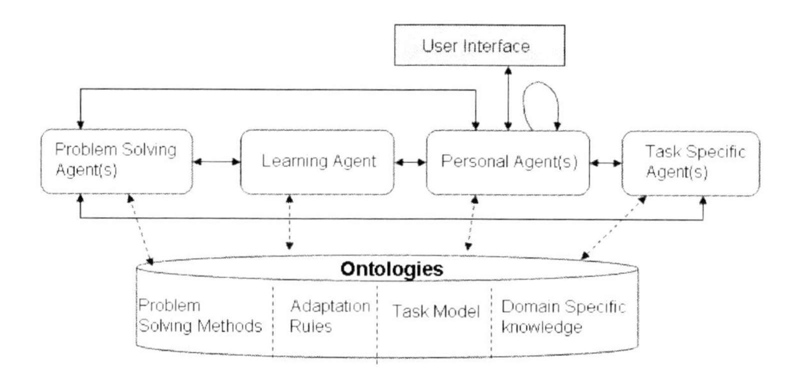

Fig. 9.1. Abstract architecture for mixed-initiative multi-agent systems

in different classes of problems. In its simplest form, a PSA can be a library of procedural methods (or plans) that describe how problems can be solved.

Learning Agent (LA): It is important that such a complex system is able to adapt to its environment through learning. A LA will monitor interactions in the system among the various agents and based on adaptation rules will enable the system to adapt and improve its overall performance.

Task Specific Agents (TSA): Such agents are specific to the domain of application. For instance, a scheduling agent may be responsible for producing schedules for a workforce. Such agents may be making use of web services for instance to provide specific services in the system.

User Interface (UI): The User Interface in this architecture comprises a separate entity as we want to separate control from communication.

Ontologies: It is important that knowledge is represented explicitly in the system in the form of a set of ontologies. Different entities may only have access to some of the sub-ontologies. In particular, we have identified the following sub-ontologies as the minimum requirements for such a system:

- *Problem Solving Methods Ontology*: an ontology describing the problem solving methods that can be employed within the system. This may also include the description of web services or components that provide problem-solving capabilities.
- *Adaptation Rules*. Some adaptation rules may be explicitly represented so that they can be easily manipulated by the system designer/developer.
- *Task Model*: Tasks need to be represented at different levels of abstraction so that the agents can reason about these and modify or repair them when necessary. An appropriate knowledge representation scheme may include preconditions, parameters, constraints among tasks, order of execution etc. Although such a task representation is inevitably domain-specific, nevertheless it is an essential part of the system.

- *Domain Specific Knowledge*: other domain specific knowledge that needs to be explicitly represented to facilitate unambiguous communication exchange among the various entities.

Some of the most common interactions among the entities in the abstract architecture have been sketched. PAs interact with PSAs as well as the LA and other TSAs in the system. PAs also interact with each other (hence the return arrow), as in order to solve a complex problem they may have to collaborate. The LA interacts with the PSAs as well as the PAs in the system as the latter provide the input (in terms of the user interactions as well as information about the environment) while the adaptation and learning process may have to be performed on the PSAs as well as the PAs. TSAs interact with PAs, and though to simplify the diagram we have placed no connections in Figure 9.1 between these agents and any others, TSAs will be interacting with other components.

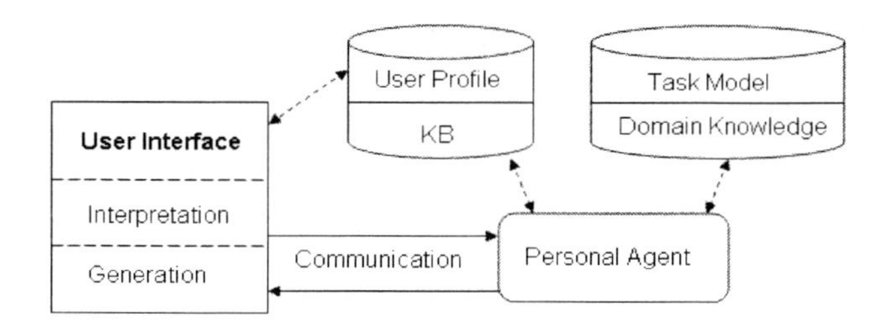

Fig. 9.2. The personal agent and the user interface components

Although the abstract architecture of Figure 9.1 illustrates the main entities in the system, we need to take a closer look at two of these entities which are central in any mixed-initiative system: the PA and the UI that enables the interaction/communication between the user and his agent. Figure 9.2 describes in more detail the essential sub-components of these two entities.

The UI provides two functionalities: *interpretation* of the user's input (action performed, goal delegated) which is passed on in an appropriate way to the agent; and *generation* of the response based on the output provided by the agent in an appropriate form for the user to understand [16]. The user profile contains the preferences regarding the appearance of the interface and viewable services. As the user will be manipulating his interface thereby personalizing his digital space, his actions are recorded and stored in his user profile. This information is used to tailor-make the interface the next time that the user logs into the system. The PA has its own knowledge base in which it maintains a representation of the environment including the user's goals and plans about how goals can be achieved. The agent also has access to the task model as well as the domain knowledge ontology as it has to be able to reason over these.

The issue of communication between the user and the agent is a particularly interesting one. Communication needs to be looked at from two perspectives: the user's and the agent's. Some researchers have advocated a natural language interaction, where the user is able to enter into a dialogue with the system [16]. But a natural language interface may not be appropriate for some application domains such as for instance the one considered here where mobile knowledge workers need to communicate with their agents through a mobile device which inherently limits how much and how information can be presented and inputted into the system. From the agent's perspective, whatever form the user is providing input into the system will have to be transformed in a form that the agent will be able to understand and process. Interpreting the user's input correctly is crucial as the agent needs to recognize the intention behind the user's communication. Software agents communicate with each other using communication languages such as FIPA ACL [17] and KQML [28] which are based on speech act theory. The performatives in these languages enable the sender and recipient of a message to communicate in an unambiguous way: the sender's intention is made clear through the performative used, while the recipient uses the performative to interpret the sender's intention. The use of such a language for communicating the user's intention to the agent is highly desirable.

The high level architecture that we have proposed takes into account the recommendations of [45] as it: (a) separates control from communication; (b) facilitates reasoning over tasks and goals; and (c) the various entities in the system perform asynchronously and autonomously. Code re-use is enabled since all PAs for instance may start with the same initial code and use the same interface configuration at start up.

9.5.2 Architecture of the System

Based on the abstract architecture discussed in the previous section, we propose to develop a multi-agent system to demonstrate how BDI agent technology can provide the underlying intelligence in the form of personal agents to achieve flexibility in process execution. The problem of how to handle exceptions during process execution will translate into the problem of handling exceptions when an agent executes its plans. Learning is also an important aspect and we propose to explore solutions whereby agents learn by observing users and making associations. Based on BT's workforce problem we will be modelling a number of entities in the system. Figure 9.3 illustrates the architecture of the system which comprises of the following entities:

- Engineer's Personal Agent (EPA): Knowledge workers such as engineers, are provided with their own personal agents who will be supporting them in their daily operations. The EPA knows the engineers' schedule of jobs and also has information about what each job involves in terms of basic tasks and plans. The EPA agent will be communicating with the user through a graphical user interface which will be designed so that information can be displayed to the user in a flexible way. There may be multiple EPAs instantiated in the system at any point in time. Such entities are only active while the user requires their assistance and is online. When a knowledge worker goes offline or decides that he no longer needs the help of his agent, the corresponding

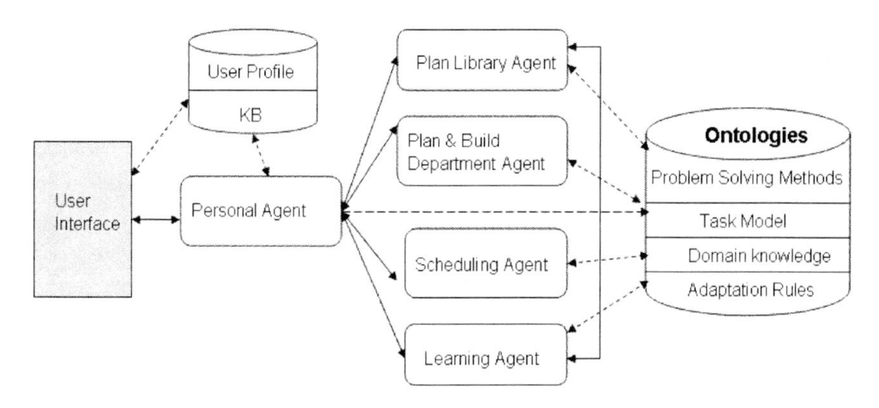

Fig. 9.3. The proposed architecture of the system

agent is de-activated. It is important to note though that the agent's state and beliefs are persistent and they are captured and stored so that the agent's belief base captures any changes as the agent executes.

- The User Interface (UI): The UI and the User in essence constitute the environment where the agent's percepts emanate from. An important issue to be addressed is to decide on the form of the percepts that the agent receives from its environment including information about exceptions. Another issue is displaying information to the engineer in an easily-accessible way.
- Plan Library Agent (PLA): The remit of this agent is, as its name suggests, to store plans that describe how goals can be achieved. In large-scale applications, "expert agents" that specialize in collecting and storing plans for specific jobs from multiple agents can be a real asset as the expertise and load in the system can be distributed and bottlenecks are avoided. Facilitator agents can be used to record information about the identities of the various PLAs and their respective areas of expertise.
- Scheduling Agent (SA): This agent is responsible for producing the schedule of jobs (or appointments) for each knowledge worker in the system. As the focus is not on how this schedule is produced, we are not concerned with the details of this agent, but assume that it generates the schedule in some way. However, as the schedule takes into account time, if there is a deviation from the prescribed job that perhaps has arisen as the result of an exception, the EPA needs to request authorization from the SA in order to proceed.
- Plan and Build Department Agent (PBDA): In addition to other domain specific operations, the PBDA is responsible for providing authorization to knowledge workers to perform various tasks based on their job descriptions. For instance, the PBDA may turn down a request from an EPA to lay cable in an area, on the grounds that a major operation is scheduled to take place in the same area the following week.
- Learning Agent (LA): The remit of this agent is to enable the system to adapt and improve. It records and analyses information on the use and

construction of plans within the system and using adaptation rules may modify the behavior of other entities in the system such as the PLA for instance.

We have implemented the system using the Jason implementation of AgentSpeak [3] as a fully distributed system using the SACI architecture. Hence, the various entities are situated in different hosts across the network and communicate with each other via the exchange of KQML messages.

Individual Agents. AgentSpeak[3] is a high level logical language based on first order logic. Agents consist of a set of beliefs and a set of plans. An agent's belief base consists of a set of belief literals (first order predicates or their negation) which describe the agent's information about the environment. For instance, salary(maria, 30000) and mother(gill, peter) are belief literals.

Agents written in AgentSpeak can have two types of goals: achievement goals which represent what the agent would like to bring about and are predicates prefixed by the '!' operator; and test goals which attempt to retrieve information from the belief base and are predicates prefixed by the '?' operator. Examples of an achievement goal and a test goal respectively are !buy(Car) and ?bank_balance(X). The test goal ?bank_balance(X) will attempt to retrieve the bank balance of the agent and unify it with the variable X (terms starting with capital letters are variables in Jason). An agent in Jason is endowed with a set of plans which describe courses of action that may lead to the achievement of the respective goals. A plan has the form te: c <- pb. te is the event that the plan is meant to handle and it can be an addition '+' or a deletion '-' event. In Jason there are 6 types of such triggering events: belief addition (deletion); achievement goal addition (deletion); and test goal addition (deletion). c is a conjunction of belief literals which describe the conditions (context) that need to be satisfied for the plan to execute. pb is a sequence of actions (internal or external) or subgoals that the agent has to achieve for the plan to complete successfully. For instance, the following plan describes how an agent can go shopping provided that it has enough money in the bank:

```
+!goShopping(Budget): bank_balance(X) & X>=Budget
<- goto(supermarket);
   ?retrieve(ShoppingList);
   !fillInTrolley(ShoppingList);
   payBill.
```

Apart from beliefs, goals and plans, the Jason interpreter manages a set of events (external and internal events) and a set of intentions. External events are generated by the environment and other agents, whereas internal events are triggered by the execution of plans by the agent itself. An intention in Jason is defined as a stack of partially instantiated plans and the agent's intentional structure includes all those intentions that the agent is currently attempting to

[3] Please note that for the rest of the text when we refer to AgentSpeak, we mean its implementation in Jason.

bring about. The agent's execution cycle in Jason follows the typical execution cycle of BDI-style agents:

- perceive the environment (includes receipt of messages and their filtering);
- update the agent's beliefs;
- update the event queue;
- select an event to process;
- retrieve relevant plans that can deal with the event;
- determine the plans whose context is satisfied in this particular situation;
- select a plan to be executed and update the intentions structure;
- select an intention to be executed and perform the next step in the intention;

The Jason interpreter uses three selection functions. The S_E function selects an event from the event queue to process. The S_O function selects one plan from the set of applicable plans for the particular event under consideration. The S_I selects the intention to be executed in the agent's current reasoning cycle.

One very useful feature of AgentSpeak as implemented in Jason is the use of internal actions. Internal actions are prefixed with the '.' operator and are different from external (or environment) actions which affect the environment. In essence, an internal action is code written in Java by the programmer which performs an operation. Internal actions have been used for instance to implement agent communication using KQML performatives.

Knowledge Representation. A knowledge worker[4] has a job description and as part of this he can perform a number of generic jobs (e.g. install new line, lay cable etc.). We assume that these are specified as part of the knowledge worker's contract and if new jobs are added to the job description or he is trained to do new jobs, these are added to his repertoire of jobs.

Each generic job that an engineer can perform has its own id, a set of parameters, a set of tasks associated with it and a textual description. Each generic job may have variations, i.e. there may be perhaps a number of different ways in which a job can be performed depending on some conditions or constraints. A particular job variation may not include all the tasks that are associated with a generic job, but only a subset of these. Each task has an id, a set of parameters, and a set of constraints associated with this task.

In AgentSpeak each job is associated with a high level achievement goal which requires the execution of a plan. There may be a number of ways to achieve the high level achievement goal and these different ways are in essence the various plans that can be used. Although such plans have the same triggering event they may differ along the following dimensions:

- each plan may be applicable in different cases (the context may differ);
- the tasks that are executed as part of the plan may vary;
- the order of execution of the various tasks may vary.

Each task (or subtask) can be an internal action, an action or a subgoal in AgentSpeak. For instance the job to install a new line (*installNewLine*) has a

[4] We are considering a particular type of mobile knowledge worker who is an engineer.

Table 9.1. Notation for expressing conditions on tasks

Notation	Meaning
!n	the task must always come last
!1	the task must always come first
this >! k	*this* task must come after the kth task
this >? k;l;m	*this* task must come after task k, l or m
this <! n	*this* must come before the last task

number of associated tasks: *consult customer*, *perform installation*, *perform line tests*, *bill customer* and *sign off job*. Each one of these tasks may have a number of associated conditions with it which in essence specify ordering conditions. For instance, the *sign off job* task can only come last as it signals the successful completion of the job. We have used a simple notation to express conditions on tasks which is illustrated in Table 9.1.

This representation scheme of jobs and tasks can be translated into a relational database and also into the appropriate representation in AgentSpeak. In AgentSpeak, a particular table is represented as a set of predicates with the same name. The agent needs to have access to the task model ontology as it may have to reason with it in order to formulate a new plan or repair an existing plan.

User Interface and Communication. Achieving a fluid collaboration and interaction requires efficient signalling between users and agents. Mixed-initiative systems need to understand what the user wants to achieve, therefore it is essential that a common ground is established between the user and his agent.

Although some authors have advocated the use of natural language to engage in a dialogue with the user [16], this is not indicated in the domain under consideration since a knowledge worker will have access to his agent through a mobile device such as a mobile phone or a PDA. Such devices have limited screen size, the average being 320×240, and therefore how much and how information is presented is inevitably limited by the physical dimensions of the device. The mode of interaction with such devices also rules out the use of natural language; if a knowledge worker is using a mobile phone, he will most likely have to use the arrow buttons to navigate in the interface and select where to go to next. PDA users in general, consider scrolling in two axes irritating. So although we can assume that the length of the screen is infinite, the width is fixed. However, preferably the user should not have to do a lot of scrolling. The interface developed for the system was quite simple using buttons to demonstrate the various options to the user. The user may have to scroll down to see all his appointments or tasks of a job, but not across the screen. The same basic design was used for all screens shown to the knowledge worker providing a uniform interface.

The user interface is in essence a web application which was developed with JSP pages and runs on an application server. In order to facilitate communication between the user and its agent through the web application we need to allow the application to send and receive KQML messages. Such messages describe the user's input and what it is that the user would like to do in the system. To achieve this, the application makes use of a Java bean component which is created when the application starts and is shared by all the pages [3]. This Java bean component is a SACI mailbox that allows the web pages to send messages to an agent whose name is specified in the construction of the mailbox.

To be more specific, the login process is taken care of by an auxiliary agent whose task is to validate the user. When a user logs into the system via the login.jsp page, an appropriate KQML message is send to the login agent to first check that such a user exists and then create the user's corresponding agent. As part of the validation process, the login agent also checks if the user's agent is already active in which case it simply logs in the user without creating his agent. There can be many EPAs running concurrently. Once a user has logged in successfully the rest of the pages that comprise the user interface send messages to the user's EPA and not the login agent. For instance, when the user presses the "Retrieve Appointments" button on the interface, a KQML message is sent to the EPA requesting that all the appointment information is retrieved and displayed. When the user selects an appointment to be executed, another KQML message signals to the agent that the user wants to execute that particular appointment. In this way, the user communicates with the agent in a clear, unambiguous way.

9.5.3 Mapping of BDI Concepts

The purpose of a personal agent is to assist the user in the achievement of his goals. We have chosen the BDI model of agency to model the agents and in particular we used the Jason implementation of the AgentSpeak language for developing the agents. However, as discussed in section 9.4, the BDI concepts are not interpreted in a uniform way. In Jason, there are no explicit language constructs for desires and intentions. Beliefs represent the agent's information about the world and goals are states that the agent wishes to bring about. The developers of Jason describe how the BDI concepts map into programming constructs in Jason as follows [3]. Desires are those goals that are currently in the set of intentions for which no relevant and applicable plans have been found yet. This interpretation of desires becomes more confusing by the introduction of an internal action .desire() in Jason which returns as desires all the achievement goals that appear in the set of events or appear in intentions (including intentions that have been suspended). This mapping is not very intuitive, and is not in par with the usual interpretation of desires in the philosophical literature. Desires represent states of affairs that the agent would ideally like to bring about and they may not necessarily be consistent with each other. Desires need not be triggered by a particular event, in contrast to Jason where all desires are triggered by events. Intentions are more straightforward in Jason as there is an explicit intention structure which is a set of partially instantiated stacks of plans.

Although we have developed the agents using Jason, the mapping of the BDI concepts as this is currently described by the Jason developers is not entirely satisfactory and fails to capture important aspects of the mixed-initiative interaction and its effects on both the user's as well as the agent's cognitive models.

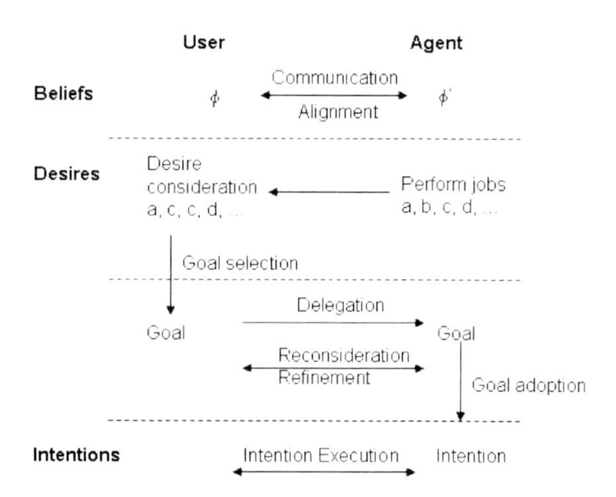

Fig. 9.4. The cognitive models of the user and the agent in mixed-initiative interaction

Both the user and the agent have beliefs about the world, but in order to be able to work together they need to establish a common ground. This is achieved through communication exchange which may lead to one of the agents to change its beliefs. The agent's desire is to help the user achieve his goals. Initially the agent retrieves the user's desires when the user requests his list of appointments which is provided by another agent in the system (SA). The agent's desires thus take the form of the appointments or jobs that the user has been assigned to perform. This representation of desires is implicit in the sense that there is no specific structure in AgentSpeak to identify desires (though when retrieving the list of appointments we could add a beliefs of the form desire(perform, Appointment) in the belief base to capture desires in a more explicit way). These desires are communicated to the user who considers them and chooses one of them to become now his goal. This is indicated by the user through selecting the appropriate job which is now delegated to the agent as the user's adopted goal. However, this goal may require refinement or reconsideration (exceptions may arise and the goal may have to be modified). Once an appropriate plan has been selected to bring about the goal, this now becomes the agent's intention and the associated with the plan tasks are revoked and executed in collaboration with the user who is now executing his part of the intention. How the agent's and user's cognitive states change through their mixed-initiative interaction is illustrated in Figure 9.4.

9.5.4 Understanding Exceptions

Our objective in using a mixed-initiative BDI multi-agent system is to provide flexibility and control to the knowledge worker with regard to performing tasks, but also handling exceptional circumstances that may arise during task execution. One of the fundamental issues that arises is how one can deal with exceptions when attempting to complete business processes. Using BDI agents, the problem translates into how to handle exceptions when an agent attempts to achieve its goals. Goals are achieved through the execution of plans. Currently, if a plan fails an agent can try to re-execute the plan or execute alternative plans until it achieves the goal. What are known as *commitment strategies* [40, 8, 51] describe an agent's commitment to achieving a goal and this may vary in different agents. A *blindly-committed* agent will only drop a goal if it believes that the goal has been achieved. A *single-minded* agent will drop a goal if it believes that the goal is no longer achievable. An *open-minded* agent will drop a goal if it believes it has been achieved, the goal is no longer achievable or the motivation for pursuing the goal no longer exists. We have implemented our agents with a mixture of commitment strategies depending on the importance of the goal. However, a commitment strategy may not necessarily account for how to deal with possible exceptional circumstances. In the following we identify the sources of exceptions and potential ways of dealing with them. In particular, we would like to make use of the other agents present in the system and more specifically the PLA as well as the user himself. The idea is that if an agent does not know how to deal with a problem, it can attempt to find solutions through other experts in the system. The role of the PLA is exactly that to provide solutions to problems when requested by other agents. In its simplest form the PLA is an agent that assembles and stores plans. However, the PLA may use a traditional planning system to construct a new plan, if it does not contain any plans that satisfy the agent's request. The user is another expert in the system whose operational knowledge can be harnessed. The following describes how the agent attempts to deal with possible exceptional circumstances as these arise:

A. The agent does not know how to deal with a situation:
There are no relevant (no plans match the triggering event) or applicable plans (i.e. although the agent has plans with the appropriate triggering event, the context of those plans does not match the current situation):

- Initiate a plan that requests plans from the Plan Library Agent.
- If there are such plans, the agent incorporates them into its own plans and attempts to re-execute the original goal.
- If no relevant plans can be found, the agent asks the user to put together a new plan by assembling tasks-actions. The task model ontology is used to guide the user in putting new plans together by providing information on constraints in the order of execution of tasks etc.

B. An exception has arisen during the process of executing a plan:

1. An exception has arisen due to the fact that the context is no longer applicable. This is captured by modifying the agent's commitment strategy. Hence, apart from motivation conditions, other essential conditions for the execution of the plan that become false during its execution are captured and a new plan that aborts the original one is invoked which also notifies the user who is given the option to re-attempt the original goal.

2. The goal that was selected for execution was the wrong one. The agent was given the wrong triggering event (i.e. goal) and the user has realized in the middle of the operation that the incorrect goal is executing. The old plan needs to be aborted and redefined in collaboration with the user.
3. A plan has failed in the classical sense, i.e. an internal action has failed or an instantiation (through a test goal or condition) has returned false. This is captured by the open-commitment strategy that we are using, though instead of automatically attempting to find an alternative plan to achieve the goal the agent prompts the user whether the goal should be retried.

The following fragment of code (please note that for simplicity some of the implementation details have been omitted) shows how an agent can deal with the first type of exceptions in a generic way:

```
-!AnyJob: not planException(1, AnyJob)
<- +planException(1,AnyJob);
   !retrievePlan(AnyJob);
   !AnyJob.
-!AnyJob: not planException(2, AnyJob)
<- +planException(2,AnyJob);
   !constructUserPlan(AnyJob);
   !AnyJob.
+abort(AnyJob)
<- .abolish(planException(_, AnyJob));
   .drop_intention(AnyJob);
   .send(SA, tell, abortedJob(MyName, JobID)).
```

The first plan enables the agent to request plans from the PLA by invoking the !retrievePlan(AnyJob) subgoal. If the list of returned plans by the PLA is not empty the goal is re-attempted. Alternatively, the agent requests the user to put a new plan together. If the user at any point decides that the goal needs to be aborted, then the third plan cleans any potential beliefs from the agent's belief base, drops the intention and also notifies the SA that the job has been aborted. The aborted job may have to be rescheduled the following day or delegated to another knowledge worker who has spare capacity.

9.5.5 Constructing New Plans

The usefulness of a task model is in that an agent can reason explicitly over tasks and jobs and can enable the knowledge worker to tailor-make how he deals with jobs to suit his particular work style. Furthermore, putting together alternative plans is one possible way to handle exceptions that perhaps no PLA knows how to deal with in the system. When prompting the knowledge worker to put together a new plan, the agent needs to display the right information in terms of the sort of tasks that can be put together to formulate a new plan. The information that needs to be displayed to the knowledge worker includes the current job, the possible associated tasks with this job and the conditions that

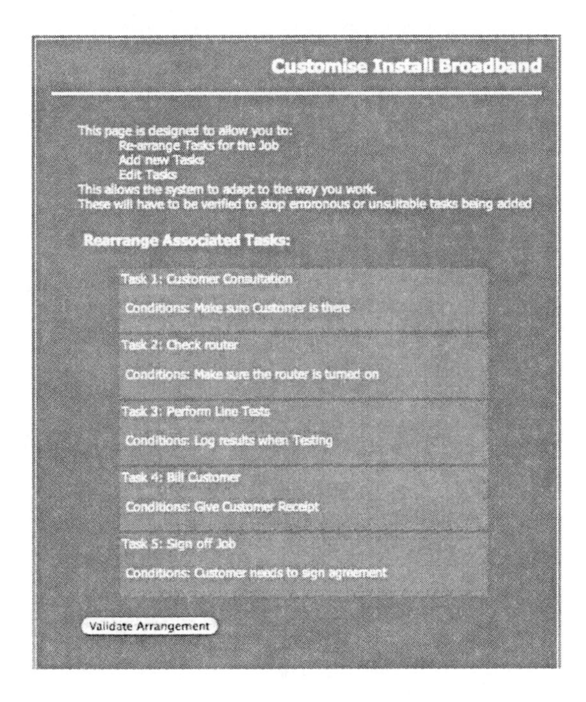

Fig. 9.5. The screen enabling the rearrangement of tasks

could be applied in this new job variation (plan). Information on the conditions can be retrieved from the task model.

We have built a flexible interface to allow the knowledge worker to put together new plans quickly by choosing tasks and then dragging and dropping them in the preferred order, Figure 9.5. However, in doing so the system checks on whether any of the conditions associated with the order of execution are violated and if this is the case the user is not allowed to place a task in a position that violates a constraint. Once the user has put a new plan together to achieve a goal, he selects the "Validate Rearrangement" button which sends a message to the agent with the plan representation as a string. This is stored in the agent's local plan library and goal execution is re-attempted (see next section).

9.6 Use Case Scenarios

In this section we will describe a couple of characteristic use case scenarios that illustrate the functionality of the system and in particular the interaction of the knowledge worker with his EPA. When the multi-agent system starts executing, the PLA contains a set of plans that can be used to execute various jobs. At the beginning of every day the SA produces a schedule of appointments for knowledge workers. The PBDA retrieves information about the various knowledge workers in the system as well as their job descriptions (i.e. information about the kind of tasks that each knowledge worker is allowed to perform). Once a knowledge worker logs into the system, his EPA initializes and requests its daily schedule from the SA. The EPA updates the user interface (Environment) and displays the list of jobs in an appropriate way for the engineer to see.

Use case (A). The engineer selects the job to lay cable (this is a new capability that has just been added to the engineer's job description). The EPA attempts to retrieve an appropriate plan to deal with this job, but has none. The EPA notifies the knowledge worker that it lacks plans to deal with this job and will be requesting appropriate plans from another agent. The EPA requests relevant plans from the PLA. The PLA returns a set of plans and the EPA attempts to re-execute the job. The EPA works in collaboration with the user and retrieves appropriate services needed for this job, such as technical maps.

Use case (B). The knowledge worker selects to execute a job. EPA attempts to retrieve an appropriate plan, but it has none. EPA agent informs the knowledge worker that has no relevant plans to execute this job and prompts the user what to do: abort job or attempt to put a new plan together. If the user selects the latter, the EPA retrieves the set of associated tasks with this particular job and displays them prompting the knowledge worker put a new plan together. The knowledge worker can selects task and formulate a new plan. Any pre-specified order in the execution of tasks needs to be preserved as captured by the task model and the EPA monitors this. Once the user finalizes the plan, this needs to be approved by the PBDA, so the EPA sends it for authorization. Authorization is received from the PBDA. The EPA attempts to re-execute the job. If the job is executed successfully, then the plan is forwarded for storing to the PLA. The LA is also notified of the acquisition of a new plan.

Use case (C). The knowledge worker selects the job to install a new line for a customer. The EPA retrieves an appropriate plan to deal with this job and starts its execution. While executing the plan, the EPA gets feedback from the knowledge worker that an exception has arisen: the currently selected job is not the correct one. Although the engineer has been told to install a new line, during the customer consultation it has emerged that the customer requires his old line to be reconnected. This is an exception that arises at the task level, but in fact affects the overall job execution. The EPA gives the engineer the following options: (i) abort job altogether and (ii) modify the job. The engineer selects the latter and the EPA provides a list of alternative jobs to the engineer who chooses one (or can abort the job at this stage). As the nature of the job to be executed has changed this has to be authorized. The EPA asks for authorization from the PBDA. The PBDA grants authorization, so the EPA then requests authorization from the SA (as the scheduling has now been affected). The SA provides authorization and the engineer is notified. The EPA attempts to perform the job and the interface is updated with associated services.

9.7 Related Work

In some instances, the criticism regarding the BDI model has been that it is not well suited to certain types of behaviors. In particular, the basic BDI model appears to be inappropriate for building complex systems that must learn and adapt their behaviors and such systems are becoming increasingly important in today's context in business applications.

Introducing learning into a BDI agent is not a trivial problem as current learning techniques tend to require large computational power while agents are small and time-critical processes. It involves analyzing different AI techniques, determining which are the most useful and incorporating them into existing architectures [18]. Moreover, a problem arises from the fact that BDI agents exhibit open-loop decision making. That is, a BDI agent requires an event from the environment to generate a goal (either directly or by changing its beliefs) so that it can search through its plans to form an intention on how to act. When the plan has finished executing (by succeeding or failing), the decision process finishes. What happens afterwards depends on the environment because the agent can have several plans executing concurrently corresponding to different goals. In addition, BDI agents are optimized for a specific scenario and hence an agent written for a particular environment will not easily translate to another.

These shortcomings can be removed by introducing a learning component into the agent. The feedback caused by learning introduces a closed loop decision process analogous to Boyd's Observe-Orient-Decide-Act (OODA) loop [37]. As a result, a new generation of coordination/cooperation architectures is starting to emerge. Consoli et al. [9] show how the BDI and OODA architectures can affect coordination and cooperation within a multi-agent system. They propose a recommendation on how these concepts can be designed and implemented in multi-agent system. This means that the decision process can proceed more rapidly, and decisions can be evaluated before they are committed.

One possibility in improving the learning capability is to use a hybrid BDI agent model. Lokuge and Alahakoon [29] propose a new extended BDI framework with an intelligent module for handling complex situations which essentially extends the learning and adaptability features of BDI agents. The Neural Network and Adaptive Neuro Fuzzy Inference system (ANFIS) in a hybrid BDI framework has shown improvements in learning and decision-making processes in a complex, dynamic environment. The change rate of beliefs and expected cost of reaching the final goal state from different states in the plan hierarchy have been considered by the agent in the proposed architecture. This would enable agents to identify the alternative plans in the intention structure with the change of the environment. Dynamic selection of plans and expected cost of achieving the final goal state from various plan paths are modeled with the use of a supervised neural network. ANFIS has been incorporated in making the final rational decisions in the agent model.

Another approach is to customize the BDI model by adding adaptation components to it. Houari and Far [24] present an approach that customizes the BDI model to define a so-called "DIBRA: Desire Intention-Belief-Rapport-Adaptation" as a generic method to support progress from individual autonomous agent concept towards interactive multiple agents. Rapport here refers to the component that connects an agent to its environment, whereas the Adaptation module incorporates mechanisms of learning. It is claimed that the five proposed tiers for multi-agent system development serves for mastering the complexity and the difficulty of setting up effective autonomous interactive systems.

A further option would be to combine the BDI approach with learning capabilities offered by Case Based Reasoning (CBR) techniques. Corchado et al. present [10] an application of a wireless tourist guide that combines the BDI approach with learning capabilities of CBR techniques. Their application demonstrates how to develop adaptive agents with a goal driven design and a decision process built on a CBR architecture. They claim that the resulting agent architecture can be generalized for the development of other personalized services.

A viable option would be to incorporate the proposed BDI architecture within a larger semantic web services semantic framework. Chepegin et al. [7] argue that the future of adaptive web-based systems lies in the modularity of the architecture and the openness to interoperate with other applications or components. They propose a service-oriented framework for adaptive web-based systems where the main goal is to help the semantic enrichment of the information search and usage process and to allow for adaptive support of user activities. To achieve this they adopt the concept of the UPML framework for semantic web service integration. In their work an important part of the modeling of user tasks, goals, roles and cognitive processes, is played by learning theories. Two of the most applied ones which they consider are ACT-R [2] and Constraint-Based Modeling (CBM) [36]. They are both based on the distinction between declarative and procedural knowledge, and the view that learning consists of two main phases. In the first phase the declarative knowledge is encoded and in the second it is turned into more efficient procedural knowledge.

To summarize, in order to facilitate adaptation and learning we will need to enhance the BDI framework by incorporating Advanced Information Processing (AIP) supporting technologies such as knowledge based systems (rule and case based), distributed blackboard systems (e.g. neural networks) or semantic web technologies (e.g. web services frameworks such the Internet Reasoning Service [34]). In this way we would enhance the capabilities offered by the BDI approach which appears to be well suited to the problem space and in the mean time overcome the shortcomings described above.

9.8 Conclusions and Future Work

In this chapter, we argued that agents and multi-agent systems offer the potential for enhancing the user experience and providing flexible and personalized support. In particular, we considered the problem of providing decision support to mobile knowledge workers who interact with their agents through their digital service space. We proposed the use of the Belief Desire Intention model of agency for developing agents to whom goals can be delegated and who can work collaboratively with the users. Such agents have the ability to represent the user's information state and objectives and execute plans to bring them about. Users and agents interact in a mixed-initiative mode to establish a common ground and the shared goals to be achieved. Plans as the means to bring about the user's goals, were utilized to enable the refinement and fine-tuning of business processes to better conform to the work style of the individual knowledge worker. In addition, we considered possible ways of dealing with exceptional circumstances in this context and how an agent can essentially learn how to deal with such situations by interacting with other agents in the system as well as the user who is considered to be an the expert.

The contribution of our work is four-fold. First, we proposed an abstract architecture for developing mixed-initiative multi-agent systems and identified the main generic components and underlying ontologies that are required to support automated reasoning, problem solving and adaptation. Second, we considered the sources of exceptions and when these can arise and studied possible ways in which exceptions can be handled in a mixed-initiative BDI multi-agent system by utilizing the other agents in the system including the user. Thirdly, the mixed-initiative approach that we advocate, provides users with the ability to tailor-make the system (in terms of the underlying business processes used, i.e. plans) to their individual work style within a well-specified organizational framework; hence empowering them. This mixed-initiative approach in combination with the BDI model of agency enables the provision of flexible, personalized and adaptable support to the users. Finally, we have demonstrated the potential of our approaches and the proposed architecture through developing a mixed-initiative BDI multi-agent system. The system which was developed as a proof of concept satisfies the key requirements for mixed initiative systems as discussed in section 9.3, namely: *control* as the user has a high degree of control over the system; *personalization* as the system provides flexibility and conforms to the individual user's work style; *teachability/adaptation* as the system can learn from other agents as well as the user; and *transparency* as the system provides guidance on what the user should do or what it requires through a simple interface which enables the user to give the system clear instructions. Even though a more detailed evaluation of the system would have been desirable this is beyond the scope of the current work and will be undertaken as part of future work.

Although, the agents as well as the overall system learns and adapts in the ways mentioned above and the system behavior can improve, we would like to explore other ways in which the BDI agents and the system overall can learn and adapt. In particular, we would like to explore a utility-based approach to choosing plans to deal with exceptional circumstances as well as other methods as described in section 9.7. Our work on dealing with exceptions bears some similarities with the work of [46] on aborting and failing goals and plans. Another direction that we are currently exploring is that of exploiting synergies between plans as in [47]. In particular, when one plan needs to be abandoned and an alternative needs to be executed, we are considering exploiting common tasks/subgoals to avoid repetition of tasks. The use of an explicit task model in our approach provides us with the means to address this issue in a systematic way. A further option for future research is to expand the problem solving capabilities of the system through deploying web services (web service orchestration) as well as incorporate a planning service to enable planning from first principles.

Acknowledgments

Aspects of the system described in section 9.5.2 have been implemented by William S. Allen, Department of Computing and Electronic Systems, University of Essex.

References

1. Ancona, D., Mascardi, V., Hübner, J.F., Bordini, R.H.: Coo-AgentSpeak: Cooperation in AgentSpeak through plan exchange. In: Proceedings of the Third International Joint Conference on Autonomous Agents and Multi-Agent Systems (AAMAS 2004), New York, NY, pp. 696–705 (2004)
2. Anderson, J.R. (ed.): Cognitive Psychology and its Implications. Studies in Fuzziness and Soft Computing, vol. 153. Worth Publishers, New York (2000)
3. Bordini, R.H., Hübner, J.F., Wooldridge, M.: Programming Multi-Agent Systems in AgentSpeak using Jason. John Wiley and Sons, Chichester (2007)
4. Bratman, M.E.: Intentions, Plans, and Practical Reason. Harvard University Press, Cambridge (1987)
5. Bylander, T.: The computational complexity of propositional strips planning. Artificial Intelligence 69, 165–204 (1994)
6. Chapman, D.: Planning for conjuctive goals. Artificial Intelligence 32, 333–377 (1987)
7. Chepegin, V., Aroyo, L., De Bra, P., Houben, G.-J.: Chime: service-oriented framework for adaptive web-based systems. In: Proceedings of Dutch National Conference InfWet, pp. 29–35 (2003)
8. Cohen, P.R., Levesque, H.J.: Intention is choice with commitment. Artificial Intelligence 42, 213–261 (1990)
9. Consoli, A., Tweedale, J., Jain, L.C.: An architecture for agent coordination and cooperation. In: Apolloni, B., Howlett, R.J., Jain, L. (eds.) KES 2007, Part III. LNCS (LNAI), vol. 4694, pp. 934–940. Springer, Heidelberg (2007)
10. Corchado, J.M., Pavon, J., Corchado, E.S., Castillo, L.F.: Decelopment of CBR-BDI agents: A tourist guide application. In: Funk, P., González Calero, P.A. (eds.) ECCBR 2004. LNCS (LNAI), vol. 3155, pp. 547–559. Springer, Heidelberg (2004)
11. Dastani, M., Dignum, F., Meyer, J.-J.: 3APL: A programming language for cognitive agents. RCIM News. In: European Research Consortium for Informatics and Mathematics (53) (2003)
12. Dennett, D.C.: The Intentional Stance. MIT Press, Cambridge (1987)
13. d'Inverno, M., Kinny, D., Luck, M., Wooldridge, M.: A formal specification of dMARS. In: Rao, A., Singh, M.P., Wooldridge, M.J. (eds.) ATAL 1997. LNCS, vol. 1365, pp. 155–176. Springer, Heidelberg (1998)

14. Duff, S., Harland, J., Thangarajah, J.: On proactivity and maintenance goals. In: Proceedings of the Fifth International Joint Conference on Autonomous Agents and Multi-Agent Systems (AAMAS 2006), Hakodate, Japan, pp. 1033–1040 (2006)
15. Fasli, M.: Heterogeneous BDI agents. Cognitive Systems Research 4(1), 1–22 (2003)
16. Ferguson, G., Allen, J.: Mixed-initiative systems for collaborative problem solving. AI Magazine 28(2), 23–32 (2007)
17. FIPA. Foundation for Intelligent Physical Agents (2007), http://www.fipa.org
18. Fulcher, J., Jain, L.C. (eds.): Applied Intelligent Systems: New Directions. Studies in Fuzziness and Soft Computing, vol. 153. Springer, Berlin (2004)
19. Georgeff, M.P., Lansky, A.L.: Reactive reasoning and planning. In: Proceedings of the Sixth National Conference on Artificial Intelligence (AAAI 1987), Seattle, WA, pp. 677–682 (1987)
20. Hanser, M.: Intention and teleology. Mind (107), 381–402 (1998)
21. Hartrum, T., DeLoach, S.: Design issues for mixed-initiative agent systems. In: Cox, M. (ed.) Proceedings of the AAAI Workshop on Mixed-Initiative Intelligence, Orlando, Florida, July 18-19 (1999)
22. Hayashi, H., Tokura, S., Ozaki, F., Tetsuo Hasegawa, T.: On-line interruption planning using dynagent: Integrating deliberation and emergency deliberation. In: Proceedings of the International Workshop on Moving Planning and Scheduling Systems into the Real World, Providence, Rhode Island (2007)
23. Horty, J.F., Pollack, M.E.: Evaluating new options in the context of existing plans. Artificial Intelligence 127(2), 199–220 (2001)
24. Houari, N., Far, B.H.: An architecture for agent coordination and cooperation. In: Proceedings of 10th IEEE Conference on Emerging Technologies and Factory Automation (EFTA 2005) (2005)
25. Howden, N., Rönnquist, R., Hodgson, A., Lucas, A.: JACK Intelligent Agents - Summary of an agent infrastructure (2001), http://www.agent-software.com/shared/resources/reports.html
26. Huber, M.J.: JAM: a BDI-theoretic mobile agent architecture. In: Proceedings of the Third Annual Conference on Autonomous Agents (AGENTS 1999), Seattle, WA, pp. 236–243 (1999)
27. Jason (2006), http://jason.sourceforge.net/
28. Labrou, Y., Finin, T., Mayfield, J.: KQML as an agent communication language. In: Bradshaw, J. (ed.) Software Agents, pp. 291–316. MIT Press, Cambridge (1997)
29. Lokuge, P., Alahakoon, D.: Hybrid BDI agents with improved learning capabilities for adaptive planning in a container terminal application. In: Proceedings of the Intelligent Agent Technology, IEEE/WIC/ACM International Conference (IAT 2004), pp. 120–126 (2004)
30. Mascardi, V., Demergasso, D., Ancona, D.: Languages for programming BDI-style agents: an overview. In: Proceedings of the 6th AI*IA/TABOO Joint Workshop "From Objects to Agents": Simulation and Formal Analysis of Complex Systems, WOA 2005, Camarino, Italy, pp. 9–15 (2005)
31. Meneguzzi, F.R., Luck, M.: Composing high-level plans for declarative agent programming. In: Baldoni, M., Son, T.C., van Riemsdijk, M.B., Winikoff, M. (eds.) DALT 2007. LNCS (LNAI), vol. 4897, pp. 69–85. Springer, Heidelberg (2008)
32. Meneguzzi, F.R., Zorzo, A.F., da Costa Mora, M., Luck, M.: Incorporating planning into BDI agents. Scalable Computing: Practice and Experience 8(1), 15–28 (2007)

33. Morley, D., Myers, K.: The SPARK agent framework. In: Proceedings of the Third International Joint Conference on Autonomous Agents and Multi-Agent Systems (AAMAS 2004), New York, NY, pp. 714–721(2004)
34. Motta, E., Domingue, J., Cabral, L., Gaspari, M.: Irs-ii: A framework and infrastructure for semantic web services. In: Fensel, D., Sycara, K.P., Mylopoulos, J. (eds.) ISWC 2003. LNCS, vol. 2870, pp. 306–318. Springer, Heidelberg (2003)
35. Myers, K., Berry, P., Blythe, J., Conley, K., Gervasio, M., McGuinness, D., Morley, D., Pfeffer, A., Pollack, M., Tambe, M.: An intelligent personal assistant for task and time management. AI Magazine 28(2), 47–61 (2007)
36. Ohlsson, S.: Constraint-based student modeling. In: Greer, J.E., McCalla, G.I. (eds.) Student Modelling: The Key to Individualized Knowledge-Based Instruction. NATO ASI Series, vol. 1488, pp. 127–146. Springer, Berlin (1994)
37. Osinga, F.P.B. (ed.): Strategy and War: The Strategic Theory of John Boyd. Routledge, UK (2006)
38. Pokahr, A., Braubach, L., Lamersdorf, W.: Jadex: Implementing infrastructure for JADE agents. Exp - In Search of Innovation 3(3), 76–85 (2003)
39. Rao, A.S.: AgentSpeak(L): BDI agents speak out in logical computable language. In: Perram, J., Van de Velde, W. (eds.) MAAMAW 1996. LNCS, vol. 1038, pp. 42–55. Springer, Heidelberg (1996)
40. Rao, A.S., Georgeff, M.P.: Modeling rational agents within a BDI-architecture. In: Proceedings of the Second International Conference on Principles of Knowledge Representation and Reasoning (KR 1991), Cambridge, MA, pp. 473–484 (1991)
41. Rao, A.S., Georgeff, M.P.: Decision procedures for BDI logics. Journal of Logic and Computation 8(3), 293–343 (1998)
42. Rieder, B.: Agent technology and the delegation paradigm in a networked society. New Media, Technology and Everyday Life in Europe Conference 2003 (2003), `http://www.lse.ac.uk/collections/EMTEL/Conference/papers/Rieder.pdf`
43. Sardina, S., de Silva, L., Padgham, L.: Hierarchical planning in BDI agent programming languages: a formal approach. In: Proceedings of the Fifth International Joint Conference on Autonomous Agents and Multi-Agent Systems (AAMAS 2006), Hakodate, Japan, pp. 1001–1008 (2006)
44. Sardina, S., Padgham, L.: Goals in the context of BDI plan failure and planning. In: Proceedings of the Sixth International Joint Conference on Autonomous Agents and Multi-Agent Systems (AAMAS 2007), Honolulu, Hawaii, pp. 1–8 (2007)
45. Tecuci, G., Boicu, M., Cox, M.T.: Seven aspects of mixed-initiative reasoning: An introduction to this special issue on mixed-initiative assistants. AI Magazine 28(2), 11–18 (2007)
46. Thangarajah, J., Harlan, J., Morley, D., Yorke-Smith, N.: Aborting tasks in BDI agents. In: Proceedings of the Sixth International Joint Conference on Autonomous Agents and Multi-Agent Systems (AAMAS 2007), pp. 1–8 (2007)
47. Thangarajah, J., Padgham, L., Winikoff, M.: Detecting & exploiting positive goal interaction in intelligent agents. In: Proceedings of the Second International Joint Conference on Autonomous Agents and Multi-Agent Systems (AAMAS 2003), Melbourne, Australia, pp. 401–408 (2003)
48. Thangarajah, J., Padgham, L., Winikoff, M.: Detecting and avoiding interference between goals in intelligent agents. In: Proceedings of the 18th International Joint Conference on Artificial Intelligence (IJCAI 2003), Acapulco, Mexico, pp. 721–726 (2003)

49. Weiser, M., Brown, J.S.: The coming age of calm technology. In: Denning, P.J., Metcalfe, R.M. (eds.), Beyond calculation: The next fifty years, Copernicus, New York, NY, pp. 75–85 (1997), `http://www.ubiq.com/hypertext/weiser/acmfuture2endnote.htm`
50. Wilkinson, A.: Empowerment: Theory and practice. Personnel Review 27(1), 40–56 (1998)
51. Winikoff, M., Padgham, L., Harland, J., Thangarajah, J.: Declarative & procedural goals in intelligent agent systems. In: Proceedings of the Eights International Conference on Principles and Knowledge Representation and Reasoning (KR 2002), pp. 470–481. Morgan Kaufmann, San Francisco (2002)

10

Ontology for Agents and the Web

Ronald L. Hartung

Division of Computer and Information Sciences, Franklin University,
201 South Grant Ave.,
Columbus, Ohio 43224
hartung@franklin.edu

Abstract. Ontology has caught on as a major tool in both the Web 2.0 and in multi-agent systems. In Web 2.0, ontology is being used to extend the search capability; for multi-agent systems, ontology becomes a basis for communication and negotiation. In support of both these areas of application, the designer needs to apply ontology with care to realize the advantages of ontology. The coverage in this chapter is to clarify some of the issues around the use of ontology, especially in the case of multiple ontologies from multiple sources and/or authors. The issues include types (uses) of ontologies, good design principles, mapping and grounding. The approach taken is designed to provide the practitioner to find guidelines and techniques to apply in problem solving.

10.1 Introduction

The aim of this chapter is to provide an overview of ontology as used in multi-agent systems and the Web. The approach is to concentrate on the definition of ontology as found in software applications. Indeed some may argue that computer scientists have used ontology as something other than the traditional meaning, which may disappoint those with strong opinions about ontology derived from philosophy. The purpose here is to serve the practitioner that will use ontology as representation structure.

There are a number of uses for ontology in current applications. Some use the ontology to support purely human activities. In this work, the use of ontologies in computer applications, especially in multi-agent systems and/or the Internet, are of interest. There are many systems/applications where domain knowledge can be represented in an ontology and this chapter provides some guidance for building an ontology. The major focus here is in applications where multiple ontologies are required. In the agent domain, the focus will be systems of agents that interact with other agents where each agent has its own ontology. Ontology-based agents can be built to provide autonomous functioning for negotiation between systems, distributed search, or intelligent exploration of the web. Likewise, searching in the web 2.0 will become dependent on working with multiple ontologies to evaluate semantic content.

Ontologies, found in the Internet, can be of use for both computer applications and for human interpretation. So far, there are a large number of ontologies, well over 13,000 on the Internet alone. However, there are no standards for building them and the differences between the ontologies are significant. The differences make it difficult to

N.T. Nguyen, L.C. Jain (Eds.): Intel. Agents in the Evol. of Web & Appl., SCI 167, pp. 223–248.
springerlink.com

find and search the ontologies. Moreover, ontologies can come in many sizes. Some examples include the Adult Mouse Anatomy that has 2744 classes and the NCI Thesaurus that has 3304 classes describing human anatomy. CYC is another classic example, which has a very large number of atomic terms, over 100K [35]. However, one will find many examples with smaller ontologies as well. With this diversity, dealing with multiple ontologies becomes a considerable problem.

It is important to differentiate the uses and the kinds of ontology. Also, it makes good sense to provide some principles for ontology design. This becomes especially critical when dealing with systems that must interact with multiple ontologies designed by different authors. In the work on multiple ontologies, we examine some of the techniques to map or merge those ontologies. As the reader can surmise, the discussion of the topics about the applications and ontologies cover a large range. The approach will be to survey the topics and provide guidance to detailed primary sources.

10.2 Definition

Ontology in philosophy is generally defined as a study of reality, the nature of being and the structured relationships between things that exist. Gruber [30] provides a commonly referenced definition for computer applications as: "An ontology is a specification of a conceptualization". Ontology as defined in philosophy is singular and the issue of "what exists" is global within the world; it is not divided into parts. This leads to the expectation that all the ontology implementations should merge into one master ontology. Unfortunately, this is not the reality of ontology as used in computer applications. Moreover, it is arguable that the study of ontology has lots of complicated issues and that no philosopher has yet produced an acceptable single ontology. The practitioner that studies the philosophers' views of ontology will find the discussions very instructive. Indeed, as an AI researcher, the issues found in ontology may tell us much about the human mind.

Ontology should not be confused with epistemology. Epistemology is the study knowledge and belief. Nonetheless, epistemology does have a role in systems that use ontologies in their interaction with external systems. A useful approach to understanding the interaction between ontology-based systems is that the ontologies can express an internal worldview of each system. For a system to "understand" another system, epistemological reasoning, i.e., reasoning about what the other system "believes", is a useful approach. When this reasoning is done, modal logic becomes a required tool.

When we apply ontology in applications, ontology becomes a formal representation. The formal representation will give a syntactic definition, and in some cases semantic and pragmatic definition. For the applications, there are a number of formal ontology languages where many of the languages are based on logic representations.

The way the ontology is used has a large impact on how it is designed. The actual use will determine the form and structure of the ontology as well as what kind of objects are being defined. This can be obscured by the use of a single name for all of the kinds of ontology. As the field matures, an ontology of ontology types may become an important tool, but as this is written, this level of maturity has not been achieved. There are some basic usage distinctions that can be offered here.

It has become common to refer to several kinds of ontology. These are really distinctions based on the use of the ontology. However, while ontologies appear in these forms, not all of the members of these types are ontologies. For example, taxonomy is not necessarily ontology. Only when the taxonomy is expressing a description of things in the world it becomes an ontology.

The simplest type of ontology is a taxonomy. This is a structure with a basic hierarchy, that is, basic parent child relationships. Taxonomy is beneficial for classification tasks; it also provides a very simple worldview.

A thesaurus extends a basic taxonomy of words by adding synonyms and possibly other linguistic data. When constructing a thesaurus, it is easy to end up with something that is not really an ontology.

A conceptual model is a construction that represents knowledge about a domain. It has s much richer structure than either taxonomy or a thesaurus. The conceptual model has a set of objects and the relationships between them. This is closer to the use of ontology within philosophy.

Another form, the logical theory, is a more complex and rich structure that implements a logical model of a domain. A logical theory adds full reasoning capability to the ontology. The other forms provide a basis for reasoning by supplying objects to reason about, but logic is external to them.

Another significant variation is a populated ontology. A populated ontology is one that contains the objects as well as the class that contain the objects. In the purest sense, ontology is classification of objects. The objects themselves are not part of the ontology. However, some users and systems find it convenient to place the objects in the ontology. Also, a populated ontology can be easier to align or map with other ontologies by matching on the objects.

10.3 Formal Definitions and Languages

There are a number of approaches to formal ontology languages and representation systems. These are based on frames, first order logic, description logic, conceptual graphs, or semantic nets[1]. For agents, and especially Internet agents, the main ontology languages are RDF, RDFS, and OWL. RDF is a semantic net-based representation for which RDFS is a schema model, and OWL is description logic. These form a family of ontology languages on the Internet and are integral to Web 2.0 or the semantic web. OKBC is a frame based knowledge model. It can used as a meta-ontology in the FIPA (Foundation for Intelligent Physical Agents) specifications. [15] In addition to the language based ontology representations, there are programming language-based approaches, for example in Java classes. Here, the intent is to provide an overview of these languages; the practitioner using one of these will be well advised to study them carefully.

The basis of many of the formal languages is description logic [1]. While normal first order logic is about sentences of predicates, description logic[2] is used to define,

[1] A good reference for this material is found in Gruber [30].

[2] The description logic is sometimes called a description language.

or describe, categories and instances. Description logic, and ontologies, provide the objects used in first order propositional logics.

Following the definition found in Brachman and Levesque [1] there are three classes of non-logical symbols. These are atomic concepts, roles and constants. Concepts and constants are names, where concepts are capitalized names. Roles are written with a leading ":" applied to a constant. For the concepts there are the forming operators: ALL, EXISTS, FILLS and AND. This description allows the following definition of a concept.

Concept is defined as:

- An atomic concept
- If r is a role and d is a concept, then [ALL r d] is a concept.
- If r is a role and n is a positive integer, then [EXISTS n r] is a concept.
- If r is a role and c is a constant, then [FILLS r c] is a concept.
- If d_1 ... d_n are concepts then [AND d_1 .... d_n] is a concept.

The [ALL r d] defines concept d where all members of d have a role r.

The [EXISTS n r] stands for the class of object that have role r with cardinality n or greater.

The [FILLS r c] is the class of objects that have role r related to c. [FILLS :workFor boss] is the class of employees who work for the boss.

Description logic also has notation for "isa" so that a hierarchy can be constructed. Description logic provides a model for extension of ontology into reasoning systems. The description logic is a formal model of the ontology contents and can be used to extract predicate expressions from the ontology to augment a reasoning system. Even though this topic is beyond the current scope, a simple example is the following. Suppose we have a rule that states: A manager is required to approve a purchase. The ontology can be queried to find a list of persons that "FILLS" the role manager.

In the Internet, there are two major ontology languages, RDFS and OWL. These have emerged as components for the semantic web. RDFS is a direct extension of RDF, resource description framework. OWL is built on RDF plus DAML. These are the current standard for the Internet, and will have considerable impact on ontology design and specification for near future.

10.3.1 RDFS

Resource Description Framework, RDF, and RDF Schema, RDFS, are basic resource description frameworks[3]. RDF makes statements about web resources. RDFS uses the same basic language form as RDF, but is used to define ontologies. Since RDF and RDFS are used to describe resources that occur on the Internet, a Uniform Resource Identifier (URI) defines resources. Properties are characteristics, attributes, or relations used to describe the resources. Sentences in RDF are triples: subject, predicate, object. Subject, predicate and object are written as a URI-based vocabulary. For example, to express Ron wrote this chapter, "Ron" is the object, the predicate is

[3] The definitions for RDF and RDFS can be found at the W3C web site.

"wrote" and "this chapter" is the subject. RDF uses XML encoding so the triple is written as:

```
<?xml version="1.0"?>
<ref:RDF xmlns:rdf=" http://www.w3c.org/1999/02/22-rdf-
syntax-ns#"
xmlns:myns="http://www.franklin.edu/hartung" >
<rdf:RDF>
        <RDF:Description RDF:about="this chapter">
        <myns:wrote> Ron </myns:wrote>
        </RDF:Description>
</rdf:RDF>
```

Note, that namespaces need to be handled with some care. In the above XML sample, a local namespace is used for the locally defined vocabulary. The term "wrote" will be in a namespace for this XML document ("myns"), while elements prefixed with "RDF" are from the W3C namespace for RDF[4]. This example shows only a single triple. Additional triples for the same object can be added by nesting them inside the parent element, "Description" in this case. For example, to add a date, insert the line "<myns:DateWritten>March 2008</myns:DateWritten>" after the line defining who wrote the chapter, i.e., "<myns:wrote> Ron </myns:wrote>".

Another way to view RDF is using diagrams. The diagrams are simple directed graphs. The subject and object are drawn as nodes and the predicate is a directed arc.

RDF resources can be classes. A triple can asset that an object is a member of a class or that classes are members of another class. However, this is not semantically enforced, it is only by knowing a class is inferred by a URI and that a relationship entails class membership. RDF classes include containers that collect resources and instances with various semantics. This allows RDF to express sequences, bags (unordered) and alt. An alt implies that the intension is to select one of a set of alternatives. The implementation of all three is all the same, a simple list. The semantic differences are implied, but not explicit.

A deficiency with RDF is the lack of the capability of enforcing the semantics. Moreover, RDF limits the structure of what can be expressed. For example, the graph view is a tree of nodes, corresponding to the triples. Also, RDF is designed to be a description language for instances.

RDF Schema (RDFS) is a schema for ontologies. RDFS adds the semantics to explicitly define classes and relationships that is lacking in RDF. RDFS uses triples just like RDF and allows the definition of classes that are types or categories, have properties and are placed in a hierarchy. Thus, the triple: author, rdf:type, ref:class, defines author as a class. The "isa" hierarchy is constructed using the "ref:subClassOf" as the relation of a triple. The example below shows the XML form for an RDFS.

---

[4] More information about the RDF vocabulary is found at http://www.w3c.org/1999/02/22-rdf-syntax-ns#

```
<?xml version="1.0"?>

<!DOCTYPE rdf:RDF [<!entity xsd
http://www.w3c.ord/2001/XMLSchema#>]>

<rdf:RDF

         xmlns:rdf="http://www.w3c.org/1999/02/22-rdf-
syntax-ns#"

         xmlns:rdfs="http://www.w3c.org/2000/01-rdf-
schema#"

         xml:base="http://franklin.edu/hartung" >

<:rdf:Description rdf:ID="author">

         <rdf:type
rdf:resource=http://www.w3c.org/2000/01/rdf-
schema#Class/>

</rdf:Description>

</rdf:RDF>
```

As mentioned above, classes in RDFS can have properties, for example the triple: age, rdf:type, rdf:property defines age as a property. Properties can be described by a domain, and range, as well as cardinality. The range is used to specify the values the property can take. The domain identifies the classes that have the property. Properties are defined globally, and then included in a class definition. This is not like the typical class semantics of object-oriented systems, where properties are defined locally to a class.

10.3.2 OWL

OWL[5] is built on RDF and RDFS. It is intended to facilitate machine interpretation of the information. OWL extends RDF with 35 keywords, resulting in a more complete description logic. OWL adds complex class constructors using union and intersection operators to RDF. It extends the description of classes to include inverse and transitive relationships and allows local property descriptions for cardinality and type.

OWL has three sub-languages, with different semantics and axioms. OWL-Lite supports simple hierarchy and some cardinality constraints. OWL-DL has maximum functionality, while insuring computational completeness and decidability. OWL-DL is a description logic. OWL-Full allows the full range of features, but it will not be guaranteed to support reasoning, as it is not complete or decidable. One example of the difference is that terms in OWL-DL are not allowed to be overloaded, e.g. names that are both a class and a property.

The added constructs in OWL include owl:Class that is defined to be a sub-class of owl:thing. Note the namespace identifier owl is used here for exemplify and clarity, but can be defined to any identifier in the namespace declarations. There is also an empty class owl:nothing. That is something ontologiests might find a bit troubling. Ontologies are about things in the domain, and nothing really a not a thing.

---

[5] Details about OWL is found at http://www.w3.org/TR/owl-features/

A simple class definition is:

```
<owl:Class rdf:ID="aClass" >
        <rdf:subClassOf ref:resource="parent" />
        <rdfs:label
xml:lang="en">className</rdfs:label>
        ….. .
</owl:Class>
```

This example shows a class with an ID used to refer to the class, a human readable name (label) and a sub class relation.

Properties in OWL are of two types. Object properties are a relation between two instances of classes. Data type properties define a relation between a class and a literal value.

```
<owl:ObjectProperty rdf:ID="realtionID" >
        <rf:domain rdf:resoursce="#from"/>
        <rdf:range rdf:resource="#to"/>
</owl:ObjectProperty>
```

The above defines a relation between two resources. The "#" indicates it is a relative ID found in the XML document. The full form is the namespace called name, followed by the "#", which is followed by the ID. In the example, from and to would indicate classes in the ontology.

To add the property to a class:

```
<owl:Class rdf:ID="aClass" >
        <rdf:subClassOf ref:resource="parent" />
        <rdfs:label
xml:lang="en">className</rdfs:label>
        <rdf:subClassOf>
          <owl:Restriction>
            <owl:OnProperty
ref:resource="#realtionID"/>
<owl:minCardinality
ref:datatype="&xsd;nonNegativeInteger"> 1
</owl:minCardinality>
          </owl:Restriction>
        </rdf:subClassOf>
</owl:Class>
```

This defines a restriction on the class requiring the relation to exist.

Data type properties have a similar syntax:

```
<owl:DatatypeProperty rdf:ID="realtionID" >
        <rf:domain rdf:resoursce="#from"/>
```

```
            <rdf:range rdf:resource="&xsd;string"/>
</owl:DatatypeProperty>
```

The range can use any of the data types defined in XML xschema documents (xsd).

The properties can also have characteristics like TransitivePorperty, SymmetricProperty, FunctionalProperty, InverseProperty, etc. These define the reasoning axioms that will restrict the semantics of the property.

OWL includes class definition by set operations, namely union, complement, and intersection. For example:

```
<owl:Class rdf:ID="example" >
            <owl:UnionOf  rdf:parseType="Collection">
              <owl:Class rdf:about="#class1"/>
              <owl:Class rdf:about="#class2"/>
            </owl:UnionOf>
</owl:Class>
```

This defines the example class as the union of two classes. Similarly, intersection and complement can be used and mixed into a complex set operation to define a class. As is discussed in the section on design or ontologies, not everyone believes these operations should be used in ontology construction.

10.3.3 OKBC

The Open Knowledge Base Connectivity (OKBC) [31] model was developed to connect knowledge bases. Some would argue it never achieved that end. FIPA describes using it as a meta-ontology to provide a common view of ontologies. It is a frame-based representation and it is capable as an ontology definition language, although not commonly used for this purpose. The primitives in OKBC are constants, frames, slots, facets, classes, and individuals. In this model, classes are represented as class frames; frames are more primitive than classes. Into these class frames, ontological objects are mapped. Properties become slot-value pairs in the class frame. If individuals are parts of the ontology, these are likewise mapped into frames.

Relations in OKBC exist in two ways. First of all, some primitive relations are defined directly in OKBC. Thus, instance-of, type-of, subclass-of, superClass-of are part of the language of OKBC. Secondly, the relations are defined as needed. These are defined as a frame, where the slots of the frame define the properties of the relation.

OKBC is a representation that is more general than ontology, but can be used to represent ontology. If it is used to design ontologies, the ontologist should be careful not to allow the richness of the representation since it can lead to construction of bad ontologies.

10.4 Ontology Design

Given the varied uses of ontology within computer science and applications, as well as the complexity of achieving a design with a broad range of application, ontology design is a difficult topic. Some systems do not need to interact with external systems, and their ontology design does could be considered less critical. Although, even within a single system, an inconsistent design will lead to poor results. However, systems that must interact with external systems, especially those designed by diverse communities, the design is very critical. Such systems often face a wide range of design practice, and will need to adapt. This topic will also arise in ontology alignment methods. For any developer of agent systems using ontology, it is worthwhile to have a view of good design practice. The design practice can also serve as an evaluation criteria applied to ontologies encountered by our systems. Ontologies encountered by an agent system could be evaluated using some of these guidelines, and the quality of the ontology inferred.

A very good discussion of ontology design can be found in Barry Smith's recorded lecture[6][2]. The focus is the construction of scientific ontologies, but the principles are sound enough to extend beyond scientific ontologies. The discussion has some very nice thoughts on what an ontology should be. Barry Smith likes to think of ontology as a window onto the world and that seems a very good analogy indeed [Smith]. Ontology does indeed present a theory of what exists in the world. And it becomes the lens used by the system as it interacts with the rest of the environment.

The notion of scientific ontologies uses the definition of ontology as a hierarchy defining entities that are universals found in the domain. The primary relationship is "isa", but this definition does not exclude other relationships. The nodes of ontology are universals and denoted by terms. The definition allows a node to have an identifier, and the node may have synonyms, definitions, glosses and comments.

In our ontology-based systems, the domain is a subset of the world. As mentioned above, the philosopher may wish to think of ontology as a singular thing. To our best knowledge, no philosopher has yet found that singular ontology that is accepted widely. The best achievement to date in philosophy is an upper level ontology. In computer science applications, multiple ontologies are the rule. The domain is a partition of reality into a subject area. Defining the domain is important for the formation of the ontology. Mixing domains or confusing domains can cause some difficult problems in design. In Barry Smith's work, he offers the chemical periodic table as a great example of ontology [2]. The table works because it captures a well-defined set of properties. It does not mix the chemical domain with some other domain. For example, the table does not try to describe sub atomic particles as part of the structure.

The universals have been defined as entities in the domain. A universal is a category defining a collection of entities in the domain that share a universal defining property. Universals in ontology are labeled with a term. Understanding what is and what is not, a universal is central to successful design. As will be seen labeling is harder to discern that one might think. To expose the problem, consider the difference between a term (universal), a class and a concept. The argument is that these are different things, and that they are positioned in a subset hierarchy. A universal is a classification of entities in the domain. This description is similar to a class and indeed it is also a class.

[6] The lecture of Barry Smith can be found at the web page http://ontology.buffalo.edu/ smith/ articles/ontologies.htm, or at his home page at the University of Buffalo.

However, a class is a more general object than a universal. Any universal will have an extension that is a class of individuals from the domain and the universal is the definition of this class. To observe the problem with classes, consider the following example. Cat is a universal – that is cats exist in the world. Likewise dog is a universal. Now the class of dogs and, i.e., union, cats is clearly a class. However, there is no dog-cat entity in the world. This example is rather trivial, however in real world ontologies, this kind of problem can become much harder to detect. This is one of the possible issues with OWL. The semantics of OWL allow easy formation of classes by union and intersection. This will violate the requirement that forming classes like "cat union dog" forms an ontology from universals.

The issue with concepts is likewise a subtle problem. Unless the ontology is to deal with the domain of concepts, concepts do not go into ontology. The ontology is concerned with entities in a domain but to understand why, examine the following concept: trees on the moon. This is a perfectly fine concept. But it does not appear to be an entity in the world, at least not at this time. There is another subtlety here as well; ontologies are not universal in time.

There are several other examples of problems in defining universal terms. Consider terms like "fist". A fist does not always exist; it must be formed from the hand. So sometimes the person has a fist and sometime, they do not. Likewise, whole/part relationships may not be universal. A man without legs is still a man. Humans do not have a lot of trouble with this kind of distinction. However when encoded into an ontology and operated on by reasoning systems the situation can become much more brittle.

Additional problems are introduced by natural language itself. Words have multiple definitions and all shades of meaning. A noun can refer to a universal and a collection can be a class. For example: "influenza is a viral infection" versus "influenza is spreading". In the first case we have a universal, while in the second we are describing a collection of instances that are occurring. Such distinctions can be subtle and hard to detect.

In forming an ontology "is" is the backbone. Other relations can exist, but "isa" is the central organizing relation. That is to say, the hierarchy is always the core of the ontology. When designing an ontology, principle attention is focused on the "isa" relations. There are a number of common terms for this relation, specialization and generalization that are used, depending on the direction. The use of terms does not relieve the care that should be exercised with the other relationships, but in our ideas on design, the hierarchy is the main consideration.

In defining or constructing an ontology, the core issue is deciding the universals and relations to be used. The domain, under consideration, is part of the key to determining the universals. Things universal in one domain are not universal in another.

Rules

The rules proposed for design are based on the lecture by Barry Smith [2]. They constitute an excellent general design guideline. The rules may appear quite simple; however applying them is a real challenge.

1) A term naming a concept in an ontology should be a singular noun. Terms in an ontology are universals. These terms define a class, but they require the additional property of universality. For example, cat is a universal and it is term that defines a set of cats. Staying with only singular terms helps to keep the ontology clean. The universals must be determined with respect to a domain. Thus, whether or not it is a universal for a given ontology will depend on the domain. Since an ontology in a system generally has a domain, it also has a purpose. The purpose will often define what universals are included. For example, cat may be useful for some purposes, while other systems may chose to subsume all animals in a single term animal. The argument is not so much that mapping and other uses can not deal with plurals, rather that using singular forms is a cleaner design practice.

2) Avoid use-mention confusion. Be careful when you use terms that are concept names (e.g. use), and, moreover, use them to refer to the word itself (e.g. mention). For example, cat is an animal and "cat" is a three-letter word. The easy way to look for this problem is when the term should be put in quotes for proper grammar. The second cat sentence is properly: "Cat" has three letters.

3) Avoid confusing words and things and concepts in minds with entities in reality. The work of an ontologist is patient and careful. Dealing with word sense and usage is difficult. The ontologist needs to capture just entities with in a domain.

4) Avoid special meanings. This is difficult since domains often use common terms with specialized meanings. When using commonly occurring terms within a domain, use the commonly accepted meaning. Using common terminology is especially helpful for reasoning systems and ontology mapping. However, if the domain is not specified as part of ontology, this will not help in mapping. Mapping needs to find ways to define correspondence between terms in diverse ontologies.

5) Terms have one meaning – ONLY. This should be obvious, but as ontology grows in size or if there is more than one author, the name usage problems can be a little harder to detect.

6) Don't allow circular definition. One cannot define everything and a circular definition is useless in an ontology. There will be a set of primitive undefined terms, the rest of the definitions are grounded in those primitive terms. If this were well defined in the ontology, then mapping might become easier by focusing on the fundamental undefined terms as the focus of the mapping.

7) Univocity. Terms have the same meaning in all uses in the ontology. This includes relations.

8) Universality. Relationships in an ontology should only include universal relations. This is univocity for relations. Note, ordering the relation, or direction of the relation, can fix some problems. Not every child becomes an adult, but every adult was a child.

9) Complements require care. The complement of a universal is not necessarily a universal. For example, if dog is a term in an ontology, a non-dog is probably not a universal term. Non-dog is a class, but it does not have a singular defining concept as a universal. One probably can find a good exception, for example a dog ontology may be quite useful with a term that includes all dogs and a concept for all things not dog. However, as can be seen, when mapping ontologies, complement terms like this can and will cause bad problems. In general, reasoning systems have problems with complement terms.

10) Supply definitions where possible to help humans, but remember some terms have to be primitive. While the definitions are mostly useful to humans, mapping tools can access definitions using natural language processing techniques.

11) Don't give definitions more complex than the term itself.

12) Use Aristotelian definitions. Aristotelian definition is a form: An A is a B which C. Man is an animal that thinks. Aristotelian definition is easy to use, since it supports the hierarchy formation. It also works well with reasoning systems.

13) Issues with logical expression (Positively). Be careful of negatives. Do not use of conjunctions and disjunctions. These lead to classes rather than universals.

14) Avoid unknown. Do not use unknown things. "Unknown" belongs to epistemology, not ontology.

15) Single inheritance. Multiple inheritance is complex and can lead to a lot of problems. "Isa" can come to mean two different kinds of things in the class hierarchy. A solution is to use separate ontolgies for the different "isa", each having only single inheritance. Example: vehicle hierarchy and a color hierarchy.

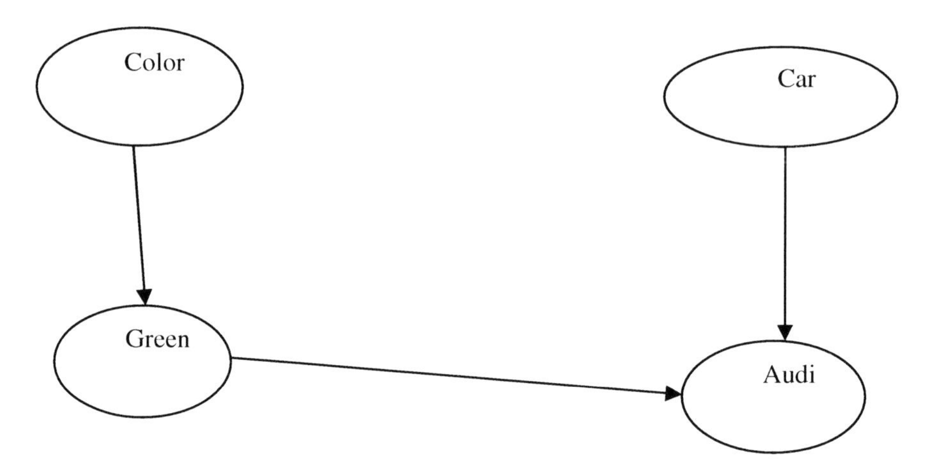

Fig. 10.1. Multiple Inheritance

The Fig 10.1 shows an example of multiple inheritance with "isa" relations. The term Audi inherits from green and car. This would in fact also say that an Audi is a kind of color. Clearly, this does not what makes sense in a normal universe. The use of multiple inheritance can hamper statistical search tools.

16) Compound terms. The meaning of a compound term is composition of the individual terms.

The rules above are a good starting set to avoid problems in ontology design. These rules are not perfect. Good ontology design remains an art, not a science. An additional problem is how to proceed when you are the ontologist and not also a domain expert. A very good approach is incremental development like agile methods in software development. Start with a few terms near the top of the ontology and check with the domain experts, and then add terms a little at a time. This is much better than massive rework efforts when a bad choice is made early in the design process.

10.5 Mapping

The web has become a source of multiple ontologies. They are heterogeneous in multiple dimensions; in implementation language; in the source natural language; in design principles; in intended use; and in type. While the implementation language can be a serious issue, it is not too hard to deal with. There are a number of frameworks that have been successful in providing a common interface layer over multiple ontologies, for example, the Ontology Mapping Specification Language API[7]. The JENA project[8] provides a framework, plus already developed interfaces for OWL and RDF. The issue of natural language brings in the typical translation problems between languages. The worst problems arise as a result of intended use, the design principles or practices used by the author, and the type of ontology. Even in agent applications, which do not access the web, multiple ontologies will also occur when agents, implemented by different designers, interact.

A critical approach for multiple ontology systems is mapping or merging. There has been a great deal of work in this mapping area and there is a wide range of approaches. This includes approaches ranging from manual to purely algorithmatic approaches. Mapping is not a simple problem and even manual schemes are hard. Given the explosion of ontologies on the web and the possible use of autonomous agents for a variety of purposes, automatic mapping is a real need.

Even though the practitioner will find a survey of mapping methods and a guide to finding the source documents for these methods, it is impractical to provide more than a brief overview of the methods here; a full treatment would require a full-length text. Since so many methods exist, it is a good indicator that the state of the art falls short of a complete solution.

There are a number of mapping approaches found in the literature and the references include some good surveys that are an excellent source to find the algorithm details [32; 8; 14].

At the outset there are three possible ways to deal with multiple ontologies. One can try for a single global ontology, and not allow multiple ontologies. But that clearly has not happened, nor is this an expected solution. The opposite approach is to allow multiple ontologies and provide mapping between them. The third possibility is a hybrid approach, which is an ontology with a global shared vocabulary used by and extended by the multiple ontologies. This third approach may be possible in some domains and could be a great help when the mapping persists over long periods of

[7] Information about Ontology Mapping Specification Language API is found at http://dome.sourceforge.net/sheets/lang.html

[8] http://jena.sourceforge.net/

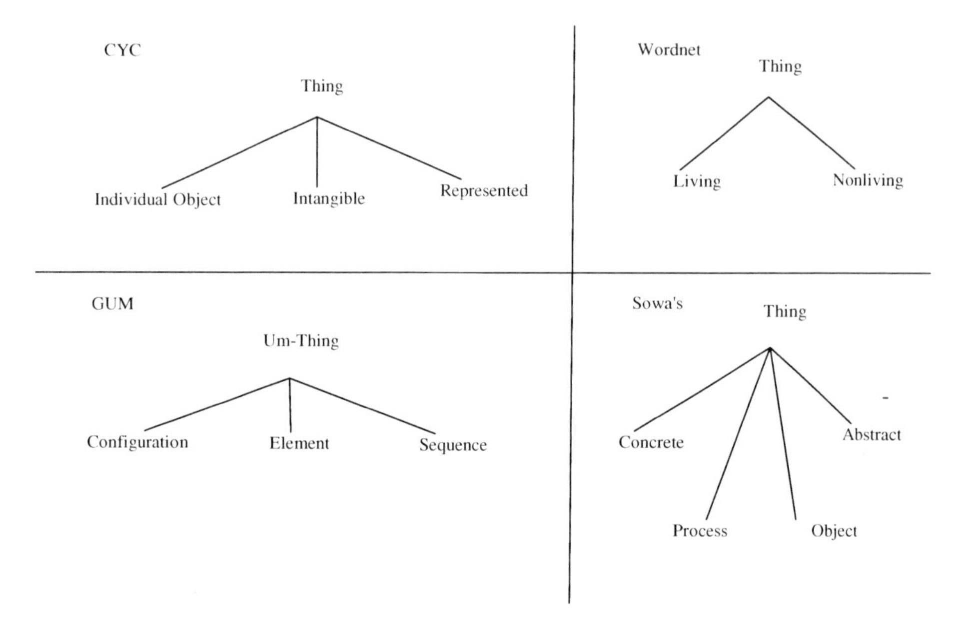

Fig. 10.2. Sample top-level ontologies, from Chandrasekaran *el al.* [5]

time and is contained in a single domain. Although, some experience observed makes this still appear problematic. The caution is to approach the hybrid approach with great care in the design of the global ontology. This is illustrated by the multiple upper level ontologies found in practice. The figure below (Fig. 10.2) shows four upper level ontologies reported in Chandrasekaran *el al.* [5].

Ontology mapping can be found under a number of names in the literature, alignment, morphism, etc. Also, the terms merging, fusion and integration are found. Merging is to form a single ontology from the ones being mapped. Sometimes this is a desired approach, especially if the subject ontologies are used frequently. However, in many applications, a maintaining a temporary translation between them is sufficient. Ontology mapping has been formally defined by Ehrig and Staab [12]: "Given two ontologies O_1 and O_2, mapping one ontology into another means that for each entity (concept C, relation R, or instance I) in ontology O_1, we try to find a corresponding entity, which has the same intended meaning O_2."

An ontology mapping can be a total mapping, a total mapping between the two ontologies captures the complete structure and the axioms of the ontology. However, it is also possible to have a partial mapping. In this case there is a subset of the ontology were the mapping preserves structure and axiom content. A different possibility is to measure the degree of heterogeneity between ontologies by examining the mismatches between them [33].

Another issue in mapping is the difference between a populated and an unpopulated ontology. With a populated ontology, the instances can be used to help identify matches. With an unpopulated ontology only the terms are available for determining the mappings. The information flow based mapping techniques use the instance of populated ontologues as a fundamental feature.

The term alignment is often used to define a more complex and richer process. In alignment we have not just a set of identity mappings as defined earlier, instead we can have a richer set of binary relations between the elements of the two ontolgies. For example, the term in one ontology may be a subset of a term in the second ontology. This is very useful when the subset relation holds, but useless if an equivalent term does not exist between the two ontologies. This can lead to an approach to construct intermediate source ontology between the two subject ontologies. The approach can be partial or total as well. Examples of this technique see the works of Maedche [26], Madhavan [25], and Compatangelo [10].

At the base of mapping approaches are some fundamental matching techniques. These are divided into element and structural mappings. While these can be applied singly in a mapping method, they are often combined.

10.5.1 Element Level Mapping

The element mapping methods seek to define a mapping between two elements from different ontologies [13] In this case, "element" refers to entities, relations and attributes in the ontology. A wide range of techniques are employed and often combined. These techniques include the following.

String based methods use similarity measures on strings. The strings can be simple names or any descriptive text associated with an element in an ontology. The similarity measures use variations of string distance measures. Some of these are as simple as the number of changed character positions between the two strings.

Natural language and linguistic techniques provide more sophistication over simple string matches. Tokenization and morphological transformations are used to reduce the words to root forms. Parsing techniques and dictionaries can be applied to eliminate words with less content that are unlikely to define concepts, for example articles, and prepositions. Synonyms, perhaps from a thesaurus, may be used with advantage. Wordnet is a semantic net that can be used to provide matches and similarity measures. A semantic net is stronger than a thesaurus, as it contains more complex linguistic data that can be used to measure semantic distance, approximated by distance in the net. If the ontology has definitions in natural language, they can also be used to gain improved performance.

Constraint based methods use properties like cardinality and data types to constrain the possible matches. When trying to match two entities, i.e., the terms with matching data types, can be possible matches even if the labels are not exact matches. The data types can include range restrictions, uniqueness constraints, and cardinalities, as well as, basic types. The data type approach is used to strengthen other methods.

Existing mappings are a very useful technique when they are available. If a pair of terms from two ontologies is matched and a third ontology is available that is mapped together with both of the ontolohgies being mapped, then a mapping may exist using transitive reasoning through the third ontology.

Upper level ontologies are sometimes available. When a common upper level is shared between two ontologies, then it may be used to limit possible mappings.

10.5.2 Structure Level Mappings

The structural mappings use the ontology structure defined in terms of the relations between terms. As mentioned above, structure can also be defined as the attributes defined for a term, using the data types and number of attributes.

Graph based techniques use a labeled graph representation of the ontologies to apply graph-matching algorithms. Because of the cost of these algorithms, approximation methods may be used. Approximation methods can be developed by reducing the types of graph features considered, for example, children, leaves, or reducing relations to types of relations. Graph matching, while computationally expensive, could be used to extend or augment other techniques.

A taxonomy-based method is a graph technique were only the specialization ("isa") relations are used. Using only specialization relations can reduce the computational cost by reducing the graph. Since specialization is a universally appearing relation in ontoligies and is arguably the most well defined relation, it is a reasonable reduction.

Repository of Structures is a technique that keeps a set of fragments of previously mapped ontologies. The similarity measures are also retained. The fragments are related by similarity measures, which provides a set of already established mappings to simplify future mappings.

Model based methods use model theoretic tools like Propositional Satiability and Description Logics to model semantics.

10.5.3 Survey of Methods

This section gives a survey of some of the more successful mapping methods. While this section is clearly not detailed enough to implement from, it does give an idea of the range of solution techniques that have been developed.

MAFRA [26] uses a normalization process called LIFT to bring the ontologies into a uniform representation and applies a semantic process to detect similarities. MAFRA constructs the mapping in a structure called a semantic bridge, which contains the mappings between elements of the ontologies.

LOM [24] is Lexicon-based Ontology Mapping. This technique uses the whole term, word constituent, synset and type matching to find mappings between ontology terms. LOM was designed to assist a human in ontology mapping by providing potential mappings.

QOM [12] is a dynamic algorithm that identifies mappings. The algorithm rates the likelihood of possible mappings and discards the less likely candidates. QOM is designed to produce an ad-hoc mapping for lightweight ontologies of large size.

CTXMATCH [3] is a logic algorithm that is based on the hierarchical classifications (isa relations). The CTXMATCH algorithm examines the terms in the hierarchy and produces binary relationships between them, specifically more general, less general, equivalent, and incompatible relations. To map terms, CTXMATCH depends on three kinds of knowledge, i.e., lexical knowledge about synonyms, world knowledge about the terms used, and the structural relations in the ontology.

ONION [27] uses a linguistic matcher to determine a similarity measure between all pairs of terms in the two ontologies. Scores above a threshold trigger define an articulation rule between the two terms. Then a structural matcher extends the set of matches. Finally, an inference-based matcher generates the matches. The user provides the rules for the inference matcher. Generally, ONION is iteratively run with human input after each cycle until a suitable match is generated. While this is not applicable for an agent system, due to the human input, it is a worthy technique that might work in some environments, without the user input. It seems like a good research problem.

OMEN [28] uses a probabilistic approach to mapping. The approach uses a Bayesian Net with meta rules. Moreover, it can use existing mappings to infer other mappings.

CMS [21] depends on the rich structure of OWL for mapping. CMS is designed to consult external linguistic resources and uses a multiple strategy similarity aggregator and similarity evaluator.

OntoMorph [7] uses a rule language for mapping ontologies. OntoMorph uses both semantic and syntactic rewriting rules. The syntactic rewriting rules are pattern directed whereas the semantic rules use semantic models and logical inference rules.

HMatch [4] uses a linear combination of linguistic based affinity and structural affinity. The linguistic approach uses a thesaurus derived from WordNet. The HMatch has four relations where synonymy denotes terms with the same meaning. There is a pair of relations, one for broader terms and one for narrower terms. These define more general or less general meanings. There is also a relation denoting terms that have some degree of relatedness to each other. The range of these relations is weight over [0,1], and a product of the four weights is used to combine them. The structural affinity is defined on the union of two sets. One of the sets contains properties of a concept. The other set contains other concepts that have a semantic relation with the concept. The value of relationships and properties are given a weight-based value on the strength of the relationship or property.

ASMOV [20] analyzes four features: textual description, external structure, internal structure and individual. Textual description is defined as name, id, and comments and external structure is defined as parent child relationships. Internal structure is defined as property restrictions, type, domains and range for properties, and individual similarly of the entries. These four are combined into a confidence measure by a weight sum. The weighted sum is hand tuned by the test cases.

LILY [37] uses a structure called a semantic sub graph. The semantic sub graph is extracted from the ontology; the algorithm used extracts a sub graph that is the best representation of the connection between a pair of edges or nodes. The semantic graph is viewed as a definition of the terms. After defining terms, a similarity measure is computed on the sub graphs. In some cases the graph produces too few mappings, so a propagation algorithm is used to extend the mapping. The final step is to threshold the similarity measure to select the final mappings. The threshold is dynamically adjusted.

RiMON [39] uses a six-step process for mapping. First the similarity measures for structure and for label are computed. The next step is to select a strategy, if label similarity id is higher then use natural language techniques. Otherwise, if structural similarity is higher, then use similarity propagations methods. In some cases, they selected the methods manually. Once selected, they run the strategies separately to obtain alignment. The fourth step combines the alignments (i.e., if more than one is used) and the fifth step uses similarity propagation if the structural similarity is high. Finally, a set of heuristic is used to refine the alignment. RiMON has a set of six strategies to apply.

An interesting approach to reasoning about mapping and alignment is found in Laera [23]. The authors apply a technique called Argument Framework to record the reasoning behind alignments. The argument framework keeps arguments both for and against the possible mappings. The arguments are maintained in a positive base format and a second parameter records whether this is an argument that supports or disproves a mapping. The technique is strong enough to allow multiple degrees or types of mapping relations. In Hartung and Hakansson [19], this technique is extended to allow partial mappings over multiple ontologies. The algorithm works by maintaining partitions of consistent ontologies and the mapping and partitioning is accomplished as the ontologies are accessed and searched.

There are a number of ontology mapping competitions and they are a good source of results[9]. The results of the 2007 Ontology matching initiative [37] summarized the results on 13 matching systems. The evaluation criteria are precision and recall. Three of the systems show a precision in the range around 95% and recall in the range around 90%. While precision stays up close to the 90% range, the recall on some drops to around 50%. The three best systems were ASMOV, LILY, and RiMON.

The conclusion on mapping is that approaches are being improved and progress is clearly being made. On the other hand, the benchmarks used in the tests are limited to a few examples. Moreover, it is unclear if the quality of these ontologies is comparable to the quality of the ontologies found on the web. It would be unreasonable to assume the quality of the ontology design will not affect the results of mapping. The author's thesis is that the methods will under perform when used on the multiple ontologies on the web. There are, however, many good examples contained in the reference to allow the designer to select some possible mapping methods. Unfortunately, there is little real guidance on how to pick one to fit a domain.

10.6 Communications with Ontologies

A fundamental approach to inter-agent communication is using ontology, or ontologies, to express the communication vocabulary. The issue of agent communication has a long history. Michael Genesereth and Steven Ketchpel [17] did some work on agent communication back in 1994. They identified three issues for agent communication, what language to use, how to build the agent and what the communication architecture to apply. For this work, we consider a possible communications language. In particular, the use of ACL, Agent Communications Language, has been widely adopted. ACL was designed by researchers in the ARPA Knowledge Sharing Effort (NFFGPSS 1991). One approach to agent communication is to combine the ontology with ACL. In order to explore this approach, we first need a quick overview of ACL.

The format of the ACL messages includes a performance term based on speech acts. There are twenty-two of these acts found in the specification FIPA00037.

Act	Description
Accept Proposal	Accept an offered proposal The content is a repeat of the action for the proposal.
Agree	A positive response to a request The content is the action agreed to.

[9] The link http://ontologymapping.org/ is one source of ontology mapping competitions.

Cancel	Informs an agent that a previous request is not needed. Content is the action
Call for Proposal	Ask agents to propose executing an action. Content is the action
Confirm	Sender confirms the truth of a proposition. Content is a proposition
Disconfirm	Sender denies the truth of a proposition. Content is the proposition
Inform	Inform indicated the truth of a proposition. Content is the proposition
Inform If	This is a macro operation. It actually sends an inform. The if condition is based on the truth or falsity of a proposition. If the agent believe the proposition is true, it send an inform on the proposition, otherwise, the inform will send the proposition is false. The content is the proposition.
Inform Ref	This is a macro operation. It sends an inform on an object referenced by a name. Inform Ref can be sent as an action in a request. Content is the object
Not Understood	Sent to inform an agent that a message was not understood. Content is the message not understood.
Propagate	Used to pass messages on to other agents. The content identifies recipients plus the message to send.
Propose	Proposes to perform an action and may place preconditions on the action. The content must define an action.
Proxy	Like propagate, forwards a message. Content identifies the agents to send to by using a description of the agents to select plus the message to send
Query If	Ask an agent to inform the sender of the truth of a proposition. Content is the proposition
Query ref	Ask for the object referred to by a referent. Content is the referent.

Refuse	Refuse to perform an action. Content is the action plus a reason.
Reject Proposal	Refuse a proposal. Content is the action plus the reason for the refusal.
Request	Request a agent to perform a action. Content is the action.
Request When	Request an action to be preformed when a proposition becomes true. Content is the action and the proposition.
Request Whenever	Like Request When, but will request the action be taken every time the proposition becomes true.
Subscribe	Request the receiver to notify the sender every time the value of a proposition changes. Content is a proposition
Failure	Inform an agent that an attempted action has failed. Content is the action Followed by a proposition to indicate the cause of the failure.

The other terms in the message that are of direct interest is the content, which is the object for the verb (performance) of the message. The other parameters that go with the content are language, encoding and ontology. The ontology parameter specifies the ontology used to give meaning to the symbols in the content. The proposition referred to in the above table is a proposition from first order logic that uses the ontology as the source of base terms.

In examining this level of communication, it is important to find out what can be determined about the needs of ontologies in agent systems that employ multiple ontologies. The needs are dependent on the application. Some of the use, implied in the speech acts, is an exchange of information about propositions. The simplest end of the spectrum can be simple properties with values that are booleans or even integers. The only requirement is that names of the properties are shared. The names can be identified by identical names or by correct mapping through the ontlogies. Even in this simplest end of the spectrum, mapping can fail.

However, more interesting applications can specify operations as part of the message. This approach can be more complex. The mapping of the names for the operations and inferring the correct nature of the operation is critical. For example, consider building agents to be sellers and buyers for an automated commercial system. If a buyer tries to acquire a widget and sends out request for a proposal to sell widgets, what has to be assured? First of all, the widgets being sold had better be the widget the buyer wants. This can be an issue if some names of widgets are specialized by domain and the same name can mean very different items in different domains. Second, since commerce is being attempted, the issue of country of origin may be an issue, which can affect shipping, money exchange and political restrictions. This task may require a whole cascade of ontologies and operations to fully enact the solution.

In some cases, understanding the operation itself can be a complex problem. Mapping objects can be difficult, however, the name of an operation does not guarantee the actual semantics of the operation. Indeed ontology was designed to capture the definitions of object types, operations are objects, but the nuances of the semantics of an operation can be subtle and a definition beyond simple "isa" relations would seem to be necessary.

The final summary of these issues is that ontology based communication requires mapping. Furthermore, defining operations looks more difficult than defining physical entities.

10.7 Symbol Grounding

The final important aspect of ontologies is the issue of symbol grounding. The reader should not anticipate a simple answer. The basic issue here is the same one that has plagued symbolic AI from the earliest time and is still remains unanswered. This has been summarized in two questions [16]. First, can AI work with symbolic systems or do we need sub-symbolic representation? Second, do we really need representation? The first question is often framed in the debate between neural nets and symbolic systems. There are, indeed, hybrid approaches with a symbolic layer over the neural components. The second question is crystallized by the writings of Rodney Brooks but there is no consensus on this debate. There have been limited success with all the approaches and, yet, no one will argue that AI is a solved problem.

Symbol grounding is attributed to Harnad [18]. How do humans (or artificial systems) form a link between a real world entity and an internal categorical representation? This is reflected in the famous Chinese Room argument of Searle. Demonstration systems have been built that seem to show the formation of language and, hence, symbol associations with real world objects. Still these systems do not show a full-scale intelligence and that fact requires careful consideration about how fully capable their solution really is.

Symbol grounding and ontology mapping should not be confused even though they are complementary in some ways. Grounding is how to relate symbols to objects in the world. It is also about how two systems can try to relate how they are using terms to ensure they are communicating the same concepts. Mapping is a process of relating terms between two ontologies by examination of the structure and terms used in the two ontologies. Grounding could enhance mapping. Likewise a mapping can be a starting point to apply grounding.

Ontology is clearly a symbolic tool. There are several cases to consider. First of all, systems that use a single ontology are not the main issue here. Such systems may or may not achieve results of an intelligent behavior but they have a single shared vocabulary embodied in the shared ontology. More interesting are systems with multiple ontologies. One case is represented by a pair of agents communicating, each agent with it's own ontology. The other case is a single agent working against multiple ontologies. The first case is the typical agent communication problem; the second case is more typified by Web 2.0 search. Both are examples of achieving an understanding between entities. Note, that in the web search example one entity is active while the ontolgies on the web may be passive.

Before turning to how symbol grounding may be achieved, it is instructive to consider communication in natural language as an issue. Studying the problems of natural language communication has always been a key subfield of AI. A number of approaches have been tried and all have given problems. One issue raised by Winograd is learning [38]. Language is a dynamic space and the grammar and vocabulary of human speech is always changing. Winograd argued that without learning, no natural language system would really be ale to succeed [38]. Another commonly ascribed cause is pragmatics without world knowledge that sets context; nuances of the conversation are lost. This is sometimes stated in terms of real world knowledge. Another complication that arises is the use of metaphor. Language is argued by some to be based on metaphor from the first times and most especially movement in the physical domain [9].

A complaint about the nouvelle AI faction would raise is that the real mechanisms are more stimulus-response and less representational. That is, the basic level of language and real world knowledge is built on a strong reaction based layer. How this arises into a full theory of language in humans has not yet been expressed. It is clear that the underlying mechanisms in the brain have a functional organization that supports language [9].

Is agent communication like human communication? In both, there are two entities exchanging messages and hopefully exchanging information. In human communication there is a sense of motive. We view our communications with others in light of motives we ascribe to them. Likewise agents will have goals embodied in their programming. When humans converse, they cannot be absolutely sure they use the words with the same sense and meaning. When different authors build the ontologies, the vocabulary is not guaranteed to have a shared meaning. The concept of dishonesty can have just a much meaning between agents as between humans. There is no reason to view a society of agents as being different than a society of humans. At the present state of the art, the language used by agents is clearly simpler than that used by humans. In fact, the agents' language is probably comparable to the one used between humans in our far distant past, sometime referred to as "me Tarzan, you Jane". However, the need to ground the terms, even in a simple language, is still great.

There is one very significant difference between human and agent communication. In human communication, the human has access to a universal constant, the world itself. This assumes the world we experience is the same, even if an individual's senses may vary; the aspect being experienced is the same. Thus, a concrete source of grounding is available. Take color as an example, the way one individual senses blue may be different from another individual, they will both reach a point of agreeing on blue as a shared concept. But this external world is missing for an agent. Since the agents may vary widely, even if they have the same external world, they may still have trouble reaching agreement on a concept.

When we turn to the semantic web, the same issues arise [11]. The stated intent of the semantic web is to leverage human knowledge to extend the ability to find and use information. To improve search, the thesis is having users annotate their pages with semantic information and using tools to generate such semantic marking. This is yet to be proven. In examining the problem, Pollock and Hodgson [34] have analyzed possible mismatches into seven types. These are shown in the figure 10.3.

Terminology	Same word with different meanings Or Different words meaning the same
Representation at the Instance level	Same information is represented differently
Representation at the Concept level	Concepts abstracted differently
Representation at the Structural level	Different choices about division of the domain into instances, classes or properties Or Different axioms
Representation at the Superstructure level	Different modeling constructs Or Different representational methodologies
Granularity	Same information at different level of granularity
Perspective or context	Point of view issues
Underlying Conceptualization	Differences in conceptualization, theory or ideology
Origin	Purpose or method of collection of the data creating bias

Fig. 10.3. Mismatch Types

The mismatch types can be viewed in light of the section on ontology design. The principles discussed are designed to help eliminate the mismatch problems. The solutions for these problems are proposed, but not so easy to implement. Terminology could be solved by alignment or mapping of the ontologies. Representational problems can be solved provided a transformation can be identified. Granularity can be achieved if a mapping into a common ontology is possible, but this presupposes a global master ontology to be available. Perspective, underlying conceptualization, and origin require that these aspects need to be made explicit. Perhaps this calls of a meta-ontology to define these aspects in some computable form. This is clearly an unsolved problem.

Another problem brought to light in the web ontologies is the use of URL's to help resolve mapping or alignment. Since the instances in web ontology are all object named by a URL, it is possible to use the instances for a concept as a key to mapping. That is, if the extensions of two classes are the same, then the classes are identical. The problem with this is that the classes in an ontology is are more than classes, they are conceptualizations. It is very easy to form identical classes, with respect to membership, with different definitions.

These observations should give concern in the ability to use ontology as a communication mechanism. After all, the use of ontology is to define some vocabulary. The vocabulary from an ontology can take several forms. First of all the vocabulary can be set of entities in a domain, which is the closest to the sense of philosophical ontology. At the other extreme the ontology can be taxonomy for some domain. While these two definitions can be identical, they do not need to be. In some cases, the vocabulary can be a set of nouns that do not fit into a well-defined ontology. This can be the result when the hierarchy has different senses of "isa". It can also result if the ontology is not well defined or contains errors.

One proposed solution to communication between ontologies is to develop ways of viewing symbol grounding. The whole topic of mapping is one approach of grounding the ontologies by reducing terminological mismatch.

Another proposed approach to grounding is found in Katarzyniak *et al.* [22]. The approach uses modal logic as tool for grounding and since the only recourse to grounding is by observation of the ontologies exposed by other agents, it seems a good approach. This approach uses the observation of properties of objects to determine modal predicates about the properties. It is one approach that could be applied to grounding since it includes time as parameter, enabling ontolgies to change over time and allowing agents to collect expertise over time. The semantics uses in the approach are based on the formula p(o) and its complement ~p(o). These are interpreted as object o exhibits, or does not exhibit, property p. The interpretation of the modal predicates: Pos() meaning it is possible, Bel() meaning it is believable, and Know() meaning that it is known.

The model uses observations. An observation is defined as a set of objects and properties. The observation is the collective observations of the state of the world at a time *t*. The observations can be represented by sets, where each set is the exhibiting of a property or the non-exhibiting of a property. The set members are the objects that are observed to exhibit, or not to exhibit, the property.

For an object o and a property p, two sets can be defined. The set of observations of the object exhibiting the property is named A1 and the complement set is named A2. The cardinality of these sets can be used to define two metrics:

$$\lambda(t, p(o)) = Card(A1) / (Card(A1) + Card(A2))$$

$$\lambda(t, \sim p(o)) = Card(A2) / (Card(A1) + Card(A2))$$

In order to use these values to establish modal predicates a set of threshold is defined:

$$0 < \lambda_{minPos} < \lambda_{maxPos} <= \lambda_{minBel} < \lambda_{maxBel} < 1$$

Now the model predicate can be evaluated by applying the threshold to the values of the metrics.

The next steps are to look at strategies for handling the observations over time. The very simplest of these is to remove profiles over certain time distance. In simplest terms this is to allow the agent to forget.

This approach to grounding does offer some ways to refine a shared understanding for ontologies. However, the basic mapping problem such as getting the mapping between names of the objects still remains. The grounding approach to properties can be used to check the mapping by comparing properties between objects and check if they are mapped as identical.

The better possibility offered by this grounding approach is the communication with multiple ontologies and the establishing of belief and possibility modalities. In communicating between agents, forming hypothesis around the belief about what other agents think, leads to deeper abilities to function. In human negotiation, each person makes assumptions about beliefs and motivations of others. This example and the others work with modal approaches offer the future for agents that negotiate on behave of their owners.

10.8 Conclusion

The subject of ontology has become very active. The Semantic Web and Web 2.0 have major plans for the use of ontology to enhance search and access to information. In the domain of agent applications ontology is proposed for communication. As the use of agents reaches greater autonomous and intelligent action, ontology can provide the basis for rich and profitable applications. However, design of ontology is difficult and interpretation of ontology remains a challenge. Mapping and grounding both need more development to realize the full potential of ontology-based systems. There is much work remaining before the goals of agent based systems and Web 2.0 are realized.

References

1. Brachman, R.J., Levesque, H.J.: Structured Representations. Knowledge Representation and Reasoning, Ch. 9. Elsevier, Amsterdam (2004)
2. Smith, B.: Ontology: An introduction. Video: How to build Ontology, http://ontology.buffalo.edu/smith/articles/ontologies.htm
3. Bouquet, P., Serafini, L., Zanobini, S.: Peer-to-Peer Semantic Coordination. Journal of Web Semantics 2(10) (2005)
4. Castano, S., Ferrara, A., Montanelli, S.: Matching Ontologies in open networked systems: techniques and applications. Journal on Data Semantics (JoDS) V (2006)
5. Chandrasekaran, B., Josephson, J.R., Benjamins, V.R.: What Are Ontologies and Why Do We Need Them. IEEE Expert Systems, 1094–7167 (1999)
6. Chaudhri, V.K., Farquhar, A., Fikes, R., Karp, P.D., Rice, J.P.: Open Knowledge Base Connectivity 2.0.3 (proposed) (1998), http://www.ai.sri.com/~okbc/
7. Chalupsky, H.: Ontomorph: A translation System for Symbolic Knowledge in Principles of Knowledge and reasoning (2000)
8. Choi, N., Song, I., Han, H.: A Survey on Ontology Mapping. SIGMOD Record 35(3) (September 2006)
9. Calvin, W.H.,, Bickerton, D.: Lingua ex Machina: Reconciling Darwin and Chomsky with the Human Brain, 1st edn. MIT Press, Cambridge (2001)
10. Compatangelo, E., Meisel, H.: Intelligent support to knowledge sharing through articulation of class schemas. In: Proceedings of the 6th International Conference on Knowledge-Based Information and Engineering Systems (KES 2002), Crema, Italy (September 2002)
11. Cregan, A.M.: Symbol Grounding for the Semantic Web
12. Ehrig, M., Staab, S.: QOM – Quick Ontology Mapping. In: McIlraith, S.A., Plexousakis, D., van Harmelen, F. (eds.) ISWC 2004. LNCS, vol. 3298, pp. 683–697. Springer, Heidelberg (2004)
13. Euzenat, J., Issac, A., Meilicke, C., Shvaiko, P., Stuckenschmidt, H., Svab, O., Svatek, V., van Hage, W.R., Yatskevich, M.: Results of the Ontology Alignment Evaluation Initiative. In: The 6'th International Semantic Web Conference and the 2'nd Asian Semantic Web Conference (2007), http://www.ontologmatching.org
14. Falconer, S.M., Noy, N.F., Sorey, M.A.: Ontology Mapping – A User Survey
15. FIPA Ontological Service Specification, Document number XC00086D, http://fipa.org
16. Franklin, S.: Artificial Minds. MIT Press, Cambridge (1997)
17. Genesereth, M., Ketchpel, S.: Software Agents. Communications of the ACM 37(7), 48–53 (1994)
18. Harnad, S.: The symbol grounding problem. Physica D 42, 335–346

19. Hartung, R., Hakansson, A.: Using Meta-agents to Reason with Multiple Ontologies. In: Nguyen, N.T., Jo, G.S., Howlett, R.J., Jain, L.C. (eds.) KES-AMSTA 2008. LNCS (LNAI), vol. 4953, pp. 261–271. Springer, Heidelberg (2008)
20. Jean-Mary, Y.R., Kabuka, M.R.: ASMOV Results for OAEI. In: The 6th International Semantic Web Conference and the 2nd Asian Semantic Web Conference (2007), http://www.ontologmatching.org
21. Kalfoglou, Y., Hu, B.: CROSI Mapping System (CMS) Results of the Ontology Alignment Contest. In: K-CAP Integrating Ontologies Workshop 2005 Banff, Alberta, Canada (2005)
22. Katarzyniak, R.P., Nguyen, N.T., Jain, L.C.: Soft Computing Approach to Conceptual Determination of Grounding Sets for Simple Modalities. In: Apolloni, B., Howlett, R.J., Jain, L. (eds.) KES 2007, Part I. LNCS (LNAI), vol. 4692, pp. 230–237. Springer, Heidelberg (2007)
23. Laera, L., Tamma, V., Euzenat, J., Bench-Capon, T.J.M., Payne, T.: Arguing over ontology alignments. In: Proceedings of International Workshop on Ontology Matching (OM 2006) (2006)
24. li, J.: LOM: A Lexicon-based Ontology Mapping Tool Proceedings of the Performance Metrics for Intelligent Systems (PerMIS 2004) (2004)
25. Madhavan, J., Bernstein, P.A., Domingos, P., Halevy, A.: Representing and reasoning about mappings between domain models. In: Proceedings of the 18th National Conference on Artificial Intelligence (AAAI 2002) Edmonton, Alberta, Canada (August 2002)
26. Maedche, A., Staab, S.: Semi-automatic engineering of ontologies from texts. In: Proceedings of the 12th International Conference on Software Engineering and Knowledge Engineering (SEKE 2000) Chicago IL USA, pp. 231–239 (July 2000)
27. Mitra, P., and Wiederhold, G.: Resolving Terminological Heterogeneity in Ontologies. In: Proceedings of the ECAI 2002 workshop on Ontologies and Semantic Interoperability (2002)
28. Mitra, P., Noy, N.F.: OMEN A probabilistic Ontology Mapping Tool International Semantic Web Conferences, pp. 537–547 (2005)
29. Gomez-Perez, A., Fernandez-Lopez, M., Corcho, O.: Ontological Engineering. Springer, Heidelberg (2003)
30. Gruber, T.R.: A translation approach to portable ontologies. Knowledge Acquisition 5(2), 199–220 (1993)
31. OKBC Open Knowledge Base Connectivity Project, http://www.ai.sri.com/~okbc/
32. Kalfoglou, Y., Schorlemmer, M.: Ontology Mapping: the State of the Art
33. Visser, P.R.S., Jones, D.M., Bench-Capon, T.J.M., Shave, M.J.R.: Assessing Heterogeneity by Classifying Ontology Mismatches. In: Working Notes of the AAAI Spring Symposium on Ontological Engineering
34. Pollock, J.T., Hodgson, R.: Adaptive Information: Improving business through semantic interoperability, grid computing and enterprise integration. Wiley Series in Systems Engineering Management. John Wiley and Sons, Inc., Chichester (2004)
35. Reed, S.L., Lenat, D.B.: Mapping Ontologies into CYC (2002)
36. Sowa, J.F.: Knowledge Representation Logical, Philosophical and Computational Foundations, Brooks/Cole (2000)
37. Wang, P.,, P.: Xu, B.: LILY: The Results for the Ontology Alignment Contest OAEI. In: The 6'Th International Semantic Web Conference and the 2'nd Asian Semantic Web Conference (2007), http://www.ontologmatching.org
38. Winograd, T.: Language a Cognitive Process: Syntax. Addison Wesley, Reading (1982)
39. Zhong, Y.L.Q., Li, J., Tang, J.: Result of Ontology Alignment with RiMON at OAEI. In: The 6'th International Semantic Web Conference and the 2'nd Asian Semantic Web Conference (2007), http://www.ontologmatching.org

11

Ontology Agents and Their Applications in the Web-Based Education Systems: Towards an Adaptive and Intelligent Service

Alejandro Peña Ayala

WOLNM, UPIICSA-IPN, ESIME-Z-IPN
31 Julio 1859 # 1099B Leyes Reforma, DF 09310, Mexico
apenaa@wolnm.org, apenaa@ipn.mx

Abstract. The ontology agent (OA) has become a necessary approach to enable and manage the acquisition, query and reuse of knowledge in distributed artificial intelligence (DAI) approaches and applications, such as Multi-agent Systems (MAS) and Web-based Education Systems (WBES) respectively. In this chapter, the aim is to point the reader at what I perceive to be the main theoretical and methodological issues regarding definition, design and implementation of an OA. Hence, the underlying concepts about agents, ontologies, OA and WBES are stated firstly. Afterwards, some agents and ontologies applications in the WBES arena are introduced. Next, the formal model for agents and ontologies is set. Also, a method for building agents and ontologies is outlined. With this baseline, the development of an OA for delivering adaptive and intelligent teaching-learning experiences through a WBES is shown. Based on the trial's outcome of the approach, it is concluded that a WBES is able to improve the apprenticeship of students. This chapter ends with a resume of the content, a discussion about the usefulness of the OA on a WBES, and a declaration of further work to be done in order to encourage research in this field.

11.1 Introduction

Essentially, an OA encapsulates the functionality for playing the role of a knowledge provider in DAI environments. The OA accomplishes the tasks for encoding and storing ontologies. Also, it takes over the requests for translating, interpreting and executing incoming queries [1]. In WBES, an OA is able to manage the semantic content of repositories as teaching material and student profile. So, in order to define the concepts about an OA and its application in WBES, this section is arranged as follows: firstly, the underlying concepts about agents are stated, such as definition, properties, kind of systems and types of agents. Later, the concepts about ontologies, as definitions and types, are pointed out. Then, the profile and functionality of the OA are depicted. Next, background and development stages of the WBES are outlined. Following this, how an OA supports the functional modules of a WBES in order to provide adaptive and intelligent services for the students is explained.

11.1.1 Agent's Underline Concepts

According to Kay, the idea of an agent was conceived by John McCarthy in the mid 1950's, and the term was stated by Oliver Selfridge a few years later [2]. They had in

N.T. Nguyen, L.C. Jain (Eds.): Intel. Agents in the Evol. of Web & Appl., SCI 167, pp. 249–278.
springerlink.com

view a system that given a goal, the system could carry out the details of the appropriate computer operations, and could ask for and receive advice offered in human terms when it was struck. An agent would be a soft robot living and doing its business within the computer's world. Since then, a wide variety of agents has emerged as biological, robotic, artificial life and software agents [3]. The last variety is defined by Shoham as: "A software entity which functions continuously and autonomously in a particular environment, often inhabited by other agents and processes" [4]. Among the versions of software agents appears one that is called *intelligent agent*, which is described by Gilbert et al. in terms of a space characterized by three properties [5]:

- *agency* is the degree of autonomy and authority vested in the agent. As a minimum, an agent must run asynchronously,
- *intelligence* is the degree of faculty for reasoning and learning that the agent owns. Also, it is the agent's ability to accept the user's statements of goals and fulfill the tasks delegated to it. As a minimum, there are some statements of preferences,
- *mobility* is the degree to which an agent travels through the network from one machine to another during the execution, while it carries accumulated state data.

In addition, Etzioni and Weld consider other kinds of attributes for agents as [6]:

- *communicative* is the facility that an agent owns for sharing knowledge,
- *capable* is the faculty held by an agent to take actions in some sort of world,
- *reactivity* is the ability to selectively sense and act,
- *autonomy* is the agent's self-starting, goal-directedness and proactive behavior,
- *adaptive* is the faculty of the agent to be able to adapt its behavior based on a combination of user feedback and environmental factors,
- *collaborative* is the aim of the agents to work together to achieve a goal,
- *inferential* is the agent's capability for acting on abstract tasks specifications using prior knowledge about general goals and preferred methods to achieve flexibility,
- *temporal continuity* is the persistency of identity and state during a long period.

Beside these properties, the agents are stated by their information attitudes and pro-attitudes. *Information attitudes*, as belief and knowledge, are related to the information that an agent holds about the world it occupies. *Pro-attitudes*, as desires and intentions, guide the agent's actions [7].

Based on the properties and attitudes that depict agents, an agent can be a sort of reactive, intentional or social system, where: a *reactive agent* reacts before the changes in its environment or to messages from other peers; an *intentional agent* is able to reason about its intentions and beliefs to define plans and execute them; a *social agent* retains explicit models about other agents [8].

Finally, MAS is a computational system, where two or more agents interact and work together in order to perform several tasks and accomplish a set of goals [9].

11.1.2 Ontology's Underlying Concepts

According to the Foundation for Intelligent Physical Agents (FIPA) [10] an ontology is: "A philosophical term to refer a particular system of categories accounting for a certain vision of the world". Essentially, the ontology does not depend on a particular language; hence it is always the same. However, from the DAI's view, an ontology represents an engineering artifact constituted by a specific vocabulary used to describe a certain reality, plus a set of explicit assumptions regarding the intended meaning of the vocabulary words. Thereby, in order to be compliant with most of the work in DAI, in this chapter the term *conceptualization* is adopted in reference to the philosophical reading, whereas the *ontology* term holds its traditional DAI meaning.

Generally, the ontology's content concerns an agent is related to several agents, or is regarding the whole MAS. Hence, *particular ontologies* are devoted to set the knowledge about expertise, goals, plans and tasks attached to an agent. However, *common ontologies* are oriented to depict communication protocols, behavior's constraints, global goals and knowledge domain for a community of agents.

An ontology is a repository that is composed of a taxonomy of concepts. The ontology states the properties that depict concepts. Also, it holds the constraints that define the values assigned to properties. Moreover, the relations among concepts are acknowledged as part of the ontology. In addition, an ontology contains the axioms that establish the ontology's meaning and reasoning. Furthermore, the content of the ontology is stated in a language that is understandable by humans and machines.

Usually, the ontologies are classified through several dimensions such as: 1) degree of dependence on a particular domain; 2) level of detail of their axioms; 3) nature of their domain. The *degree of dependence* embraces three instances: *top-level ontologies* that depict general concepts; *domain ontologies* that state vocabularies; *application ontologies* that set concepts about a domain. As regards *level of detail*, there are two degrees: *fine-grained*, an ontology very rich in axioms and close to set the intended meaning of a vocabulary, and *coarse ontology* that consists of a minimal set of axioms to support a limited set of services. Finally, the *ontologies about the nature of their domain* define a classification of primitive terms used by a knowledge representation language like concepts, attributes and relations [10].

Too often, in MAS several ontologies meet to provide gray-levels of content scope, degree of dependence, detail or nature of the domain. As a result, six tiers of *ontology relations* appear such as: 1) *extension*: whether ontology *A* includes ontology *B*; 2) *identical*: if the vocabulary, axioms and representation used by ontology *A* are physically identical to those managed by ontology *B*; 3) *equivalent*: the vocabulary and axioms used by ontologies *A* and *B* are physically identical, but their statements are different; 4) *weakly translatable*: it is possible to translate from ontology *A* to ontology *B*, even with a possible loss of information, *B* shares a subset of vocabulary axioms of *A*; 5) *strongly translatable*: although the language used by ontologies *A* and *B* is different, the vocabulary of *A* is totally translated to the vocabulary of *B*, the axioms of *A* holds in *B*, there is no loss of information from *A* to *B*, and there is no appearance of inconsistency; 6) *approximately translatable*: when two ontologies are weakly translatable with the occurrence of possible inconsistencies [10].

11.1.3 Ontology Agent

In MAS, the agent communication is based on the assumption that collaborative agents share a common ontology in the domain of discourse. This means that agents ascribe the same meaning to symbols used in messages. Therefore, a design issue arises: how to state the ontology? When the designer chooses an *implicit encoded*, then the ontology content is embedded, or linked, in the software implementation of the agents. But, when the selection is an *explicit declaration*, the ontology is stored somewhere and is formally published as an ontology service. This kind of service is federated in MAS, thereby a community of agents can request actions such as: query, update, inference and translate. Hence, given the amount and complexity of tasks encapsulated in an ontology service this role is carried out by an OA.

In short, an OA is: a specialized agent that provides access to one or more ontologies with the aim to offer ontology services for an agent community.

An OA federates services as the following: ontology meta-definition; content maintenance through insert, delete and update items; content integrity verification; discovery of public ontologies; publication of the available ontologies; query answers for relationships between terms; translation of expressions between ontologies; identification of shared ontologies for communication and collaborative work among agents; provision of the context meaning for the job to be accomplished by agents, and recovery of the ontology's items requested by agents.

In a sense, an OA is a middleware between agents and ontologies, because it takes over the requests sent by agents that ask for ontology services. However, the OA is not alone in responding to the agents' demands. In MAS there is a set of OA that shares the workload according to some kind of criteria. Also, between the OA and the ontology physical storage there is an *ontology engine*. This software component is the real ontology manager, like a data base engine. The ontology engine seeks, recovers and modifies the ontology items according to the ontology language used for encoding the content as: eXtended Markup Language (XML) [11], Resource Description Framework (RDF) [12], Defense Advanced Research Projects Agency Agent Markup Language (DAML) [13], and Web Ontology Language (OWL) [14].

In MAS, an OA is a kind of wrapper for an ontology manager with the aim to provide the sort of functionalities for the administration of knowledge stored in specific ontologies. Therefore, like a front-end, each OA federates the ontology services for some ontology that it is able to manage. Hence, every agent can request the services to the appropriate OA that is able to satisfy its requirements. The communication is achieved by messages encoded through an *Agent Communication Language* (ACL), as Knowledge Query and Manipulation Language (KQML) [15] and FIPA-ACL [16]. Once the agent identifies the OA that meets its requirements, it includes a communicative interface to express the nature of its requests as part of the message. Such interface is depicted through an *internal language* to state facts about ontologies as: concepts, relations, parameters and commands. This kind of language, as the Open Knowledge Base Connectivity (OKBC) [17], is neutral in respect to the language used to store and represent the ontology.

11.1.4 Web-Based Education Systems: Roots, Evolution and Trends

WBES is the result of the research and application work fulfilled in the Computer-Assisted Education (CAE) arena [18]. In the mid 60's the earliest approach for Computer-Assisted Instruction (CAI) was a set of systems that generated problems in arithmetic and in vocabulary recall that was built by Leonard Uhr [19]. Since then, broad instances of research, applications, outcomes and approaches have been fulfilled, such as: Computer-Assisted Learning (CAL) provides content ad-hoc to the nature of teaching domain and the requirements of the students [20]; Computer-Assisted Instruction (CAI) is a teaching system which tries to adapt to the particular needs of each student [21]; Intelligent Computer-Aided Instruction (ICAI) introduces the Artificial Intelligence (AI) to the education [22]; Intelligent Tutoring System (ITS) plays the tutor role for helping the student to resolve problems [23]; Hypermedia Systems set content through a mixture of hypertext and multimedia [24]; Intelligent Learning Environments (ILE) try to adapt themselves for improving the learning of students [25]; Computer Supported Collaborative Learning (CSCL) promotes the workgroup between peers to learn a subject [26]; Learning Management Systems (LMS) and Instructional Management Systems (IMS) provide teaching content in multi-user environments [27].

Later, as a result of the invention Internet, in the 80's, the scope of the CAE paradigms has been enhanced world wide. In addition, a new version for CAE applications emerged: the WBES. This kind of paradigm fixes some attributes, goals and components of its ancestors. Besides, a WBES includes the Web technologies in order to deliver contents and facilitate learning wherever and whenever the users like. In spite of their short life, the WBES have already experienced four generations: the first one is characterized by static content; the second introduces the eReading modality by interactive communication between student and WBES; the third generation aims at Web Lecturing through real-streaming [28]. Nowadays, WBES are grounded on an interdisciplinary frame that includes: cognitive, psychology, philosophy, pedagogy and communication sciences. Furthermore, WBES take advantage of DAI approaches, as agents and ontologies, and the advances of the intelligent Web, as semantic Web and ubiquitous computing. With this baseline, WBES focus on education paradigms as: Student Centered Education, Situated Learning, Blended Education, Collaborative Learning and Learning by Reflection.

Among the trends of WEBS appears the characterization of an adaptive and intelligent behavior. The WBES attempt to be more adaptive by building a model of goals, preferences and knowledge of each individual. This goal is fulfilled through the interaction with the user. So a WBES adapts itself to the requirements of the student. Such WBES tries to be more intelligent by performing tasks executed by people who coach students or diagnose their misconceptions.

The core of WBES is fixed by the following branches: adaptive hypermedia, such as adaptive navigation and adaptive presentation; adaptive information filtering, as content-based filtering and collaborative filtering; intelligent collaborative learning, by adaptive collaboration, adaptive group formation and virtual students; intelligent tutoring, through curriculum sequencing, intelligent solution analysis and problem solving support; and intelligent class monitoring, like a remote teacher [29].

11.1.5 Ontology Agent Support to Web-Based Education Systems

In order to identify the targets where an OA is willing to support a WBES to provide adaptive and intelligent services it is necessary to take into account a WBES model, as the learning technology standard architecture [30]. Thus, a statement is given next.

The role for each functional module is: *Content Provider* delivers the teaching material regarded to the learning experience; *Sequencing* plans the learning experiences for the student; *Manager* administrates the teaching-learning sessions; *Evaluator* estimates the apprenticeship achieved by the student; *Assessment* tracks the behavior and outcomes fulfilled by the student; *User-System Interface* shows the content and provides the communication facilities between user and WBES; *Student Model* carries out a mental model about the goals, preferences and knowledge of the student.

Besides the content stored in the repositories of a WBES, some ontologies can be attached to each repository in order to depict the knowledge domain that corresponds to: curricula, teaching material, student record and student profile. Therefore, the administration of such ontologies is accomplished by several OA, which provide ontology services to agents that accomplish the tasks of the functional modules.

11.2 Related Work

In order to explain the nature of an OA; in this section some approaches that provide ontology services in MAS are depicted. Also, a survey of agents and ontologies applications in WBES is stated. Based on these works, the support that an OA is able to provide to WBES for achieving an adaptive and intelligent behavior is grounded.

11.2.1 Ontology Agents Approaches

An interesting approach for delivering ontology services is the Hybrid Local-Global Model for handling ontologies proposed by Brena and Ceballos [1]. They set part of the common ontology for MAS in each agent, through a Client Ontology Component (COC), whereas the whole ontological knowledge is handled by an OA. The COC takes over ontology handling capabilities. Every time that agent takes up its work, its respective COC is loaded with a subset of a common ontology. When additional knowledge is required, the COC requests it from the OA. The OA answers the queries, achieves the tasks required, and translates the results in a way understandable for the COC. The approach pursues efficiency through the use of a local ontology and a personal COC for every agent. So it reduces the overhead in agent's communication.

Another useful approach corresponds to the framework for Agent Coordination through Ontology Management stated by Bai and Zhang [31]. The OA plays the role of a coordinator to manage and model agent cooperation. It holds three elements: ontology base, ontology indexer and ontology distributor. The *ontology base* contains the ontologies required by MAS. The *ontology indexer* is a database that collects and stores abstracted information of the ontologies in MAS. The *ontology distributor* is a software component that is in charge of distributing ontology services.

When an item of knowledge needs to be updated in MAS, the OA locates the ontology through the ontology indexer. Next, it refreshes the ontology stored in the ontology base. Afterwards, the OA distributes the refreshed ontology to the user agents by the ontology distributor. This approach facilitates the achievement of insert, delete and update operations in order to meet the dynamics of MAS.

11.2.2 Agents Applications in WBES

MAS is a suitable platform to deploy WBES, where agents fulfill the functionality of the modules. For instance, the Open, Adaptive and Multi-Subject WBES, stated by Alvarez and Fernandez, holds the following kinds of agents: *selector* for sequencing, *student* for student modeling, *domain* as content provider, *planning* as manager, *student resources* for assessment, *evaluation* and *interface* [32]. The adaptability lies in WBES's self-tailoring in execution time for any subject or student.

As regards teaching content, Querrec et al., built a MAS oriented to train firemen through virtual environments [33]. They set a model to organize the interactions among agents. As a result, agents own reactive, cognitive and social abilities to simulate a fire. The fire fighters deal with such virtual situations during their training.

Related to student modeling, Limoanco and Sison developed *learner agents* that create and maintain the student profile during apprenticeship activities achieved by a person [34]. The learner agent owns a model world that contains: data about the subject that is learned by the student; a self model, which holds the student's profile; a social model that stores information about other entities as tutor agents and peers.

Regarding sequencing, Conati and Zhao built *pedagogical agents* that provide personalized instruction integrated with the entertaining nature of the games on number factorization [35]. With this application, the children develop their interest in maths.

11.2.3 Ontology Applications in WBES

Ontologies provide context support to teach content. For example, Falquet and Ziswiler set a virtual hyperbook to support collaborative learning [36]. The model holds a domain ontology for a reusable document repository. Also, the model includes a hyperbook ontology to depict relations between fragment repository and domain ontology. The hyperbook accesses the content according the user's point of view.

Ontologies help to evaluate the apprenticeship of the students, like the test ontology that Soldatova built for supporting a quiz generator system [37]. The ontology allows analyzing test characteristics, structure and process composition. The exams are characterized in the ontology through reliability, validity and precision properties.

Also, ontologies are willing to implement pedagogic models, as the ontology support model for learning in a Web classroom outlined by Mirjana and Vladan Devedzic [38]. They set two kinds of ontologies: education domain and pedagogic. For instance, the *education ontology* is devoted to demography subjects. The *pedagogic ontology* focuses on instructional design and human learning theories to ensure educational justification of learning, assessment and collaboration between students.

11.3 Agent and Ontology Formal Models

In spite of the quantity of literature about formal models for agents and ontologies, there is a lack of an integral work that sets the baseline for agents and ontologies. Thus, in order to deal with such issue, in this section two goals are pursued: to state the elements of a formal theory for MAS, and set the formal account of ontologies.

11.3.1 MAS Formalism

Based on the work fulfilled by Wooldridge [7, 39, 40], in this section a formal method for reasoning about MAS is stated. So a formal model for MAS is set, next an execution model is depicted, and temporal belief logic for modeling MAS is outlined.

Model of MAS. The MAS formal model has to confront the following questions: how to conceive agents? What are the agent's properties to take into account? How to formally represent agents? The answers to such questions are given as follows:

A weak conceptualization of agency is: An agent communicates with their peers by exchanging messages in an expressive ACL. Agents can be large entities with some sort of persistent control. This notion attaches properties to agents as autonomy, social ability, reactivity and pro-activeness. A stronger conceptualization adds mentalistic notions, such as knowledge, beliefs, desires and intentions. Also it includes human-like attributes such as emotion and animation.

The agent's properties and the formal representation are stated as follows:

- *names*: each agent from the set Ag is uniquely identified by an *agent id* like: i, j,
- *beliefs*: the agent's knowledge is depicted by a belief set, which is expressed in an internal logical language L, where: $Form(L)$ is the set of well-formed formulae of L, and BS is the set of possible belief sets that an agent could have according to the generation function ω, thus: $BS = \omega(Form(L))$,
- *actions*: agents can perform private actions. They can only perform actions which operate on their own state, where Ac is the set of all such actions,
- *messages*: agents can communicate by sending messages under the format: $<i, j, \varphi>$ where $i \in Ag$ is the sender agent, $j \in Ag$ is the recipient agent, and $\varphi \in Form(L)$ is the message content. Let $Mess$ be the set of all messages: $Mess = \{<i, j, \varphi> \mid i, j \in Ag \text{ and } \varphi \in Form(L)\}$,
- function rcv is defined with the signature: $rcv : Ag \times powerset(Mess) \rightarrow \omega(Mess)$,
- an agent is: $<B^0, A, M, N>$ where: $B^0 \in BS$ is the agent's initial belief set; $A : BS \rightarrow Ac$ is the agent's action function; $M : BS \rightarrow \omega(Mess)$ is the agent's message generation function; $N : BS * Ac * \omega(Mess) \rightarrow BS$ is the agent's next state function,
- on the basis of its initial beliefs, an agent selects an action to perform using the function A, and some message to send, using the function M. Then, the function N transforms the agent from one state to another, on the basis of the message it receives, and according to the action it has just performed,
- a MAS is an indexed set of agents: $\{<B_i^0, A_i, M_i, N_i> : i \in Ag\}$.

Execution Model. The execution model takes into account the notion of the state of a system and the changes in state being produced by transitions. A state is: a picture of the belief set of every agent in the system at some point in time. Likewise, a state change, or transition, occurs when one or more agents receive messages and perform actions. Moreover, the belief set is assumed to be in some kind of equilibrium.

At time u the next state B^{u+1} of agent i is stemmed from (11.1) based on its current state B_i^u, the action that i performs, and the messages that are sent to i. Hence, the current state u of the whole system is given by every agent i having its current belief set B_i^u. Thus, the execution of a system is a cycle process where each agent chooses an action to perform, sends messages, receives messages, and shifts into its next state.

As the MAS executes, the model traces out an execution history, which describes each agent's state, the actions performed, and the messages that it sent at each moment in time u. As a result, the set of all execution histories Σ is outcome, where σ

denotes a member of such set, s_u points the uth state of σ, t_u for the uth transition of σ. So, σ is drawn as non-terminating: σ: $s_0 - t_0 \rightarrow s_1 - t_1 \rightarrow s_2 \ldots - t_{u-1} \rightarrow s_u - t_u \rightarrow \ldots$

$$B_i^{u+1} \overset{def}{=} Ni(B_i^u, Ai(B_i^u), rcv(i, \bigcup_{j \in Ag} Mj(B_j^u))). \quad (11.1)$$

Agent Logic. A propositional logic for reasoning about MAS model is Agent Logic (AL). AL is based on a model of time that is linear, bounded in the past and infinite in the future. This means respectively: each point of time is assumed to have just one successor, there is a beginning of time, and there is no final moment of time. AL achieves reasoning about agents by their beliefs, actions and messages sent by them.

AL is characterized by the internal language *L*, whose syntax owns the next items:

- *symbols* {*true*, *Bel*, *Send*, *Do*}: corresponds respectively to: truth logical value, operator for describing the agent's beliefs, operator for depicting the messages that agent sends, operator for pointing the action that agent achieves,
- *agent constants* $Const_{Ag}$: denotes agents' identifiers,
- *action constants* $Const_{Ac}$: identifies the actions that agents perform,
- *internal language L*: holds the agents' beliefs stated as closed formulae of *L*,
- *propositional connectives* {¬, ∨}: depict respectively *negation* and *disjunction*,
- *unary temporal connectives* { ∘, • }: represent *next* and *last* respectively,
- *binary temporal connectives* {*U*, *S*}: correspond respectively to *until* and *since*,
- punctuation symbols{), (}.

The set of well-formed formulae of AL, based on *L*, is stated as follows:

- if *i*, *j* are agent ids, φ is a closed formula of *L*, and α is an action constant, then the following are atomic formulae of AL: *true*, (*Bel i* φ), (*Send i j* φ), (*Do i* α),
- if φ, ψ are formulae of AL, then the following are formulae of AL: $\neg\varphi$, $\varphi \vee \psi$, $\circ\varphi$, $\bullet\varphi$, $\varphi U \psi$, $\varphi S \psi$,
- the meaning of non-standard operators in AL is: (*Bel i* φ), agent *i* beliefs φ; (*Send i j* φ), agent *i* sends message φ to agent *j*; (*Do i* α), agent *i* performs action α; ∘φ, next φ; •φ, last φ; φ*U*ψ, φ until ψ; φ*S*ψ, φ since ψ.

The semantics for AL is set by the structure: *M* = < σ, *Ag*, *Ac*, *bel*, *action*, *sent*, *I* > where: σ ∈ Σ is an execution history; *Ag* is a set of agent ids; *Ac* is a set of agent actions; *bel* : Σ x *Ag* x *N* → *BS* is a function that takes an execution history σ, an agent id *i*, and a moment in time *u* to return the belief set B_i^u; *action* : Σ x *Ag* x *N* → *Ac* is a function that takes an execution history σ, an agent id *i*, and a point in time *u* for returning the action α performed by the agent; *sent* : Σ x *N* → *ω*(*Mess*) is a function that takes an execution history σ, a time *u*, and returns the set of messages sent in the execution history, at the time *u*; *I* : *Const* ↔ (*Ag* ∨ *Ac*) interprets constants.

There is a close relationship between the formal model of MAS and logical models for AL. In order to characterize such relation, it is necessary to set the conditions under which a model for AL is worthy to depict an execution of MAS. This means that a model M for AL, $< \sigma, Ag, Ac, bel, action, sent, I >$, represents a execution result of the system $\{<B_i^0, A_i, M_i, N_i>: i \in Ag\}$, iff the condition set by (11.2) is met:

$$\forall u \in N \circ \forall i \in Ag \circ bel(\sigma, i, u) = B_i^u \wedge action(\sigma, i, u) = Ai(B_i^u) \wedge sent(\sigma, u) = \bigcup_{j \in Ag} Mj(B_j^u). \quad (11.2)$$

The satisfaction relation $\models$ for AL holds between pairs of the form $<M, u>$ and formulae of AL, where M is a model for AL, and $u \in N$ is a point of time into M; then the following semantic rules for AL are stemmed as follows:

1. $<M, u> \models$ true,
2. $<M, u> \models$ (Bel i φ) iff $\varphi \in bel(\sigma, I(i), u)$,
3. $<M, u> \models$ (Send i j φ) iff $<I(i), I(j)\ \varphi> \in sent(\sigma, u)$,
4. $<M, u> \models$ (Do i α) iff $action(\sigma, I(i), u) = I(\alpha)$,
5. $<M, u> \models \neg \varphi$ iff $<M, u> \not\models \varphi$,
6. $<M, u> \models \varphi \vee \psi$ iff $<M, u> \models \varphi$ or $<M, u> \models \psi$,
7. $<M, u> \models \circ\varphi$ iff $<M, u+1> \models \varphi$,
8. $<M, u> \models \bullet\varphi$ iff $u > 0$ and $<M, u-1> \models \varphi$,
9. $<M, u> \models \varphi\ U\ \psi$ iff $\exists\ v \in N$ such that $v \geq u$, $<M, v> \models \psi$ and $\forall$ w $\in N$ such that $u \leq w < v$, $<M, w> \models \varphi$,
10. $<M, u> \models \varphi\ S\ \psi$ iff $\exists\ v \in \{0,\ldots, u-1\}$ such that $<M, v> \models \psi$ and $\forall$ w $\in N$ such that $v < w < u$, $<M, w> \models \varphi$.

The 1st rule is a logical constant for truth and is always satisfied. The 2nd rule means: agent i believes φ. The 3rd rule is satisfied if agent i sends a message with content φ to agent j. The 4th rule depicts that agent i executes the action α. The propositional connectives $\neg$ and $\vee$ hold standard semantics *not* and *or* respectively; whereas the remaining connectives are defined as abbreviations in the usual way.

11.3.2 Ontology Formalism

The conceptualization concerns the formal structure of reality as perceived and organized by an agent independently of the actual occurrence of a specific situation. Whereas an ontology is a set of logical axioms designed to account for the intended meaning of a vocabulary. Thus, with the purpose to state the formal meaning for both terms, the definition for conceptualization and ontology is set in this section based on the work fulfilled by FIPA, Genesereth, Gruber, Guarino and Giaretta [10, 41-44].

Model of Conceptualization. Conceptualization aims at the meaning of conceptual relations in a domain of representation independently of a state of affairs. For instance, in the puzzle domain, the meaning of the *above* relation lies in the way it refers to certain couples of blocks according to their spatial arrangement. Conceptual relations are stated on a domain space by the structure $<D, W>$, where D is a domain and W is the set of all states of affairs of domain, called *possible worlds*. For example, D is the set of blocks, and W is the set of all spatial arrangements of such blocks.

Given a domain space $<D, W>$, a conceptual relation p^n of arity n on $<D, W>$ as a total function p^n: $W \rightarrow (2^D)^n$ from W into the set of all n-ary ordinary relations on D is defined. For a generic conceptual relation p, the set $\boldsymbol{E}p=\{p(w) \mid w \in W\}$[1] contains the admittable extensions of p. Thereby, a conceptualization for D is defined as the tuple: $\boldsymbol{C}=<D, W, R>$, where R is a set of conceptual relations on $<D, W>$. Therefore, a conceptualization $\boldsymbol{C}$ is a set of conceptual relations defined on a domain space.

Based on $\boldsymbol{C}$, for each world $w \in W$, the corresponding world structure is the structure: $\boldsymbol{S}_{wC}=<D, \boldsymbol{R}_{wC}>$, where $\boldsymbol{R}_{wC}=\{p(w) | p \in R\}$ is the set of extensions of the elements of R. Hence, all the intended world structures of $\boldsymbol{C}$ is the set $\boldsymbol{S}_C=\{\mathbf{S}_{wC} \mid w \in W\}$.

Also, a logical language $\boldsymbol{L}$, with vocabulary V, a model for $\boldsymbol{L}$ is stated as the structure: $<S, I>$, where $\boldsymbol{S}=<D, \boldsymbol{R}>$ is a world structure and I: $V \rightarrow D \cup \boldsymbol{R}$ is an interpretation function, which assigns elements of D to constant symbols of V, and elements of $\boldsymbol{R}$ to predicate symbols of V. Hence, a model fixes a particular extensional interpretation of the language. In addition, an *intentional interpretation* is set by means of the structure $<\boldsymbol{C}, \zeta>$, where $\boldsymbol{C}=<D, W, R>$ is a conceptualization and ζ: $V \rightarrow D \cup R$ is a function that assigns elements of D to constant symbols of V, and elements of R to predicate symbols of V. Such intentional interpretation is called an *ontological commitment* for L. Thus, if $\boldsymbol{K}=<\boldsymbol{C}, \zeta>$ is an *ontological commitment* for $\boldsymbol{L}$, it states that $\boldsymbol{L}$ commits to $\boldsymbol{C}$ by means of $\boldsymbol{K}$, while $\boldsymbol{C}$ is the underlying conceptualization of $\boldsymbol{K}$.

Moreover, given a language $\boldsymbol{L}$, with vocabulary V, and an ontological commitment $\boldsymbol{K}=<\boldsymbol{C}, \zeta>$ for $\boldsymbol{L}$, a model $\boldsymbol{M}$, whose world structure is $<S, I>$, will be compatible with $\boldsymbol{K}$, if it meets three constraints: 1) $\boldsymbol{S} \in \boldsymbol{S}_c$; 2) for each constant c, $I(c)=\zeta(c)$; 3) for each predicate symbol p, I maps such a predicate into an admittable extension of $\zeta(p)$. This means that a conceptual relation p and a world w, such as $\zeta(p)= p \wedge p(w) = I(p)$, exist. In consequence, the set $\boldsymbol{I}_k(\boldsymbol{L})$ of all models of $\boldsymbol{L}$ that are compatible with $\boldsymbol{K}$ become the set of *intended models* of $\boldsymbol{L}$ according to $\boldsymbol{K}$.

Ontology Definition. Since a language L with ontological commitment K, an ontology O for L is: a set of axioms stated in such way that, the set of its models approximates as well as possible to the set of intended models of L according to K.

Thus, an ontology $\boldsymbol{O}$ specifies a conceptualization $\boldsymbol{C}$ in an indirect way, due to: it can only approximate a set of intended models, and such models are a weak characterization of a conceptualization. Hence, an ontology $\boldsymbol{O}$ for a language $\boldsymbol{L}$ approximates a conceptualization $\boldsymbol{C}$, if there is an ontological commitment $\boldsymbol{K}=<\boldsymbol{C}, \zeta>$ so that the intended models of $\boldsymbol{L}$, according to $\boldsymbol{K}$, are included in the models of $\boldsymbol{O}$.

These underlying concepts are sketched in Fig. 11.1 as follows: a language $\boldsymbol{L}$ commits to conceptualization $\boldsymbol{C}$ by means of an ontological commitment $\boldsymbol{K}=<\boldsymbol{C}, \zeta>$. As a result a set of models $\boldsymbol{M}$ is stemmed from the language $\boldsymbol{L}$, but the ontology $\mathbf{O}$ is just an approximation of the set of models $\boldsymbol{M}$. In spite this, the ontology $\mathbf{O}$ is able to depict the set $\boldsymbol{I}_k(\boldsymbol{L})$, which means the set of *intended models* of $\boldsymbol{L}$ according to $\boldsymbol{K}$.

In resume, an ontology $\boldsymbol{O}$ commits to $\boldsymbol{C}$ if: 1) it has been designed with the purpose of characterizing $\boldsymbol{C}$; 2) it approximates $\boldsymbol{C}$. Also, a language $\boldsymbol{L}$ commits to $\boldsymbol{O}$ if it meets some conceptualization $\boldsymbol{C}$ so that $\boldsymbol{O}$ agrees with $\mathbf{C}$. As a result, the following definition is stemmed: an ontology is a logical theory accounting for the intended meaning

[1] In section 11.3.2 symbols denoting structures and sets of sets appear in boldface.

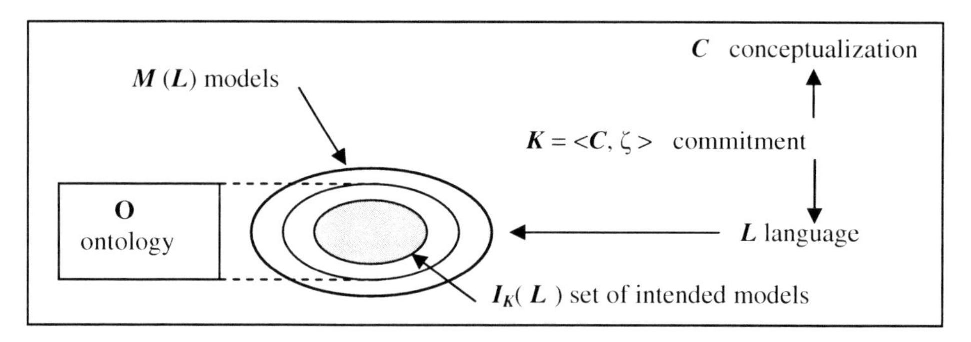

Fig. 11.1. The set of intended models $\boldsymbol{I_k(L)}$ of a logical language $\boldsymbol{L}$ reveals its commitment $\boldsymbol{K}$ to a conceptualization $\boldsymbol{C}$. Thus, an ontology $\boldsymbol{O}$ indirectly brings out such commitment by approximating the set of intended models $\boldsymbol{I_k(L)}$ stemmed from language $\boldsymbol{L}$ [10, 44].

of a formal vocabulary, i.e. its ontological commitment to a particular conceptualization of the world. The intended models of a logical language using such a vocabulary are constrained by its ontological commitment. Therefore, an ontology indirectly reflects this commitment by approximating such intended models [10].

11.4 Methodology

In order to enhance the integral view for building an OA, in this section a method for developing agents and ontologies is outlined. Firstly, the agents' life-cycle is shown. Then, an object language for setting the agent's formal specification is resumed. Finally, the design steps devoted to build an ontology are set.

11.4.1 Methodology for Building Agents

In this section a method for building MAS is set. This approach fixes the method proposed by Farias and Rodriguez [45], and the Gaia method stated by Wooldrige et al. [46]. The method's life-cycle embraces three phases: specification, implementation and verification. Each phase is carried out through several activities that outcome models. Such models reveal specific agents' issues, where one of them is pointed out through an object language that is resumed in section 11.4.2.

Specification. The specification is oriented to set the functionalities that MAS commits to cope with the problem to be solved. During this phase the following stages are achieved: analysis, design, definition of the agent model, and specification of the interaction model. A profile for each stage is introduced as follows:

Analysis. The analysis stage attempts to develop an understanding of the system and its structure. At this stage, two models are outcome: roles and interactions. The nature of the two models is described as follows.

The *roles model* identifies the key roles in the system, where a role is defined by four attributes, liveness expressions and a set of operators for such expressions:

- *responsibilities*: the functionality of the role is claimed by the responsibilities. Each responsibility embraces two properties: liveness and safety. *Liveness* asserts that "something good happens". It depicts those states of affairs that an agent must bring about, given certain environment conditions, i.e., an agent *i* that receives a *request* message from a friendly agent *j* will do the action α. *Safety* property is an invariant that the agent maintains while executing. This responsibility applies across all states. It relates to absence of an undesirable condition arising, "nothing bad happens". Safety requirements are specified by means of a list of predicates, i.e., if an agent *i* fulfils an action α, then the performance of the action α must have been preceded by a *request* message from a friendly agent *j* to do the action α,
- *permissions*: the rights granted to a role are identified by permissions attributes. The permissions are stated through the *Fusion* notation for operation schemata. The aim is to identify the resources that are available to the role for achieving its responsibilities [47], i.e., *reads* object data, *changes* content, *generates* message,
- *activities*: the attribute identifies actions accomplished by an agent without interacting with other peers,
- *protocols*: the actions to be fulfilled by an agent through the interaction with other agents are pointed out by the protocols attribute,
- *liveness expression*: the properties of infinite computations are sketched by means of a kind of *w*-regular expression. Liveness expression sets the potential execution trajectories through activities and protocols. The expression is composed by a role name and the sequence of activities and protocols, i.e. *request-Profile* = (*edit-message* . *send-message* . *receive-response* . *read-response*) where the role name appears at the beginning of the expression, the activities are underlined, the protocols are shown with normal font, and the period is the sequence operator,
- *operators for liveness expressions*: the operators set how the activities and protocols are executed. The operators are symbols that appear beside or between activities and protocols, *x*, *y*, as follows: 1) $x \,.\, y$: *x* followed by *y*; 2) $x \mid y$: *x* or *y* occurs; 3) $x \parallel y$: *x* and *y* interleaved; 4) x^*: *x* occurs 0 or more times; 5) x^+: *x* occurs 1 or more times; 6) x^w: *x* occurs infinitely often; 7) [*x*]: *x* is optional.

The interaction model depicts dependencies and relationships among roles. The interaction model contains a set of protocol definitions. A protocol definition is considered as a pattern of interaction, which has been defined and abstracted away from any particular sequence of execution. Also, a protocol definition gives rise to a number of message interchanges at run time. It owns the following attributes:

- *purpose* is the nature of the interaction like *request*, *inform*,
- *initiator* corresponds to the role responsible for starting the interaction,
- *responder* is the role with which the initiator interacts,
- *inputs* represent the information used by the role initiator,
- *outputs* correspond to information supplied by/to the protocol responder during the course of an interaction,
- *processing* is the task that the protocol initiator performs during the interaction.

Design. The design stage relates how a community of agents collaborates to achieve the system-level goals, and what is required for each individual agent in order to accomplish its specific goals. The design outcomes four models: agent, services, acquaintance and agent-class. A profile of each model is stated as follows.

The *agent model* reveals the agent type used to build MAS. The model depicts the agent instances that are stemmed from each agent type. An agent type is a set of closely related agent roles. This model is depicted as a tree, whose leaf nodes correspond to roles, and nodes with descendants reveal agent types. The number of agent instances is shown by a number, a range, or symbols *, +, that mean respectively the number of instances, the minimal and maximal number of instances, the possibility of none or several instances, and the affirmation that at least there is one instance.

The *service model* brings out the main services that are required to carry out the agent's role and specify the main properties of such services. The services are stated in the lists of protocols, activities, responsibilities and liveness properties of a role. For each service that an agent performs, the inputs, outputs, pre-conditions and post-conditions are stated. The inputs and outputs are stemmed from the interaction model, whereas pre-conditions and post-conditions are constraints on services. Such conditions are derived from the safety properties of the role.

The *acquaintance model* sketches the pathways of communication between agents. The purpose of the model is to identify any potential communication bottleneck, which may cause delays at run-time. The acquaintance model is drawn as a directed graph. Its nodes correspond to agent types, and its arcs identify communication links.

The *agent-class model* outcomes an organizational structure of MAS that is sketched by a *Class diagram* of the Unified Modeling Language (UML), where a class refers to the encapsulated unit that is achieved by an agent.

Agent Model. This is a formal model of MAS that is based on the formalism introduced in section 11.3.1. Besides the model of MAS, execution model and AL; the agent model brings out a computational interpretation that is outlined through an object language. Hence, the agent model embraces the formal definition for an agent and MAS, the execution history representation –based on (1) –, the attributes and behaviors of MAS –that are set by the Meta-language AL–, and the system description –which is stated through the programming language introduced in section 11.4.2–.

Interaction Model. The set of rules that agents must meet to interchange a sequence of messages between agents in order to achieve a goal is called: an *interaction protocol.* The interaction model contains the interaction protocols and protocol diagrams to control the dialogue maintained by agents for solving a task in a cooperative fashion.

Among the interaction protocols are: coordination, cooperation and negotiation. In the solution of a problem, the *coordination protocol* pursues to achieve the required coherence and the actions control. A *cooperation protocol* provides the mechanisms for implementing the cooperative work among agents in order to fulfill a task. A *negotiation protocol* sets the items to support a mutual agreement among agents.

The Protocol diagrams are an extension of UML Sequence diagrams, and they are part of Agent Unified Modeling Language (AUML) [48]. A Protocol diagram shows interactions among agents. Each interaction is a set of messages exchanged among different agent roles within a cooperative-work to achieve a desired behavior. The main conventions and elements of a Protocol diagram are the following:

The vertical axis points out the time, and the horizontal depicts agent roles. An interaction contains ordered messages. Each message sets communication between a sender role and a receiver role, which appear inside a box. Every agent role owns a vertical lifeline that reveals its state. When the state is *active* a wide line is drawn, otherwise a thin line is sketched. However, the destruction of the state is revealed by the symbol x. When the lifeline is split, it depicts threads that correspond to *and*, inclusive *or*, exclusive *or*. Such concurrency is pictured by a horizontal heavy bar with a non filled diamond, like the inclusive *or* with x inside the diamond. The messages are drawn by an arch with an arrowhead that is depicted for asynchronous messages, but it is filled for synchronous. Messages can be sent in parallel mode according to the conventions stated for lifetime's threads. Nested protocols are sketched in a modular way by a rectangle with rounded corners. Also, the protocols represent concurrency by the threads conventions, and they can use parameters.

Implementation. In the implementation phase the computational system that meets the specifications stated previously is built. The implementation is fulfilled through three stages: agent architecture definition, agent platform set up, MAS development. Thereby, a profile for each stage is introduced next.

Agent Architecture Definition. The agent architecture specifies how the agents can be decomposed into a set of modules, and how these modules should be built with the purpose to interact themselves. There are several approaches for building agents, as deliberative, reactive and hybrids, whose main properties are the following:

Deliberative architecture contains an explicitly symbolic model of the world. It makes decisions through logical reasoning based on pattern matching and symbolic manipulation [7]. Such architecture deals with two problems: transduction and representation-reasoning. The first one corresponds to the challenge to translate the real world into an accurate symbolic description. The second claims: how to symbolically represent information about complex real-world entities? How to get agents to reason with such information for the outcomes to be useful?

Reactive architecture is an agent architecture that does not include any kind of central symbolic world model, and does not use complex symbolic reasoning. For instance, the subsumption architecture set by Brooks is a hierarchy of tasks that accomplish behaviors. In this taxonomy, each behavior competes with others to exercise control over a robot. The lower layers depict more primitive kinds of behavior and have precedence over tiers further up the hierarchy [49].

Hybrid architecture is a paradigm that tries to marry deliberative and reactive approaches. The Procedural Reasoning System built by Georgeff and Lansky is a belief-desire-intention (BDI) architecture. It includes a plan library, as well as explicit symbolic representations of beliefs, desires and intentions. *Beliefs* are facts about the external world or the system's internal state. *Desires* are the system behaviors. *Intentions* are knowledge areas that are associated with invocation conditions that determine which areas are activated [50].

Agent Platform Set Up. The agent platform is a framework that supports the creation, execution and transference of a community of agents. The platform is built based on the elements stemmed from the specification phase. It takes into account the interactions protocols and the agent architecture. One instance of agent platform is the Multi-agent platform for Distributed Soft Computing, proposed by Bieganski et al., which aims to implement reliable and reusable system architectures [51].

MAS Development. The most common practice for deploying agents over Internet is the use of a Java platform as JADE [52] and ZEUS [53]. Both platforms are compliant with FIPA and provide support for functionality, planning and scheduling.

In addition, there is a trend to develop agents through Web services (WS). This kind of Web program offers specific functionalities to a community of users through Internet. The client applications access the service by the use of an interface that invokes a particular activity on behalf of the client. A WS provides: interoperability, Internet friendliness, typed interfaces, ability to leverage Internet standards, support for any language and distributed component infrastructure. The facilities required for federating WS are organized into five functional tiers: 1) transport supported by Hyper Text Transfer Protocol (HTTP); 2) encoding by means of XML; 3) message format edited by Simple Object Access Protocol (SOAP); 4) description of the functionality by Web Services Description Language (WSDL); 5) publication of services by Universal Discovery Description and Integration UDDI. Among the approaches for deploying a MAS through WS is the application achieved by Peña et al, who used the dot Net [54]. Such a platform offers a multi-tier architecture that supports the installation of WS quite easily through languages such as C#.

Verification. In the verification phase, the behavior is evaluated in order to identify if MAS meets the former specifications. Hence, this phase is achieved by two stages: life-cycle verification and formal verification, whose description is the following.

Life-cycle Verification. During the life-cycle, verification is determined when MAS accomplishes the specifications. Particularly, assurance of liveness and security properties require a special evaluation. Hereby, the main issues to be considered are: 1) confirmation that executed actions correspond to expected actions to be achieved by MAS; 2) validation of actions according to the restrictions set in the interaction protocols; 3) assurance that action sequences satisfy the specifications.

Formal Verification. The formal verification aims to test in a mathematical sense that MAS is right according to the formal specifications. Among the strategies for achieving the verification are the following: Modal Semantic Trees, the algorithm for determining if a formula is valid in a model that was proposed by Rao and Georgeff [55]; the SAT algorithm designed by Manna and Pnueli [56].

11.4.2 Object Language

Although AL is a language based on logic, the appearance of the formal model for MAS is not easy to interpret. Thus, it is most suitable to use a representation like computational programming. So in this section the Language of Capacity for Agent Interaction in Open Systems, proposed by Farias with the Spanish acronym LCIASA, is introduced [45, 57]. LCIASA offers the syntactic structure and the semantic interpretation to the actions stated in AL. Moreover, LCIASA provides a computational interpretation to the formal agent model represented by AL.

Object Language Profile. LCIASA contains three items: a set of constant actions that define the tasks to be achieved by an agent, a set of sentences that provides the syntactic structure, and the semantic interpretation to the rules stated by AL.

The constant actions embrace a set of primitive tasks given by KQML messages and a set of compound activities. A compound action is the result of the integration of several KQML messages with the purpose to set a predefined function. A compound action is devoted to authenticate each task involved in an interaction between agents in order to assure the security and integrity of the specifications.

The compound actions are classified into six functional groups: 1) *Actions of agents*, which correspond to activities fulfilled by an agent in response to the request sent by other agent; 2) *Authorization*, these tasks are oriented to confirm or deny subsequent actions; 3) *Encoded*, this kind of action is a mechanism devoted to avoid unauthorized access to specific information; 4) *Authentication*, such tasks confirm the agents and messages legitimacy; 5) *Verification*, these activities evaluate the consistency of the plans; 6) *Operations in the virtual knowledge base (VKB)*, this kind of action is devoted to manage the agent's knowledge base.

The sentences are stemmed from the semantic rules of AL. An action sentence involves constant actions. Each sentence identifies an action, as: execute an agent action, e.g. *Do* (i, α); send a message, e.g. *Send* (i, j, φ); storage a belief, e.g. *Bel* (i, φ). Also, a sentence provides a syntactic structure for conditions and cycles, e.g. *While – Do*(φ, μ, ψ). The syntactic structure is introduced in the following section.

As regards the semantic interpretation for rules stated by AL, the semantic representation is pointed out at the end of this section.

Object Language Syntax. In MAS the interaction among agents is carried out by an exchange of messages. Such messages are fixed to define *plans*. Thus, a plan is set as a sequence of messages devoted to achieve a task. Hence, the syntax of LCIASA based on the plan concept is outlined as follows:

Each plan owns an identifier and a block of the plan. The plan block contains: preconditions, declaration of variables, body of the plan and postconditions. The preconditions own attributes of the plan and conditions for the invocation. The body of the plan holds AL tasks –as *send, bel, do*– and ordinary actions, which are simple or structured. Simple actions are primitive activities that correspond to KQML messages –as informative: *tell, deny*; notification: *subscribe, monitor*; query: *evaluate, reply*–, also they can be compound actions –as actions of agents: *finds-solve-message, analyze-plan*; authorization: *visualize-VKB, validate-agent*; encoded: *dispose-message, protects-message*; verification: *checks-plan, verifies-order-of-message*; authentication: *authenticates-agent, authenticates-message*; operations in the VKB: *add-plan, admits-belief*–. Structured actions are conditional, as *if-then-else*, and repetitive, as *while-do*. The LCIASA syntax appears in [57].

Object Language Semantic. The semantic structure of LCIASA is outlined as an axiomatic system. It uses the following notation for an axiom: $a; b; \ldots h \vdash z$, which means that: given the premises $a, b, .. h$ is asserted z. In addition, the derivation process is carried out by Modus Ponens inference rule, e.g. $\vdash \phi \rightarrow \psi \vdash \phi \models \psi$.

As regards the semantic rules of AL, they become the premises of the axiomatic system of LCIASA. According to this consideration, a set of rules is stemmed in order to guarantee the specification correctness. Also, the LCIASA semantic description is fully detailed in [57].

11.4.3 Ontology Development Method

In this section an ontology development method (composed by the Ontology Development 101 stated by Noy and McGuinness [58], and the Ontology Development Process proposed by Gomez [59]) is introduced. The ontology development is an iterative process, whose life-cycle embraces three sets of activities: 1) *management* whose aim is to administer the life-cycle; 2) *integral* contains specialized activities for eliciting knowledge; 3) *development* is devoted to build the ontology. These activities are described in the following sections.

Management. Management activities are devoted to assure a smooth-running ontology. There are three types of activities: 1) *planning* identifies the task to be done; 2) *control* guarantees that the planned tasks are done rightly; 3) *quality assurance* guarantees each outcome is compliant with a given quality standard.

Integral. Integral activities offer support for each activity to be done during the development of the ontology. This set encompasses four activities: 1) *acquire* elicits knowledge from different sources such as: human experts, books, dictionaries, handbooks; 2) *evaluate* is a technical judgment with respect to a framework of reference; 3) *document* states the information that depicts the ontology and its resources; 4) *configuration management* keeps the records of each release issued during the development and maintenance of the ontology.

Development. The construction of an ontology is broken down into three stages: specification, construction and maintenance. Each stage accomplishes specific goals through the achievement of specialized activities that are explained next:

Specification. At this stage the purpose of the ontology and its conceptual model are set by the specification and conceptualization of the ontology to be built as follows:

The outcomes to specify an ontology are statements about: 1) *end-users* answers the question: who is going to use the ontology?; 2) *domain* is the field of knowledge that corresponds to the ontology; 3) *purpose* is the intended use of the ontology; 4) *scope* is the summary of the ontology's content; 5) *users' needs* are stemmed from the informal competency questions; 6) *degree of formality* ranges from informal natural language to a rigorous mathematical notation.

The deliverables to conceptualize an ontology are: a conceptual model of the problem and its solution, and the decision to reuse an existing ontology or build a new one. The conceptual model encompasses scenarios that represent the problem to be solved as a history of problems and intuitive solutions. The model is depicted as a set of functionalities that are sketched by UML Use Cases diagrams. Moreover, the conceptual model holds a glossary of primitive concepts, properties and relationships. These terms are drawn as a classification tree where they appear graphically correlated. In addition, the informal competency questions, stated in natural language, are translated into a formal set of competency questions using first-order logic. This formal set is used to evaluate the extensions of the ontology. Based on the conceptual model, it is worth considering the reuse of an existing ontology that meets the requirements. For instance, there are public and commercial ontologies like Ontolingua, DAML, UNSPSC, RosettaNet and DMOZ. If such ontology exists, then an evaluation is done to consider its formal model, the language of representation and the constraints for translating. Also, other issues must be taken into account such as royalties.

Construction. During the construction stage three outcomes are achieved: formal model, design and implementation. Hence, based on the formal model previously stated, the design is described next, and the implementation is set in the fifth section.

The design of an ontology requires the participations of domain specialists as linguistics and knowledge engineers. The challenge is to define a world reality model, where concepts depict physical or abstract objects, and relationships point out conceptual or physical links among objects. Hence, the design of an ontology embraces activities such as: elicit terms of a domain, define concepts, identify relations between concepts, state classes, set class hierarchy, define properties, establish facets, create instances, and codification. These activities are outlined next:

The terms of a domain are stemmed from the following process: 1) identification of nominal groups from a corpus considered as being representative of the domain; 2) the groups that can not be chosen as terms, because of morphological or semantic attributes, are excluded; 3) the nominal sequences that are held as terms are chosen.

In order to state the concepts that reveal the meaning of the terms, it is necessary to seek their formal definition according to domain. In addition, common knowledge, which is always present in any domain, is taken into account.

The relations between concepts are stemmed from the following sequence: 1) identify the co-occurrences of the terms; 2) seek a similarity between terms with respect to the contexts they share; 3) determine terms that are semantically related by semantic, proximity, causal, physical, abstract or any other kind of relations.

With the purpose of selecting the terms that play a role in a class, the words that depict objects having independent existence must be considered. Moreover, it is worth noting terms that reveal generalities and groups of objects.

A class hierarchy can be stemmed from three approaches: 1) *top-down* starts with the most general class in the domain and goes on with subsequent specialization of concepts; 2) *bottom-up* begins with the most specific classes and continues with subsequent grouping of the latest classes into more general ones; 3) *hybrid approach* commences with the more salient classes, then generalizes them and specializes them.

Properties are stemmed from terms that describe objects. The properties depict the internal structure of concepts. Moreover, they are the class attributes. The common properties are attached to the most general class. As a result, the subclasses inherit the attributes of the general class. Also, there are several types of properties such as: *intrinsic*: innate properties to the object as size; *extrinsic*: reference properties to the object as name; *parts*: physical or abstract members of the object; *relations* between individual elements of the class and other items like the brand of a product.

Regarding the assignment of properties, suitable facets must be defined to set the cardinality –as single or multiple–, the range –minimum and maximum–, the value type –numeric, string– and other features that constrain the values.

The creation of instances is achieved through the definition of objects that are stemmed from a given class. Hence, each attribute of the class is filled with values that commit its facets and reveal the nature of the object that is being evaluated.

The encoding of the ontology is done by an appropriate language such as: XML [11], RDF [12], DAML [13] and OWL [14]. Moreover, there are specialized tools that facilitate the codification of the ontology as Chimaera and Protégé [60].

Maintenance. This stage is devoted to update the structure and the content of the ontology according to new requirements that appear during the life of the ontology. Thereby, each time a new version is outcome the whole life-cycle is achieved.

11.5 Development

Based on the formal model and method for agents and ontologies, the process for building and using an OA is depicted. The trial concerns a WBES that aims to provide adaptive and intelligent lectures. So a resume of the WBES, which holds OA, is set. Next, the activities and outcomes achieved during the specification and construction for the ontology are stated. Also, the specification and implementation phases correspond to the agents' method are outlined. Finally, a resume of the trial is given.

11.5.1 Experimental Scenario

In this section a profile of the research achieved in WBES field is set. Also, the contribution of the OA to provide adaptive and intelligent services is acknowledged.

Research Profile. In order to enhance the apprenticeship of people that use a WBES, I propose an approach that suggests the best available lecture for teaching a concept [61]. So I focus my work on the *Student Centered Education* paradigm. The research question is: How to depict and anticipate the learning that the student is able to achieve when he/she takes a lecture delivered by a WBES? Thus, I set the hypothesis: Based on the student attributes, lecture properties, and representation of the cause-effect relations between characteristics of the individual and the lecture; it is possible to estimate the causal outcome that a lecture produces on the student. According to such supposition, the three premises were dealt as follows:

A student profile for the individual is outcome from his/her *learning preferences*, *personality* and *cognitive* capacities, which are stemmed from psychological tests. Thus, concepts as logical preference, maturity and intelligence depict the person.

As regards lectures, the WBES meets one requirement: for any concept to be taught there must be different lectures. Thereby, each lecture is tailored from different views, as: *learning theory*, constructivism, objectivism; type of *stimuli*, sound, visual. As a result, a description for each lecture, used to teach each concept to be learned, is stated. The description holds concepts as: complexity, interactivity and media.

The causal relations are sketched by a *Cognitive Map* (CM) [62]. A CM is a mental model, where *nodes* depict concepts about student profile and lectures; and *links* correspond to causal relations between concepts. Also, concepts and relations were qualitatively estimated by linguistic terms. Such terms reveal *levels of intensity*, as high or low, and *variations*, as bias low or inhibit high. In this way, a specific CM is automatically generated for any combination of individual-concept-lecture.

Regarding the conclusion, this is an outcome stemmed from the causal bias that a lecture produces on the student's learning. Such result is estimated through a fuzzy-causal inference. This kind of reasoning is an iterative process, where in each cycle the causal effect that occurs on the state of each concept is computed. This value reveals a level or a variation on the concept's state. So, the CM is a dynamic system that changes progressively until it meets a stable situation, where there are no more changes, or it is acknowledged that such a situation is not feasible.

Ontology Agent Role. The contribution of an OA for a WBES is: to administrate the ontologies that reveal the meaning of knowledge repositories. As in WBES the content of such repositories is diverse, the statement and management claim a high work-load. Hence, the OA fulfills tasks such as storing content, responding to queries and achieving inferences for a given ontology. In short, the OA is responsible for meeting requests sent by agents that carry out the functionalities of WBES modules.

11.5.2 Ontology Development

While the trial is progressing, a WBES fills in several knowledge repositories. For this reason, four ontologies were previously built to depict the meaning of such files. These static ontologies are tailored to reveal: the student profile, knowledge acquired by the student, content of the CM, and domain of teaching. The ontologies are outcome by the development team according to the method set in section 11.4.3. Thus, prior to starting the trial, domain literature is consulted and the advice of psychology and CM experts is sought. Hence, during the trial, the ontologies are managed by the OA to meet requests sent by agents of the WBES. Thereby, for illustrative purposes, the construction of the ontology to state the meaning of the CM content is resumed next:

Specification. As a result of the ontology specification and conceptualization, an ontology profile and a conceptual model are achieved with the following attributes:

The ontology profile sets as end-user: the student model of the WBES. The domain is: the content of the CM, whose purpose is to state the lecture and the student to be modeled. The scope includes: classes, properties and instances that represent CM's items. One of the competency questions is: Which are the attributes of a concept? The degree of formality is: a computational language, whose notation is based on XML.

The conceptual model holds UML Cases of Use diagrams for sketching the access to ontology items and the inheritance inferences through class hierarchy. In addition, a glossary is stated with terms that shape fuzzy sets such as: area, angle, longitude. Moreover, the relations between meta-classes and classes are drawn as a classification tree. Finally, a decision for building a new ontology is made.

Construction. During the construction stage the formal model for conceptualization and ontology –introduced in section 11.3.2– is taken into account. Based on such references, the design and implementation of the ontology are achieved as follows.

The elicitation of terms, definition of concepts and statement of relations among concepts are stemmed from literature and experts [61, 62]. The identification of classes and the configuration of a hierarchy are resumed in two kinds of classes: *meta-classes* that are classes with no ancestors; and *classes* that own ancestors, sub-classes and object instances. The properties and the corresponding facets are outlined for each class in order to represent the real object accurately. The instances correspond to physical and abstract objects of the domain such as: student, learning preferences, lecture and type of content. The codification of the ontology is accomplished by means of Protégé tool, and the code is stated through OWL.

An instance of ontology content is shown next. In such OWL code, the *_fuzzy_set* class is outlined through the first element *owl:Class* and the attribute *rdf:ID*. After, the description for the class appears as the element *rdfs:comment*. Also, the reference to the super class *#_identification* is set in the element *rdfs:subClassOf*.

Extract of an ontology for Cognitive Maps from Peña (2008) PhD Thesis, Mexico [61]

```
<owl:Class rdf:ID="_fuzzy_set" xmlns:rdf="rdf"…>
<rdfs:comment rdf:datatype="www.w3.org/2001/XMLSchema#s
tring" xmlns:rdfs="rdfs">fuzzy set class</rdfs:comment>
<rdfs:subClassOf xmlns:rdfs="rdfs">
   <owl:Class rdf:about="#_identification"/>
</rdfs:subClassOf></owl:Class>
```

11.5.3 Ontology Agent Development

MAS provides a suitable DAI platform to deploy the functionalities required by WBES. Due to several specialized agents autonomously fulfilling tasks to achieve roles attached to modules and repositories in an intelligent way, the specification and implementation of an OA in charge of CM's ontology are resumed next.

Specification. The analysis and design of an OA, and several modules devoted to depict an OA are outcome during the specification stage as follows.

Analysis. As a result of the analysis two models are sketched roles and interaction. For example, the role model for Query_if role, which requests ontology items, is shown in Fig. 11.2; whereas the interaction model for Query_if role is drawn in Fig. 11.3.

Design. In this stage four models are set: *agent, services, acquaintance and agent-class.* As an example of the work achieved, the *agent model* for an OA appears in Fig.11. 4[a], where several roles are linked to it; and the *acquaintance model,* composed by four agents and their respective message flows, is shown in Fig 11.4[b]. Also, the *service model* attached to *query-if* role is sketched in Fig. 11.5. Moreover, an *agent-class model* is drawn in Fig. 11.6, where inheritance relations are distinguished by triangles.

Agent Model. The agent model for an OA is stemmed from: the formal model for MAS, execution model and AL specification, which appear in section 11.3.1. Hereby, a fragment of LCIASA code is stated in this section. The code shows the message that

Role Schema: Query_if
Description: Request to access a specific term from an Ontology
Protocols: Receive-Request, Decode-Request, Encode-Response
Activities: Await-Request, Seek-Term, Generate-Response
Permissions: *read* ontology-id, ontology-term // parameters given by requester
generates response: information-term or *null* // result of the search
Responsibilities: *Liveness* = (Decode-Request.Seek-Term.Generate-Response)
Safety: (Await-Request. Encode-Response. Send-Response)

Fig. 11.2. Roles model for *query-if* role. The underlined words identify actions, the dot operator means a sequence of actions, and the slashes // mean a remark.

Purpose: Provide the information of a term		**Inputs:** ontology-id,
Initiator: Functional_model	**Responder:** Query_if	ontology-term
Processing: access the meaning of the term demanded		**Outputs:** term, null

Fig. 11.3. Interaction model for *query-if* role. It shows the data flow for communication.

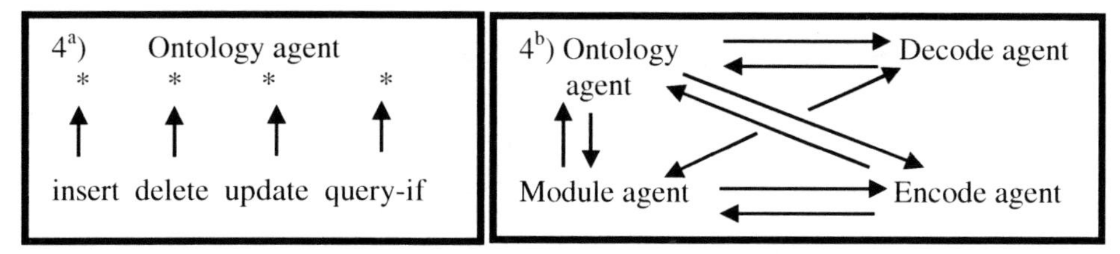

Fig. 11.4. Embraces two models: In 11.4[a] the agent model for an OA is drawn, where symbol * means zero or more instances. In 11.4[b] the acquaintance model for four agents appears.

Agent Role: Query-if class
Services: Get the information about a specific class in an ontology
Inputs: ontology-id and ontology-class-id
Outputs: response: *true* (class' information) or *false* (null)
Pre-condition: ontology-id <> *nil* and ontology-class-id <> *nil*
Post-condition: response = class' information or null

Fig. 11.5. Service model for *query-if* role shows inputs and outputs parameters and conditions

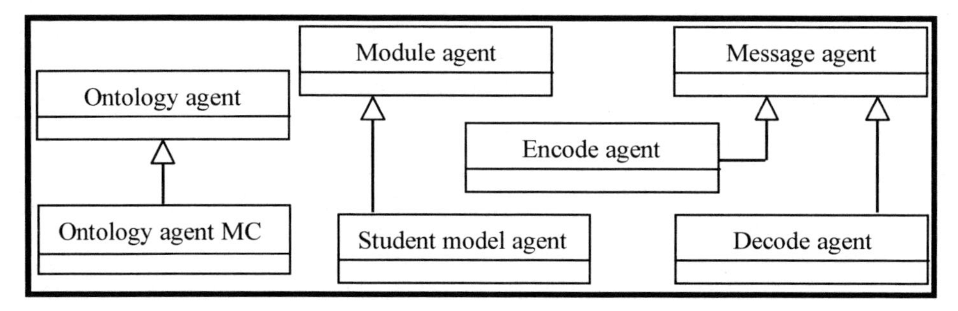

Fig. 11.6. Agent-class model for three superclasses of agents and four agent classes

a functional agent F sends the mediator agent *M* to request the support of an OA, whose id is: *OA1*. The OA1 is in charge of the ontology that depicts the meaning of the CM content. Thereby, the liveness and security properties are stated as follows:

Extract of LCIASA code for request an ontology item of a Cognitive Map

```
Plan Find Ontology Agent for Cognitive Map ontology
  Begin
    Variables Catalogue, F, M, OA1: agents
    Preconditions KQML protocol
    Begin-Plan
      While ((Catalogue = True) ∧¬Eos)
        Send(F, M, recruit-all(ask-all(x))) ∧
        Bel (M, Authenticates-Originator(F))
        Do (M, recruit-all(ask-all(x)))
        Send(OA1, M, Advertise(ask(x))) ∧
        Bel (M, Authenticates-Originator(OA1))
        Do (OA1, Advertise(ask(x)))
        Send(OA1, F, Tell(x)) ∧
        Bel (F, Authenticates-Originator(OA1))
        Do (OA1, Advertise(ask(x)))
        Send(F, OA1, Tell(x)) ∧
        Bel (OA1, Authenticates-Originator(F))
        Do (OA1, Tell(x))
      Do (Checks-Plan)
    End_Plan
End
```

Interaction Model. Agent collaboration is done through a cooperation protocol based on contracts [63]. The agent interaction is sketched as a Protocol diagram, such as the one drawn in Fig. 11.7 for a contract protocol between an OA and a functional agent.

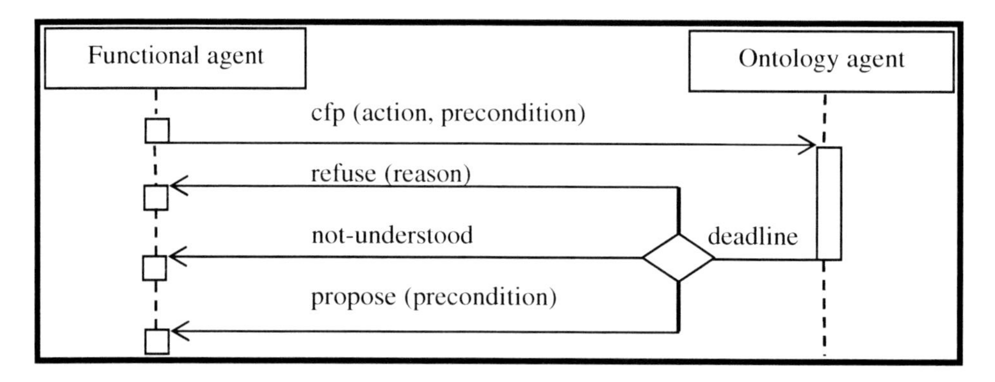

Fig. 11.7. Protocol diagram of the control net protocol between functional and ontology agents

Implementation. The agent architecture used for MAS is hybrid, as it is based on a BDI model. The agent platform is deployed by means of WS, whose definition is stated in lines 1-3 of the next code written in C#. In lines 4-7 the method to be called appears, and its reference and invocation are set in lines 8-15.

Code for Web service declaration and use from Peña (2008) PhD Thesis, Mexico [61]

```
1:  // Section 1: Web Service Class Declaration
2:  namespace     wsInterfaceAgent {
3:  public class wsInterfaceAgent :
                  System.Web.Services.WebService{
4:  // Section 2: Web Service Method Declaration
5:  [WebMethod]
6:  public int wmInterfaceAgent (Input-parameters,
           out output-parameters)
7:  { code..... return value }
8:  // Section 3: Web Service Reference
9:  namespace     waInterface
10: public class WebForm1 : System.Web.UI.Page {
11: // Namespace: host; class: wsInterfaceAgent;
12: // Instance: interface
13: host.wsInterfaceAgent interface = new
                           host.wsInterfaceAgent();
14: // Section 4: Web Service invocation;
    // Web Service method: wmInterfaceAgent
15: interface.wmInterfaceAgent(Input-parameters,
                                out output-parameters);
```

According to the arguments stated in the section "MAS development" and the code earlier set, the deployment of the functionality supported by an OA is quite easy. Indeed, the developer only has to declare the WS and the method to be called. Such method is the interface that is federated on Internet; whereas, the client agents only have to reference such interface in order to access the OA functionality. Moreover, the C# code reveals that the instructions used to state declarations and invocations are quite simple. The reason for such strength is: C# is an Object Oriented language that is devoted to develop Web applications as agents and WBES.

11.5.4 Description of the Experiment

With the aim of gathering evidence about the support that an OA offers to WBES for delivering adaptive and intelligent lectures, an experiment was carried out. The trial fulfills seven stages: statement, design, integration, measurement, selection, execution and evaluation. The trial finds out: *whether the approach accurately manages the knowledge involved in a WBES to enhance the student's apprenticeship.*

As the evaluation factor is the apprenticeship of students, an educational scenario was recreated on the Web. The scenario embraces: a WBES prototype based on MAS, knowledge domain content about "Scientific Research", three psychological tests and one knowledge domain test. Also, a universe of subjects was recruited.

Due to the experiment design being blind, the subjects were not aware of the approach, nor the outcomes of the psychological and knowledge domain tests. The trial set as *independent variable* the use of the knowledge approach for depicting students and lectures. The *dependent variable* was the difference between the average learning achieved by two groups. The *noise variable* was the student-WBES interaction.

During the integration, the promotion of the trial and the organization of a universe were achieved. The promotion was fulfilled through Internet to invite students and professors at bachelor, master and doctoral levels from several professions, universities and Mexican states. Many applications were registered on line. As a result of the analysis of electronic forms, 200 requests were chosen to set the universe.

The measurement stage included the application of psychological and knowledge domain tests. The tests were: 1) *learning preferences test* to evaluate seven kinds of preferences such as logical and visual; 2) *personality* to reveal the degree of forty personality scales such as fear and stress; 3) *cognitive* to estimate intellectual coefficients like reasoning and memorizing; 4) *knowledge domain* to identify preliminary knowledge about Scientific Research. The knowledge stemmed from the tests was set in the student profile, whose meaning was dealt by an OA. As a result, only 50 people successfully completed the tests; hence, the population was integrated by them.

During the selection, the population was trained, a sample was stemmed, and two groups were organized. Hence, the WBES delivered lectures about underlying concepts, such as science and philosophy. Afterwards, eighteen subjects were randomly chosen to set a sample. The size reveals a standard error of 0.05. Later, the sample was randomly split into two teams of nine people: *experimental* (*E*) and *control* (*C*).

The execution of the trial encompassed three activities: pre-measure, stimulus provision, and post-measure. Thus, in order to estimate the apprenticeship of the subjects ten "basic concepts", like hypothesis, law and theory, were chosen as the *reference*. Hence, for each basic concept, the preliminary knowledge of the subject was estimated according to Bloom's Taxonomy. Thereby, the number of the corresponding level, which ranged from 0 to 6, was tracked. Next, the lecture about a basic concept was delivered. Afterwards, the same test was applied to identify the level achieved. The apprenticeship for a basic concept was stemmed from the difference between the final and the preliminary levels. The *E* team was taught according to the selection of lectures that took into account the student and lecture profiles, whose meaning was managed by the OA. The *C* group received lectures that were randomly chosen.

The evaluation stage found out that: the prior knowledge of *E* team was 38 levels against 42 for *C* group. But, after the stimulus, the knowledge acquired for *E* group was enhanced up to 198 levels against 174 for *C* team. This means, that the apprenticeship of *E* team was 15% higher in the number of levels achieved than the level acquired by *C* group. Also, the analysis of variance for *E* group revealed a high regression with probability of 0.006. Therefore, the predictions' accuracy for the population was quite high. In short, the contribution of the OA was very valuable [61].

11.6 Conclusions

Knowledge is the core of cleaver behavior in humans, artificial intelligent systems and novel Internet approaches. The issues about elicitation, learning, representation and reasoning are still the focus of many research areas. Agents, ontologies and OA are paradigms for dealing with knowledge issues such as acquisition, organization, access, reuse and manipulation in open environments like Internet.

In this chapter a holistic framework for the formal definition, design, development and application of OA was outlined. Given such baseline, it was possible to appreciate underlying concepts about agents, ontologies and WBES. Also, a resume of the related work achieved in the arena was fulfilled. Alongside, the formal model for conceptualize agents and ontologies was set, in addition to their respective methods. With such grounds, how to build and use an OA in WBES were shown.

The outcome of the WBES trial showed that the OA provides a valuable support to enable WBES to deliver an adaptive and intelligent service to students. The apprenticeship of the students, whose profile was taken into account, was enhanced more than the learning achieved by students whose profile was ignored.

Regarding future work for OA, I consider the OA application to the following trends: semantic Web, Intelligent Web, Wisdom Web and Evolving WBES [64].

Acknowledgements

The author states that this work was especially inspired by his Father, his brother Jesus and his Helper as one of the projects of World Outreach Light to the Nations Ministries (WOLNM). Also this work was partially supported by IPN grants: SIP-20082973, SIP-EDI, COFAA-SIBE, CONACYT-SNI 36453. Moreover, the support for reviewing the paper achieved by Lawrence Whitehill is well appreciated.

References

1. Brena, R.F., Ceballos, H.G.: A Hybrid Local-global Approach for Handling Ontologies in Multi-agent System. In: Proc. Int. C. Intelligent Systems, Varna, Bulgaria (June 2004)
2. Kay, A.: Computer Software. Scientific American 251(3), 53–59 (1984)
3. Franklin, S., Graesser, A.: Is It an Agent or Just a Program? A Taxonomy for Autonomous Agents. In: Proc. Workshop Agent Theories, Architectures & Languages. Springer, Heidelberg (1996)
4. Shoham, Y.: An Overview of Agent-oriented Programming. In: Bradshaw, J.M. (ed.) Software Agents. AAAI Press, Menlo Park (1997)
5. Gilbert, D., Aparicio, M., Atkinson, B., Brady, S., Ciccarino, J., Grosof, B., O'Connor, P., Osisek, D., Pritko, S., Spagna, R.: Intelligent Agent Strategy. IBM Corporation (1995)
6. Etzioni, O., Weld, D.S.: Intelligent Agents on the Internet: Facts. Fictions and Forecast-IEEE Expert 10(4), 44–49 (1995)
7. Wooldridge, M., Jennings, N.R.: Intelligent Agents: Theory and Practice. Knowledge Engineering Review 10(2), 115–152 (1995)
8. Moulin, B., Chaib-draa, B.: An Overview of Distributed Artificial Intelligence. In: O'Hare, G., Jennings, N.R. (eds.) Foundations of Distributed Artificial Intelligence. Wiley, Chichester (1996)
9. Lesser, V.R.: Cooperative Multi-agent Systems: A Personal View of the State of the Art. IEEE Transactions on Knowledge and Data Engineering 11(1) (1999)
10. FIPA-OS: Ontology Service. Foundation for Intelligent Physical Agents (2001)
11. Cowan, J.: eXtended Markup Language, 1.1. World Wide Web Consortium (2000)
12. Klyne, G., Carrikkm, J.J.: Resource Description Framework (RDF): Concepts and Abstract Syntax. World Wide Web Consortium (2003)
13. Hendler, J., McGuinness, D.L.: The DARPA Agent Markup Language. IEEE Intelligent Systems 16(6), 67–73 (2000)
14. Dean, M., Schreiber, G.: Web Ontology Language Reference. In: WWWeb Consortium (2003)

15. FIPA-ACL: Agent Communication Language. Foundation for Intelligent Physical Agents (2000)
16. Stanford-OKBC: Open Knowledge Base Connectivity Specification. Stanford Uni. (1998)
17. Labrou, Y., Finin, T.: A proposal for a new Knowledge Query and Manipulation Language Specification. University of Maryland, Baltimore (1997)
18. Aroyo, L., Kommers, P.: Special Issue on Intelligent Agents for Educational Computer-Aided Systems. Int. J. Interactive Learning Research 10 (1999)
19. Uhr, L.: Teaching machine programs that generates problems as a function of interaction with students. In: Proc. 24th ACM National Conference, New York, USA, pp. 125–134 (1969)
20. Guttormsen, S., Krueger, H.: Using new Learning Technologies with Multimedia. Int. J. IEEE Multimedia (July-September 2000)
21. Sleeman, D., Brown, J.S.: Intelligent Tutoring Systems, p. vii. Academic Press, London (1982)
22. Wegner, E.: Artificial Intelligence and Tutoring Systems: Computational Approaches to the Communication of Knowledge, p. 3. Morgan Kaufmann Publishers Inc., San Francisco (1987)
23. Burton, R.B., Brown, J.S.: An Investigation of Computer Coaching for Informal Learning Activities. Int. J. Man-Machine Studies 11, 5–24 (1979)
24. Mathé, N., Chen, J.: User-centered, Indexing for Adaptive Information Access. Int. J. User Modeling and User-Adapted Interaction 6(2-3), 225–261 (1996)
25. Brusilovsky, P.: Adaptive Learning with WWW: The Moscow State University Project. In: Telematics for Education and Training, pp. 252–255. IOS Press, Amsterdam (1995)
26. Ayala, G., Yano, Y.: Collaborative Learning Environment Based on Intelligent Agents. Int. J. Elsevier, Expert Systems with Applications 14, 129–137 (1998)
27. IMS: Instructional Management Systems Global Learning Consortium Inc. (2007)
28. Sheremetov, L., Uskov, V.: Hacia la nueva generación de sistemas de aprendizaje basado en la Web. Revista Computación y Sistemas. CIC-IPN (2001)
29. Brusilovsky, P.: Adaptive and Intelligent Web-based Education Systems. Int. J. Artificial Intelligence in Education 13, 156–169 (2003)
30. IEEE-LTSA: Learning Technology Standard Architecture. Institute of Electrical and Electronic Engineer (2008)
31. Bai, Q., Zhang, M.: Agent Coordination through Ontology Managements. In: Proc. Int. C. Artificial Intelligence and Applications. Innsbruck, Austria (February 2004)
32. Alvarez, A., Fernandez, I.: An Open, Adaptive and Multi-subject Educational System for the Web. In: Proc. Artificial Intelligence in Education, AIED, Sydney, Australia (July 2003)
33. Querrec, R., Buche, C., Maffre, E., Chevailler, P.: Muti-agents Systems for Virtual Environments for Training Application to Fire-fighting. Int. J. Advanced Technology for Learning 1(1), 25–34 (2004)
34. Limoanco, T., Sison, R.: Learner Agents as Student Modeling: Design and Analysis. In: Proc. Int. C. Computers and Advanced Technology in Education, IASTED, Grece (2003)
35. Conati, C., Zhao, X.: Building and Evaluating an Intelligent Pedagogical Agent to Improve the Effectiveness of an Educational Game. In: Proc. Intelligent User Interface and Computer-Aided Design of User Interfaces, IUI-CADUI 2004, Madeira, Portugal (January 2004)
36. Falquet, G., Ziswiler, J.C.: A Virtual Hyperbooks Model to Support Collaborative Learning. Int. J. E-Learning 4(1), 39–56 (2005)

37. Soldatova, L.: Test Ontology. In: Proc. Int. C. Computers and Advanced Technology in Education, IASTED, Rhodes, Greece, pp. 175–180 (June 2003)
38. Devedzic, M., Devedzic, V.: Learning Demography in a Web Classroom: Ontological Support. In: Proc. Artificial Intelligence in Education, AIED, Sydney, Australia (July 2003)
39. Wooldridge, M.: The Logical Modeling of Computational Multi-agent Systems. PhD Thesis. Manchester Metropolitan University, Department Computation. Manchester, UK (1992)
40. Wooldridge, M.: Temporal Belief Logics for Modeling Distributed AI Systems. In: O'Hare, G.M.P., Jennings, N.R. (eds.) Foundations of DAI. Wiley Interscience, Hoboken (1995)
41. Genesereth, M.R., Nilsson, N.J.: Logic Foundation of Artificial Intelligence. Morgan Kaufmann, Los Altos (1987)
42. Gruber, T.R.: Toward Principles for the Design of Ontologies Used for Knowledge Sharing. Int. J. Human and Computer Studies 43(5/6), 907–928 (1995)
43. Guarino, N., Giaretta, P.: Ontologies and Knowledge Bases: Towards a Terminological Clarification. In: Mars, N. (ed.) Towards very large Knowledge Bases: Knowledge and Knowledge Sharing, pp. 25–32. IOS Press, Amsterdam (1995)
44. Guarino, N.: Formal Ontology in Information Systems. In: Proc. 1st C. Formal Ontology in Information Systems, Trento, Italy, pp. 3–15 (June 1998)
45. Farias, N., Rodriguez, F.: Models and Tools for Multi-agent Systems Analysis and Design. Research on Computing Science, pp. 94–106 (2003) ISBN: 970-36-0098-0
46. Wooldridge, M., Jennings, N.R., Kinn, D.: The Gaia Methodology for Agent-Oriented Analysis and Design, Autonomous Agents and Multi-Agent Systems. J. Autonomous Agents and Multi-Agent Systems 3(3), 285–312 (2000)
47. Coleman, D., Arnold, P., Bodoff, S., Dollin, C., Gilchrist, H., Hayes, F., Jeremaes, P.: Object-Oriented Development: The Fusion method. Prentice Hall, Hemel (1994)
48. FIPA-IPL: Interaction Protocol Library. Foundation for Intelligent Physical Agents (2001)
49. Brooks, R.A.: How to Build Complete Creatures rather than Isolated Cognitive Simulators. In: VanLehn, K. (ed.) Architectures for Intelligence, pp. 225–239. Lawrence Erlbaum, Mahwah (1991)
50. Georgeff, M.P., Lansky, A.L.: Reactive Reasoning and Planning. In: Proc. C. Applications of Artificial Intelligence, AAAI, Seattle, Washington, pp. 677–682 (July 1987)
51. Bieganski, P., Byrski, A., Kisiel, M.: Multi-agent Platform for Distributed Soft Computing. Int. J. Artificial Intelligence 28, 63–70 (2005)
52. Bellifemine, F., Caire, G., Grenwiid, D.: Development MAS with JADE. Wiley, Chichester (2007)
53. Nwana, H., Ndumu, D., Lee, L., Collins, J.: ZEUS: A Tool-Kit for Building Distributed Multi-Agent Systms. Int. J. Applied Artificial Intelligence 13(1), 129–186 (1999)
54. Peña, A., Sossa, H., Tornes, A.: Ontology Agent based Rule Base Fuzzy Cognitive Maps. In: Nguyen, N.T., Grzech, A., Howlett, R.J., Jain, L.C. (eds.) KES-AMSTA 2007. LNCS (LNAI), vol. 4496, pp. 328–337. Springer, Heidelberg (2007)
55. Rao, A.S., Georgeff, M.P.: A Model-theoretical Approach to the Verification of Situated Reasoning Systems. In: Proc. Int. J. C. Artificial Intelligence IJCAI, France, pp. 318–328 (1993)
56. Manna, Z., Pnueli, A.: A Hierarchy of Temporal Properties. In: Proc. ACM Symposium of Principles of Distributed Computing, Quebec, Canada, pp. 377–410 (August 1990)
57. Farias, N., Ramos, F., Larios, R.: LCIASA: A useful Language for Specification and Verification of Agent-based Systems. In: Proc. Int. C. Principles of Distributed Systems. Reims, France (December 2002)
58. Noy, N., Mc Guinness, D.L.: Ontology Development 101. Knowledge Systems Laboratory, Stanford University (2001)

59. Gomez, A.: Knowledge Sharing and Reuse. In: Liebowitz, J. (ed.) The Handbook of Applied Expert Systems. CRC Press, Boca Raton (1998)
60. Protege: The Protégé Project. Stanford University (2008)
61. Peña, A.: Student Model based on Cognitive Maps. PhD thesis. Centre of Computer Research of the National Polytechnic Institute. Mexico (2008)
62. Peña, A., Sossa, H., Gutiérrez, A.: Causal knowledge and Reasoning by Cognitive Maps: Pursuing a Holistic Approach. Int. J. Expert Systems with Applications (December 2008)
63. Farías, N., Rodríguez, F., Macias, L.G.: The Contract-Business protocol. IEEE, 14ª Reunión de Comunicaciones, Computación, Electrónica y Exposición Industrial. México (2003)
64. Liu, J.: Web Intelligence: What makes Wisdom Web. In: Proc. Int. Joint C. Artificial Intelligence, IJCAI, Acapulco, Mexico (January 2003)

12

An Evolutionary Approach for Intelligent Negotiation Agents in e-Marketplaces

Raymond Y.K. Lau

Department of Information Systems
City University of Hong Kong
Tat Chee Avenue, Kowloon
Hong Kong SAR
raylau@cityu.edu.hk

Abstract. Automated negotiation mechanisms are desirable to enhance the throughput of e-marketplaces. However, existing negotiation mechanisms are weak in supporting real-world negotiations because they assume the availability of complete information about static negotiation spaces where both negotiator's preferences and market conditions remain unchanged. This article illustrates the design and development of evolutionary negotiation agents which are able to learn from and adapt to dynamic negotiation environment. Our experimental results show that the proposed evolutionary negotiation agents outperform a Pareto optimal negotiation mechanism under dynamic negotiation conditions such as the presence of time pressure. These agents can also achieve near optimal negotiation outcomes under dynamic negotiation environment. Our research work opens the door to the development of intelligent negotiation agents to streamline real-world e-marketplaces.

Keywords: Automated Negotiations, Genetic Algorithm, Intelligent Agents, e-Market places.

12.1 Introduction

As our world is constrained by limited resources, negotiations have long been taken as the essential activities in human society. Negotiations are ubiquitous and conducted in various contexts such as the formation of virtual enterprises [19], managing labor dispute [53], resolving hostage crisis [56], streamlining logistic supply chain [54], etc. Negotiation refers to the process by which group of agents (human or software) communicate with one another in order to reach a mutually acceptable agreement on resource allocation (distribution) [34]. This article focuses on the design and development of a multi-agent system to assist human negotiators to make effective negotiation decisions in e-marketplaces. In particular, the decision making models of these autonomous negotiation agents are underpinned by a robust and efficient Genetic Algorithm (GA).

N.T. Nguyen, L.C. Jain (Eds.): Intel. Agents in the Evol. of Web & Appl., SCI 167, pp. 279–301.
springerlink.com

12.1.1 Background

Given the importance of negotiations in various settings, research into negotiation theories and techniques has attracted attention from multiple disciplines such as Distributed Artificial Intelligence (DAI) [25, 27, 53], Social Psychology [2, 44, 45], Game Theory [40, 49, 55], Operational Research [10, 20], and more recently in agent-mediated electronic commerce [19, 58]. Real-world negotiation environments (e.g., negotiating business contracts) are characterized by large negotiation spaces involving multiple parties and dozens of issues (i.e., negotiation attributes). Under such circumstance, even the most experienced human negotiators will be overwhelmed. Consequently, sub-optimal rather than optimal deals are often reached and the phenomenon of "leaving some money on the table" may occur [45]. It is believed that automated negotiation systems are more efficient than human negotiators in combinatorially complex negotiation situations [17, 28, 31, 32, 34].

In typical e-marketplaces [4, 21, 24], buyer agents and supplier agents can exchange their demands or offers of some products with the mediation of the facilitator agents which enforce the trading and negotiation protocols of the particular electronic markets. Software agents are encapsulated computer systems situated in some environments such as the Internet and are capable of flexible, *autonomous* actions in that environment to meet their design objectives [23, 57]. The notion of agency can be applied to build robust architectures of e-marketplaces where a group of software agents communicate and autonomously make negotiation decisions on behalf of their human users. Since software agents can simultaneously evaluate a large number of alternatives in resource allocations (distributions), these agents can help reduce the large cognitive load imposed on human negotiators. It is argued that software agents can incorporate experiential knowledge of past transactions to streamline the effects of volatile demand and supply conditions across multiple e-marketplaces along the electronic supply chain [52].

12.1.2 The Problems

While classical negotiation models [49, 55] provide excellent theoretical analysis of the optimal outcomes (e.g., Nash equilibrium) in limited scenarios (e.g., bilateral negotiations), these normative theories fail to advise the course of actions that a negotiator can follow to reach the optimal outcome. Another concern for the practical use of these normative theories is that the search space of considering all the possible strategies and interactions in order to identify the equilibrium solutions grows exponentially. Although some agent-based negotiation systems have been proposed [3, 9, 12, 26, 28, 36, 38, 39, 59], these systems still suffer from the problems of supporting limited types of negotiation scenarios (e.g., bilateral negotiations) [12, 26, 39, 59] and assuming the availability of preference information of the opponents [3, 26, 28, 38, 59]. Last but not least, a common weakness of existing agent-based negotiation systems is that they assume a static negotiation space where each negotiator's preferences remain the same after a negotiation process begins. In other words, the negotiation agents are not adaptive to the negotiation dynamics (e.g., changing negotiation preferences and market conditions) often encountered in e-marketplaces.

12.1.3 Justifications of the Proposed Approach

GAs have long been taken as heuristic search methods to find optimal or near optimal solutions from large search spaces [16]. In a GA-based approach, a negotiation problem is resolved by heuristically searching for an agreement (i.e., a solution) over a multi-dimensional negotiation space. As a matter of fact, GAs have been successfully applied to develop adaptive information agents [29] and negotiation systems [6, 28, 30, 32, 33]. For most real-world negotiations, only very limited information about the negotiation spaces (e.g., the opponents' preferences) is available. GAs are just able to search for feasible solutions efficiently without requiring detailed structural information about the search space as input. It has been shown that a GA can identify the same optimal solutions as that derived from a game-theoretic model under certain conditions [13, 15]. Since GAs are based on the evolution principle of "natural selection", they are effective in modelling *dynamic negotiation behavior* and other kind of social behavior where good strategies evolve and converge via the evolution process [37, 41].

12.1.4 Contributions of the Article

One of the main contributions of this article is to demonstrate how intelligent agents can be designed and developed to support real-world negotiations conducted over e-marketplaces. In particular, each negotiation agent's independent decision making mechanism is underpinned by a novel genetic algorithm which allows an agent to search for optimal or near-optimal negotiation solutions under uncertain and dynamic negotiation environments. Multi-lateral rather than bi-lateral negotiations are supported by our evolutionary negotiation agents. Moreover, the proposed GA-based negotiation system can be used as a valuable test-bed to examine emergent negotiation behavior and verify existing negotiation theories under dynamic negotiation settings.

12.1.5 Outline of the Article

The remainder of the article is organized as follows. A comparison of previous research work with ours is given in Section 12.2. In Section 12.3, the conceptualization of automated negotiations as distributed multi-objective decision making is highlighted. A base-line automated negotiation model which guarantees Pareto optimum is highlighted in Section 12.4. Section 12.5 illustrates the design and development of the GA based evolutionary negotiation agents. Section 12.6 describes the quantitative evaluation of the GA based evolutionary negotiation agents and reports our experimental results. Finally, we offer concluding remarks and describe future direction of our research work.

12.2 Related Research Work

It has been argued that the challenge of research in negotiation is to develop models that can track the shifting tactics of negotiators [28]. Accordingly, a genetic algorithm based negotiation mechanism is developed to model the dynamic *concession matching behavior* arising in bi-lateral negotiation situations [28]. The main weakness of this negotiation model is that a subjective estimation of the opponent's utility function is required to compute member fitness and to evaluate an incoming counter-offer. These negotiation agents are not adaptive since the negotiators' preferences are assumed static in this system. Our GA-based evolutionary negotiation agents do not assume that their opponents will disclose the private utility functions. Moreover, our negotiation agents can learn and adapt to the opponents' changing preferences as well as the dynamic market factors via a robust GA.

A computational system for bilateral negotiations in which artificial agents are generated by evolutionary algorithm (EA) is reported [15]. For deal evaluation, the utility of an incoming counter-offer will be compared with the user defined negotiation threshold. It is assumed that the utility of every issue as perceived by each negotiation agent is available to a centralized negotiation mechanism for the purpose of utility calculation. Another limitation of such a negotiation system is its applicability to bilateral negotiation setting only. Although an extension of the system to support multi-lateral market with competition is proposed, the outlined multi-lateral negotiation mechanism is still heavily based on the bilateral assumptions. The GA-based adaptive negotiation agents proposed in this article do not assume the availability of the opponents' negotiation preferences nor the artificial negotiation thresholds. Moreover, the system does not force an agent to negotiate with each opponent in order to compute the fitness of a chromosome. Therefore, our method is more time and space efficient. Finally, our negotiation system is truly supporting decentralized multi-lateral negotiations in e-marketplace where market conditions may change from time to time.

Rubenstein-Montano and Malaga have also reported a GA-based negotiation mechanism for searching optimal solutions for multiparty multi-objective negotiations [38]. However, the main problem of their particular GA-based negotiation mechanism is that the preferences (i.e., the utility functions) of all the negotiation parties are assumed available to a centralized negotiation mechanism. Moreover, the preferences of the negotiation parties are assumed unchanged during a negotiation process. Our agent-based evolutionary negotiation system does not assume complete knowledge about a negotiation space. Instead, the GA-based evolutionary negotiation agents can observe and gradually learn the opponents' negotiation preferences. Last but not least, the architecture of our negotiation system is underpinned by a distributed multi-objective decision making model where each negotiation agent employs their individual decision making mechanism to derive appropriate negotiation strategy.

Genetic algorithm has also been applied to learn effective rules to support the negotiation processes [36]. A chromosome represents a negotiation (classification) rule rather than an offer. The fitness of a chromosome (a rule) is measured in terms of how many times the rule has contributed to reach an agreement. In order for the system to determine if an agreement is possible, each negotiator's preferences including the reservation values of the negotiation attributes are assumed available or hypothesized by a centralized negotiation mediator. Our GA-based evolutionary negotiation system does not rely on a centralized negotiation mediator to determine negotiation solutions, instead each negotiation agent can search for the solutions over their individual search space. The emergent negotiation behavior (e.g., a consensus among the group of negotiators) is developed based on the interactions and the individual decisions made by the agents.

A GA has been used to search for the optimal negotiation tactics given a particular negotiation situation (e.g., a predefined amount of negotiation time) [35]. Even though such a theoretical analysis helps identify optimal negotiation parameters, the proposed negotiation model is not applicable to build practical negotiation system because the model is based on a centralized decision making architecture where complete information about each player is available. Similarly, a GA is also used to study the bargaining behavior of boundedly rational agents in a single issue (e.g., price only) bi-lateral negotiation situation [13]. For the negotiation model proposed in this article, a GA is applied to develop the agents' decision making mechanism. In particular, such a decision making mechanism is based on more realistic assumptions (e.g., partial information about a negotiation space) underlying real-world negotiation situations.

Oliver has applied a GA to develop the decision making and the learning mechanisms of adaptive negotiation agents [41, 42, 43]. The particular GA makes use of a binary encoding scheme with each chromosome representing a negotiation threshold. An agent's concession making process is driven by the evolution of the population of chromosomes. Standard genetic operators such as uniform cross-over and mutation are used to evolve the populations. After executing a evolution process (e.g., evolving 20 generations), a chromosome will be selected as the current solution. An incoming offer from the opponent can be evaluated based on the negotiation threshold encoded on the solution. For instance, if an incoming offer produces a utility greater than or equal to the negotiation threshold, the agent will accept the offer. The adaptive agent model is evaluated based on bi-lateral negotiation scenarios. Our work is similar to Oliver's model in the sense that a GA is used to construct the decision making and the learning mechanism of adaptive negotiation agents. However, the fitness function of our GA is developed based on ground principles (e.g., maximizing self payoffs and maximizing the chance of reaching agreements). Moreover, a decimal encoding scheme is applied to the chromosomes which represent potential offers (i.e., solutions). An artificial negotiation threshold is not required for our negotiation agents. Last but not least, our negotiation agents are evaluated under multi-lateral negotiation scenarios.

Fuzzy logic has been applied to develop intelligent negotiation agents in e-Marketplace [5]. Automated negotiation is viewed as a search process in which negotiators jointly search for a mutually acceptable contract in a multidimensional space formed by negotiable issues. In particular, nine pre-defined fuzzy rules are used to generate trade-off for quantitative issues and another nine fuzzy rules are used to generate concession for qualitative issues separately. The proposed negotiation model is somewhat limited since it is developed from the perspective of the supplier agents only. Moreover, users of the system need to provide artificial negotiation thresholds in order for the agents to decide if counter-offers from the opponents can be accepted or not. In fact, the fuzzy negotiation system is not adaptive since it cannot learn and refine the pre-defined fuzzy rules automatically. The GA-based negotiation agents proposed in this article are adaptive since they can learn the opponents' changing preferences based on previous encounters with the opponents. Moreover, our negotiation agents can take into account the dynamic market conditions to develop effective negotiation strategies on the fly.

12.3 Negotiation as Distributed Multi-objective Decision Making

Despite the variety of approaches towards the study of negotiation theory, a negotiation model consists of four main elements: *negotiation protocols*, *negotiation strategies*, *negotiation settings*, and *agent characteristics* [2, 10, 28, 34, 46, 49, 53].

- Negotiation Protocols refer to the set of rules that govern the interactions among negotiators. The rules specify the types of participants (e.g., the existence of a negotiation mediator), the valid states (e.g., waiting for bid submission, negotiation closed), the actions that cause negotiation state changes (e.g., accepting an offer or terminating a negotiation session).
- Negotiation Strategies refer to the heuristic or rules the agents employ to act in line with the negotiation protocol and the negotiation setting to achieve their objectives. For instance, negotiators should set stringent goals initially and concede first on issue of lesser importance to achieve higher payoffs in a competitive environment [2, 45].
- Agent Characteristics refer to the characteristics of the participants (agents). These characteristics include an agent's attitude (e.g., self-interested vs. benevolent), cognitive limitations (e.g., omniscient agent vs. memoryless agent), goal setting, initial offer magnitude, knowledge and experience (e.g., knowledge about the opponents), learning and adaptation capabilities, etc.
- Negotiation Settings refer to factors that are relevant to problem domain. These factors include the number of negotiation issues, number of parties, the presence or absence of time constraints, nature of domain such as purely competitive versus cooperative negotiations, and market conditions, etc.

Each agent (human or software) is characterized by its own negotiation preferences, attitudes, and negotiation strategies. These agents interact according to a pre-defined negotiation protocol (e.g., the alternate offering protocol [60]) within a particular negotiation setting (e.g., an e-Marketplace). Competition exists in an e-marketplace if two or more buyers (sellers) try to acquire (distribute) the same products or services. A *negotiation mechanism* refers to a particular negotiation protocol and the negotiation strategies (i.e., the negotiation heuristic) that operates under the protocol [34]. In other words, the negotiation mechanism is the decision making model of a negotiation agent. Generally speaking, a centralized negotiation mechanism (i.e., a centralized decision making model) is ineffective to model the real-world negotiation scenarios. The reason is that each negotiator is characterized by its own preferences and employs individual strategies to search for negotiation solutions. Figure 12.1 demonstrates one of the the client interfaces for the human buyer to specify their valuation functions. As real-world negotiations often involve multiple issues such as price, quantity ordered, shipment time, etc., an agent aims at optimizing multiple objectives at the same time and needs to make issue trade-off if it is necessary. Therefore, automated negotiations should be modeled as distributed multi-objective decision making processes.

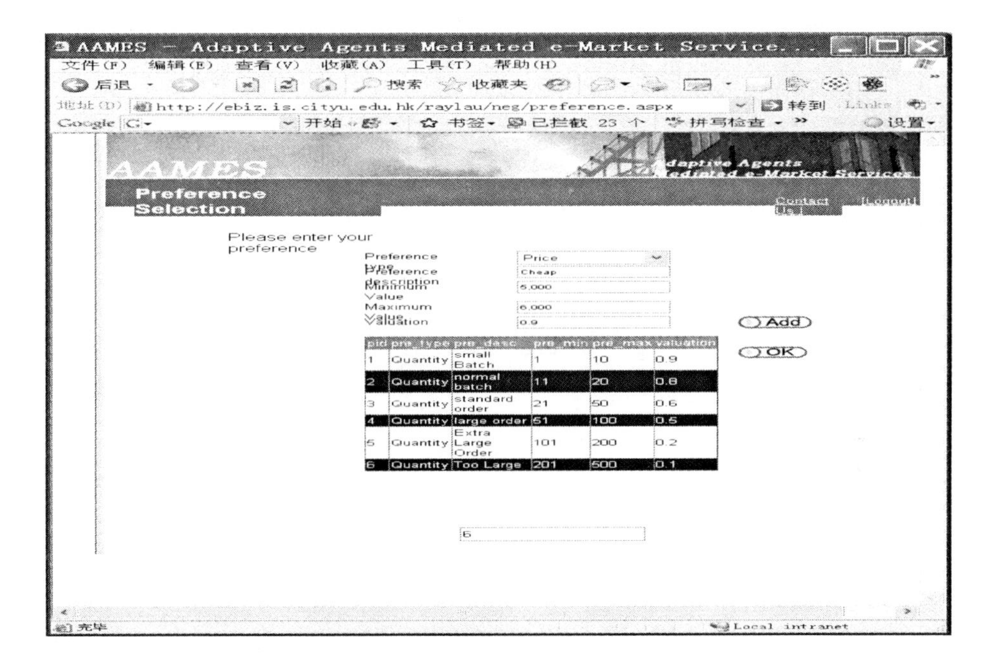

Fig. 12.1. Defining Valuation Functions Via a Client Interface

The *emergent* negotiation behavior (i.e., the distributed multi-objective decision making processes) of a group of agents is revealed due to the interactions and decisions made by the individual negotiator. Emergent behavior is the overall behavior that results from lower level factors such as individuals and their interactions [28]. It is based on the view that the nonlinear and dynamic aspects of organizational behavior are in fact the result of the interaction of the *adaptive decision makers*. To this end, the notion of autonomous software agents [22] provides an effective and robust approach for modeling the emergent behavior of a group of negotiators situated in a complex negotiation environment such as an e-marketplace. Figure 12.2 depicts an administration interface for the creation of e-marketplaces in our current prototype system. Each negotiation agent is a subrogate of a human negotiator in the e-marketplace. These agents can autonomously search for potential trading partners and negotiate with these partners according to the preferences specified by the human negotiators.

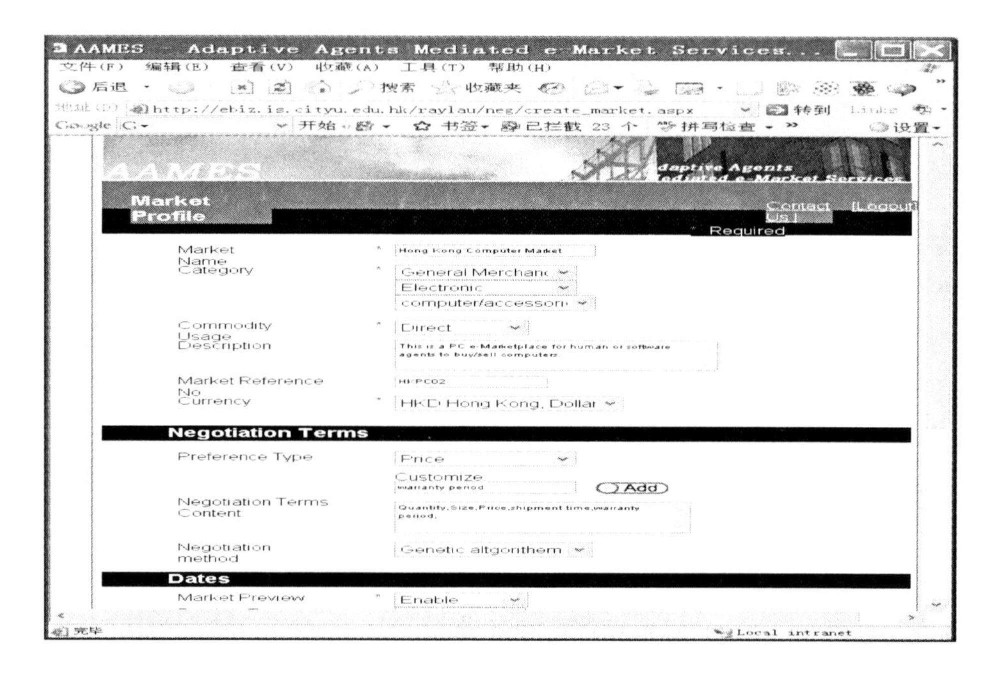

Fig. 12.2. An Administration Interface for e-Marketplace Creation

12.4 A Pareto Optimal Negotiation Mechanism

A negotiation space $Neg = \langle P, A, D, U, T \rangle$ is a 5-tuple which consists of a finite set of negotiation parties (agents) P, a set of attributes (i.e., negotiation issues) A understood by all the parties $p \in P$, a set of attribute domains D for A, and a set of utility functions U with each function $U_p^o \in U$ for an agent $p \in P$. An attribute domain is denoted D_{a_i} where $D_{a_i} \in D$ and $a_i \in A$. A utility function pertaining to an agent p is defined by: $U_p^o : D_{a_1} \times D_{a_2} \times \ldots \times D_{a_n} \mapsto [0, 1]$. Each agent p has a deadline $t_p^d \in T$. It is assumed that information about P, A, D is exchanged among the negotiation parties during the *ontology sharing* stage before negotiation actually takes place. A *multi-lateral* negotiation situation can be modelled as many one-to-one *bi-lateral* negotiations where an agent p maintains a separate negotiation dialog with each opponent. In a negotiation round, the agent will make an offer to each of its opponents in turn, and consider the most favorable counter-offer from among the set of incoming offers evaluated according to its own payoff function U_p^o. A sequential alternate-offer negotiation protocol is employed by our negotiation system in which negotiation proceeds in a discrete series of rounds. In each round, an agent can either make an offer, propose counter-offer, accept an offer, or withdraw from negotiation.

An *offer* $\overrightarrow{o} = \langle d_{a_1}, d_{a_2}, \ldots, d_{a_n} \rangle$ is a n-tuple of attribute values (intervals) pertaining to a finite set of attributes $A = \{a_1, a_2, \ldots, a_n\}$. An offer can also be viewed as a vector of attribute values in a geometric negotiation space with each dimension representing a negotiation issue. Each attribute a_i takes its value from the corresponding domain D_{a_i}. Generally speaking, a finite set of candidate offers O_p acceptable to an agent p (i.e., satisfying its hard constraints) is constructed via the Cartesian product $D_{a_1} \times D_{a_2} \times \cdots \times D_{a_n}$. As human agents tend to specify their preferences in terms of a range of values, a more general representation of an offer is a tuple of attribute value intervals such as $o_i = \langle 20 - 30(\$), 1 - 2(years), 10 - 30(days), 100 - 500(units) \rangle$.

Preference representation is concerned about rating a set of potential offers according to an agent's specific negotiation interests. The *valuations* of individual attributes and attribute values (intervals) are defined by the valuation functions $U_p^A : A \mapsto [0,1]$ and $U_p^{D_{a_i}} : D_{a_i} \mapsto [0,1]$ respectively, whereas U_p^A is an agent p's *valuation function* for each attribute $a_i \in A$, and $U_p^{D_{a_i}}$ is an agent p's valuation function for each attribute value $d_{a_i} \in D_{a_i}$. In addition, the valuations of attributes are assumed normalized, that is, $\sum_{a_i \in A} U_p^A(a_i) = 1$. One common way to quantify an agent's preference (i.e., the utility function U_p^o) for an offer o is by a linear aggregation of the *valuations* [1, 15, 50]: $U_p^o(o) = \sum_{a_i \in A, d_{a_i} \in o} U_p^A(a_i) \times U_p^{D_{a_i}}(d_{a_i})$.

If an agent's initial proposal is rejected by its opponent, it needs to propose an alternative offer with the least utility decrement (i.e., computing a concession). An agent will maintain a set $O_p^{'}$ which contains the offers it has proposed before. In a negotiation round, an alternative offer with a concession can be determined based on $\exists_{o_{counter} \in \{O_p - O_p^{'}\}} \forall_{o_x \in \{O_p - O_p^{'}\}} : [o_x \preceq_p o_{counter}]$, where $o_x \preceq_p o_y$ denotes that an offer o_y is more preferable than another offer o_x. The preference relation $\preceq_p$ is a total ordering induced by an agent p's utility function U_p^o over the set of feasible offers O_p. In each negotiation round, an alternative offer with concession is picked up from the top of the list ranked by $(\preceq_p, \{O_p - O_p^{'}\})$.

When an incoming offer o is received from an opponent, an agent p first evaluates if $o \in O_p$ is true (i.e., the offer satisfying all its hard constraints). To do this, an *equivalent offer* $o_{\simeq}$ will be computed. $o_{\simeq}$ represents agent p's interpretation about the opponent's proposal o. Once $o_{\simeq}$ for o is computed, acceptance of the incoming offer o can be determined with respect to p's own preference $(\preceq_p, O_p)$. An offer $o_{\simeq} \in O_p$ is equivalent to o iff every attribute interval of $o_{\simeq}$ *intersects* each corresponding attribute interval of o. Formally, any two attribute intervals d_x, d_y intersect if the intersection of the corresponding sets of points is not empty (i.e., $d_x \cap d_y \neq \emptyset$). The acceptance criteria for an incoming offer o (i.e., the equivalent $o_{\simeq}$) is defined by:

(1) If $\forall_{o_x \in O_p}\ o_x \preceq_p o_{\simeq}$, an agent p should accept o since it produces the maximal payoff.

(2) If $o_{\simeq} \in O_p^{'}$ is true, an agent p should accept o because $o_{\simeq}$ is one of its previous proposals.

It has been shown that if each participating agent $p \in P$ employs their preference ordering $(\preceq_p, O_p)$ to compute concessions and uses the offer acceptability criteria described above to evaluate incoming offers, *Pareto optimum* is always found if it exists in a negotiation space [1]. The advantages of this Pareto optimal negotiation model is that the preference information of the opponents is not assumed public information and artificial negotiation thresholds are not required to evaluate incoming deals.

A solution is Pareto optimum if it is impossible to find another solution such that at least one agent will be better off but no agent will be worse off [47]. It should be noted that Pareto optimum does not necessarily lead to maximal joint payoff. Joint payoff simply refers to the sum of each agent's payoff obtained at the end of a negotiation session. However, one of the problems of this optimal negotiation model is that it may take a long time to sequentially evaluate all the candidate offers before a *Pareto optimal* solution is found. Given a high dimensional multi-issue negotiation space often encountered for business negotiations over e-marketplaces, this kind of sequential search may not be feasible.

12.5 Evolutionary Negotiation Agents

The decision making models of the GA-based evolutionary negotiation agents are developed based on the basic intuition that negotiators tend to maximize their individual payoffs while ensuring that an agreement is reached [12, 28]. A population of chromosomes is used to represent a subset of feasible offers for each negotiation agent. Since an agent knows its own utility function, an offer o_{max} representing the offer with the maximal payoff can be identified. In addition, the offer $o_{opponent}$ represents the opponent's recent counter-offer. According to the basic intuition in negotiation, a feasible offer is considered *fit* if it is close to o_{max} and $o_{opponent}$. The distance from a feasible offer o_i to o_{max} and $o_{opponent}$ can be estimated based on standard distance function such as the *Weighted Euclidean distance* [8]. Evaluation of offers can be analyzed with respect to a geometric negotiation space as depicted in Figure 12.3. In Figure 12.3, offers $\overrightarrow{o}_a, \overrightarrow{o}_b \ldots, \overrightarrow{o}_n$ are the subset of feasible offers under consideration by a particular agent. In each negotiation round, the offer vectors $\overrightarrow{o}_{opponent}$ and $\overrightarrow{o}_{max}$ may change, and so are the offers considered fit by the agent. In other words, an agent is learning and adapting to the opponent's preferences gradually. For multi-lateral negotiations, the best incoming offer evaluated based on an agent's private utility function is taken as the $o_{opponent}$. Thereby, each agent only needs to maintain one population of chromosomes representing a set of relatively good offers.

Although one may consider employing an exhaustive search to identify negotiation solutions based on some criteria (e.g., o_{max} and $o_{opponent}$), such an

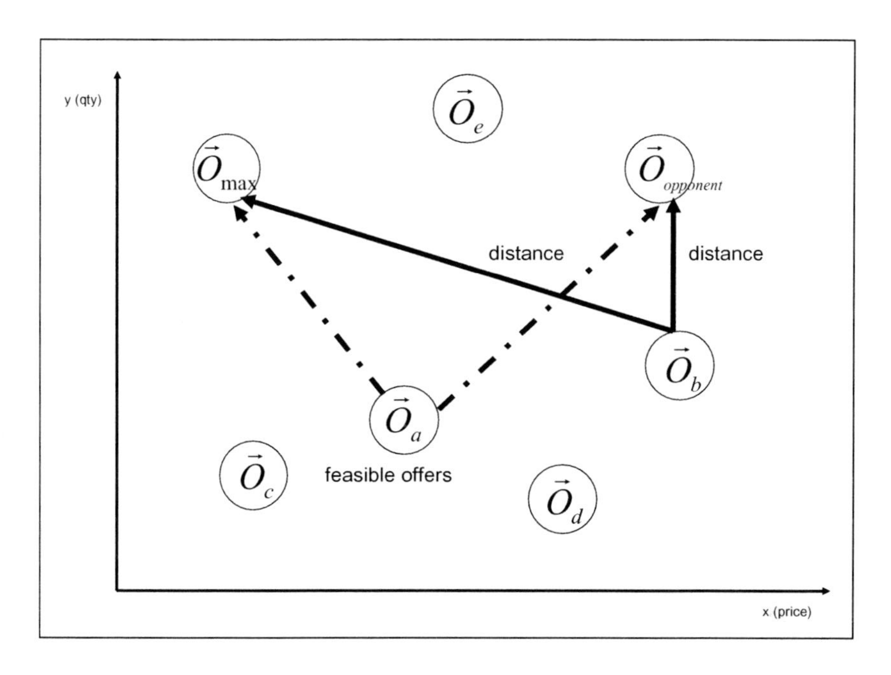

Fig. 12.3. A Geometric Negotiation Space

approach is not scalable to support real-time negotiations in e-marketplaces which may involve dozens of issues. This is the primary reason of employing the proposed GA (a heuristic search approach) to develop near optimal solutions. Although a Pareto optimal negotiation mechanism which does not assume the availability of complete information about negotiation spaces is presented in Section 12.4, such a mechanism is less applicable to e-marketplaces where the preferences of participating negotiators and the market conditions may change over time. As a result, GA-based evolutionary negotiation agents, which tend to produce near optimal negotiation outcomes under dynamic environment, seem more suitable for supporting automated negotiations in e-marketplaces. The evaluation of our evolutionary negotiation agents in simulated e-marketplaces will be presented in Section 12.6.

12.5.1 Encoding Offers in GA

Each GA-based negotiation agent p utilizes a population of chromosomes to represent a subset of feasible offers $O_p^{feas} \subseteq O_p$. Each chromosome consists of a fixed number of fields. The first field uniquely identifies a chromosome, and the second field is used to hold the fitness value of the chromosome. The other fields (genes) represent the attribute values of a candidate offer. Figure 12.4 depicts the decimal encoding (Genotype) of a chromosome (i.e., a potential offer). The genetic operators such as crossover and mutation are only applied to the genes representing the attribute values of an offer.

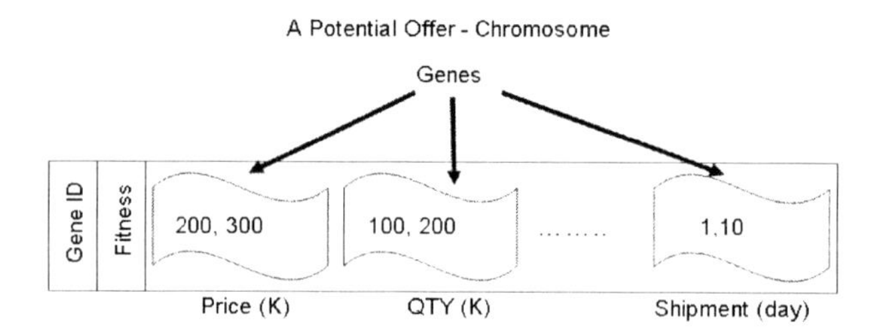

Fig. 12.4. Encoding Candidate Offers

12.5.2 The Fitness Function

One of the most important elements of a genetic algorithm is the fitness function which drives the evolution of a population of potential solutions. Ideally, the fitness function for automated negotiation should reflect the *joint payoff* of each candidate offer. Unfortunately, the utility functions of the opponents are normally not available for real-world applications. Therefore, the proposed fitness function approximates the ideal function and captures some important issues such as an agent's own payoff, the opponent's *partial preference* (e.g., the recent counter offer), and time pressure. In addition, for intelligent negotiation agents which are sensitive to the trading conditions of an e-marketplace, opportunities in trading (e.g., the number of trading partners available) and the level of competition (e.g., the number of agents competing to buy or sell the same products) will also influence the agents' concession decisions [51]. Based on aforementioned factors, the fitness function of our genetic algorithm is defined as follows:

$$\begin{aligned} fitness(o) &= \omega \times \frac{U_p^o(o)}{U_p^o(o_{max})} + \\ (1-\omega) &\times \left(1 - \frac{dist(\vec{o}, \vec{o}_{opponent})}{MaxDist(|A|)}\right) \end{aligned} \tag{12.1}$$

where $fitness(o) \in [0,1]$ defines the fitness value of an arbitrary offer o; the term o_{max} represents an offer which produces the maximal payoff based on an agent p's current utility function; $\vec{o}_{opponent}$ is the offer vector representing the recent counter-offer proposed by the opponent. The term $MaxDist(|A|)$ represents the maximal distance of a geometric negotiation space. It can be derived from the number of dimensions $|A|$ if each dimension (attribute) is normalized in the unit interval. In the very first negotiation round, the agents will use $MaxDist(|A|)$ to replace the actual distance $dist(\vec{o}, \vec{o}_{opponent})$ in Eq.(12.1) if an incoming offer has not been received and the opponent is new to the e-marketplace. This is a conservative estimation which assumes that the agents' interests are in total conflict before negotiations begin.

The parameter ω is the weight factor to control the relative importance of optimizing an agent's own utility or reaching an agreement (e.g., by considering the opponent's recent offer). The weight factor ω in Eq.(12.1) is defined according to Eq.(12.2) as follows:

$$\omega = \min(\alpha^{\omega_A} \times (\frac{|P_{partner}|}{|P_{competitor}|})^{\omega_M} \times (Time(t))^{\omega_T}, 1) \tag{12.2}$$

where the parameter $\alpha \in [0, 1]$ is used to model a wide spectrum of agent trading attitudes, from fully self-interested or aggressive ($\alpha = 1$) to fully benevolent ($\alpha = 0$). A weight factor $\omega_A \in \mathbb{N}$, where $\mathbb{N}$ is the set of natural numbers, is introduced to enable tuning the degree of influence of an agent's attitude on determining the tradeoff between maximizing the agent's own payoff and maximizing the chance of reaching an agreement. The market oriented factors are incorporated into the fitness function according to the ratio $\frac{|P_{partner}|}{|P_{competitor}|}$, where $P_{partner}$ denotes the set of trading partners and $P_{competitor}$ denotes the set of competitors in an e-marketplace respectively. In other words, if there are more trading partners available in an e-marketplace, the agent's attitude will tend to make use of such opportunities to achieve more self payoff. On the contrary, if the competition is keen (i.e., $\frac{|P_{partner}|}{|P_{competitor}|} \ll 1$), the agent's attitude will change to be more concerning about reaching an agreement [7]. The weight factor $\omega_M \in \mathbb{N}$ is used to control the degree of influence of the market conditions on the agent's negotiation attitude. In practice, information about the market condition (e.g., number of partners and competitors) is provided by the facilitator agent of an e-marketplace. Finally, the time pressure $Time(t)$ of the negotiation process will also be taken into account by the agent. Similarly, the weight factor $\omega_T \in \mathbb{N}$ is used to tune the degree of influence of the time pressure of negotiation on the agent's attitude.

When the negotiation deadline is approaching, an agent is more likely to concede in order to make a deal [18, 48]. However, different agents may have different attitudes towards deadlines. An agent may be eager to reach a deal and so it will concede quickly (*Conceder agent*). On the other hand, an agent may not give ground easily during negotiation (*Boulware agent*) [45]. Therefore, a time pressure function $Time$ is developed to approximate a wide spectrum of agents' concession attitude. Our $Time$ function is similar to the negotiation decision function referred to in the literature [11, 14].

$$Time(t) = 1 - (\frac{\min(t, t_p^d)}{t_p^d})^{\frac{1}{CA_p}} \tag{12.3}$$

where $Time(t)$ denotes the time pressure given the time t represented by the absolute time or the number of negotiation rounds; t_p^d indicates the deadline for an agent p and it is either expressed as absolute time or the maximum number of rounds allowed. The term CA_p is used to model the "concession attitude" of the agent p. An agent p will demonstrate *Boulware* behavior if the concession attitude $0 < CA_p < 1$ is set. On the other hand, an agent will exhibit *Conceder* behavior if the concession attitude $CA_p > 1$ is specified. If $CA_p = 1$ is defined, the agent holds *Neutral* attitude towards concession. The values of the $Time$ function are bounded by the unit interval $[0, 1]$. The agent concession attitude CA_p can be chosen by the human negotiator or else a system default will be applied when a negotiation process begins.

The distance between two offer vectors $dist(\vec{o}_x, \vec{o}_y)$ is defined by the weighted Euclidean distances [8]:

$$dist(\vec{o}_x, \vec{o}_y) = \sqrt{\sum_{i=1}^{|A|} w_i (d_i^x - d_i^y)^2} \quad (12.4)$$

where the weight factor $w_i = U_p^A(a_i)$ is an agent's valuation for a particular attribute $a_i \in A$. For our implementation, $\sum_{i=1}^{|A|} w_i = 1$ is enforced. An offer vector $\vec{o}_x$ contains an attribute value d_i^x along the ith dimension (issue) in a negotiation space. If an attribute interval instead of a single value is specified for an offer, the mid-point of an attribute interval is first computed. Each dimension (attribute) is scaled in the unit interval by linear scaling: $d_i^{\text{scaled}} = \frac{d_i - d_i^{min}}{d_i^{max} - d_i^{min}}$, where d_i^{min} and d_i^{max} represent the minimal and the maximal values for a domain D_{a_i}.

12.5.3 The Genetic Algorithm

The decision making model of each evolutionary negotiation agent is underpinned by the genetic algorithm depicted in Figure 12.5. The initial population P^0 is created by incorporating the member o_{max} that maximizes an agent p's payoff in the first round, and by randomly selecting the $N-1$ members from the candidate set O_p, where N is the pre-defined population size. At the beginning of every evolution process, the fitness value of each chromosome is computed according to the fitness function defined in Eq.(12.1). In other words, the fitness of a chromosome (i.e., a potential solution) is evaluated based on the most current negotiation parameters (e.g., an agent's trade-off attitude, market conditions, the opponent's counter-offer, etc.). A *Elitism* factor [16] is incorporated into the genetic algorithm to make a better balance between the functions of convergence and exploration of the proposed GA. By employing the elitism factor, certain percentage of fittest chromosomes from the current generation P^i will be transferred to the new generation P^{i+1}. *Tournament selection* [35] is employed in our GA to select chromosomes from the current generation to the mating pool MP. For tournament selection, a group of k members are selected from a population to form a tournament. The member with the highest fitness among the selected k members is placed in the mating pool. This procedure is repeated n times until the mating pool is full.

Standard genetic operators: *cloning*, *crossover*, *mutation* are applied to the mating pool to create new members according to pre-defined probabilities. These operations continue until the new generation of size N is created. An evolution cycle is triggered after every x negotiation round(s), where x is the evolution frequency defined by the human negotiator. There is another parameter $MaxE$, the maximum number of evolutions, to control the number of evolutions to be performed in an evolution cycle. When an evolution cycle is completed, the top t chromosomes (candidate offers) with the highest fitness are selected from the final population to build the solution set S. A stochastic function is then applied

Input:
Accept input parameters: U_p^o, N, $MaxE$, $Elitism$, $Clone$, $Cross$, $Mutate$
Output:
The last generation of chromosomes P^{MaxE}
The main process:

1. Set the generation number i to zero
2. Create the first population P^i which consists of o_{max} and $N-1$ individuals randomly selected from the set $O_p = D_{a1} \times D_{a2} \times \cdots \times D_{an}$
3. If the generation number i is less than or equal to the maximum number of generation $MaxE$, repeat the following steps:
   a) Develop the new generation P^{i+1} by extracting a certain percentage of best members from generation P^i) according to the elitism rate defined by $Elitism$
   b) Construct the mating pool MP by executing the tournament selection function
   c) Repeat the genetic operations until the size of the new generation P^{i+1} equals N
      i. Perform two-point crossover operation in the mating pool MP according to the pre-defined crossover rate $Cross$
      ii. Perform mutation operation in the mating pool MP according to the pre-defined mutation rate $Mutate$
      iii. Perform cloning operation in the mating pool MP according to the pre-defined cloning rate $Clone$
      iv. Add any the newly generated members to the new population P^{i+1}
   d) Increase the generation number i by 1

Fig. 12.5. The Genetic Algorithm for Evolutionary Negotiation Agents

to select a member from the solution set to form the current offer. Both of the parameters t and S are chosen by the human negotiator. After an offer with concession is computed, the agent's decision for an incoming offer can also be developed. If the incoming counter-offer produces a payoff greater than or equal to that of the current offer, a rational agent should accept the incoming offer; otherwise the incoming counter-offer should be rejected.

12.6 The Experiments

The simulated negotiation environment of our experiments was characterized by multi-lateral negotiations among two buyer agents (e.g., B1, B2) and two seller agents (e.g., S1, S2). These agents negotiated over some virtual services or products described by multiple attributes (i.e., issues) with each attribute domain containing discrete number of values represented by some natural numbers. The valuation of an attribute or a discrete attribute value fell in the unit interval of $(0, 1]$. Each negotiation case was characterized by the valuation functions U_p^A, $U_p^{D_{a_1}}, U_p^{D_{a_2}}, \ldots, U_p^{D_{a_n}}$, etc. for each agent p participating in the negotiation process. Each buyer (seller) participating in a negotiation process was assumed to have a lot of products to buy (sell). For each simulated negotiation case, an agreement zone always exists since the difference between a buyer and a seller only lies on their valuations against the same set of negotiation issues (e.g., attributes and attribute values).

As described in Section 12.4, the alternate offering protocol was adopted in our e-marketplace. At the beginning of every negotiation round, each agent would execute its genetic algorithm to generate an offer for that round. The order of out-going offer deliberation among the agents was randomly chosen by the market facilitator agent in each negotiation round. At the message exchange phase, an agent sent an offer message to each of its opponents (e.g., $S1 \rightarrow B1$ and $S1 \rightarrow B2$) in turn. After the message exchange phase, the market facilitator agent randomly assigned a sequence of trading agent invocation such as $< B2, B1, S1, S2 >$. Then, the trading agents would be activated in turn according to that particular order to evaluate their incoming offers. For instance, with reference to the above order, agent $B2$ would evaluate its incoming offers sent from $S1$ and $S2$, then agent $B1$ would evaluate its incoming offers. Each agent selected the best incoming offer (evaluated according to its private utility function) as the opponent offer $o_{opponent}$ in a negotiation round. If there was a tie, an opponent would be selected randomly by an agent. If an agreement was made between a pair, they would be removed from the e-marketplace by the market facilitator agent, and the remaining agents would continue their negotiations until either an agreement was made, the deadline was due, or no more trading partner was available.

12.6.1 Experiment One

In our first experiment, a symmetric market where 2 buyers and 2 sellers negotiated over some commercial products was constructed. The particular negotiation space was characterized by five attributes (i.e., $|A| = 5$) with each attribute domain containing 5 discrete values represented by the natural numbers $D_a = \{1, 2, 3, 4, 5\}$. For each agent p, the size of the feasible offer set is: $|O_p| = 5^5 = 3125$. Six negotiation groups were constructed with each group containing five negotiation cases (i.e., totally $6 \times 5 = 30$ simulated e-marketplaces). Each negotiation case was defined based on the respective valuation functions U_p^A, $U_p^{D_{a_1}}, U_p^{D_{a_2}}, \ldots, U_p^{D_{a_5}}$ for each agent $p \in P$ participating in the e-market. For the first negotiation group, buyers and sellers had exactly the same utility functions (i.e., no conflict). For each succeeding group, buyers and sellers were characterized by having common weighting from one (small conflict group) to five attributes (highest conflict group) respectively. Nevertheless, opposing valuations of attribute values were assigned to the buyers and the sellers. The control group consisted of the Pareto optimal negotiation agents developed according

Table 12.1. Comparison Between (GA) and (PO) Agents in e-Marketplaces Without Time Pressure

	Pareto Optimal (PO)		Genetic Algorithm (GA)	
Group	Avg. Joint-Util.	Avg. Time	Avg. Joint-Util.	Avg. Time
1	2.83	1.0	2.83	1.0
2	2.66	481.0	2.47	263.6
3	2.59	651.5	2.42	288.0
4	2.37	852.0	2.21	368.0
5	2.23	932.2	2.12	375.4
6	1.98	1212.0	1.86	464.0
Mean	2.44	688.28	2.32	293.33

to the mechanism described in Section 12.4. These agents could found Pareto optimal solutions when time constraint was not present. After running all the negotiation cases, the average joint-payoffs and the average negotiation time (in terms of rounds) were recorded for each negotiation group. For the first simulation run, no negotiation deadline was imposed for the agents. The performance of these agents was depicted under the (PO) columns in Table 12.1.

The experimental group comprised of the same number of evolutionary negotiation agents (GA) developed according to the algorithm illustrated in Section 12.5. The same set of negotiation cases was applied to the (GA) agents. Since the GA-based negotiation agents were non-deterministic, each negotiation case was executed 10 times to derive the average payoff and execution time for each agent. Then, the average figures over five negotiation cases in a particular group was computed. The performance of these agents was depicted under the (GA) columns in Table 12.1. The genetic parameters used by the (GA) system were: population size = 80, mating pool size = 40, size of solution set = 1, elitism factor = 10%, tournament size = 3, cloning rate = 0.1, crossover rate = 0.6, mutation rate = 0.05, number of evolutions per cycle = 20, and evolution frequency = 1 (i.e., one evolution cycle for every negotiation round). The agent trading attitude was $\alpha = 0.5$ for both the buyer and the seller agents. The weight factors $\omega_A = \omega_M = \omega_T = 1$ were specified for each agent. In addition, the concession attitude of an agent was set to $CA_p = 1$ (i.e., neutral) for each participating agent.

From Table 12.1, it is shown that the GA-based negotiation agents can find agreements faster than the Pareto optimal negotiation agents for all the negotiation groups where conflicts exist between the buyers and the sellers. The first negotiation group is a reference group. As the agents had exactly the same preferences in each case, they could find agreements in the first negotiation round. It is shown that both the GA-based agents and the PO-based agents behaved the same under such a special circumstance. The efficiency of the (GA) agents is compared with that of the (PO) agents by testing the hypothesis $H_{null} : \mu_{GA} - \mu_{PO} = 0$ and $H_{alternative} : \mu_{GA} - \mu_{PO} > 0$, whereas μ_{GA} is the overall mean of the negotiation time consumed by the (GA) agents and μ_{PO} is the overall mean of the negotiation time consumed by the (PO) agents. With paired one tail t-test, the null hypothesis was rejected ($t(29) = -8.67, p < .01$). Therefore, we can conclude that GA-based negotiation agents are more efficient than the PO-based agents in the simulated e-Marketplaces. The average improvement in terms of negotiation efficiency is 57.4%. In terms of negotiation effectiveness, the overall average joint-payoff of the GA-based agents is only 5.1% less than that of the PO-based agents. Based on this experiment, it is shown that GA-based negotiation agents could achieve near optimal negotiation performance when compared with their Pareto optimal counterparts.

12.6.2 Experiment Two

As real-world negotiation situations are often dynamic (e.g., the preferences of negotiators may change over time), this experiment tries to evaluate the performance of the (GA) agents under dynamic negotiation environment. Five negotiation cases of the forth negotiation group used in experiment one were modified to create bi-lateral negotiation scenarios. In each case, random changes (e.g., valuations of some attribute values) were applied at 50th and 125th negotiation round respectively. The agents' preferences were restored to their original values at 200th negotiation round. Then, the joint payoff of the agents was compared with that obtained at the Pareto optimal point for each case. After making the first lot of changes at 50th round, the (GA) agents were forced not to accept a deal until their preferences were restored at 200th negotiation round. This procedure facilitated us to observe the adaptation processes of the agents. No deadline was imposed in this experiment.

Table 12.2. Performance of GA-based Agents' Under Dynamic Negotiation Environment

Case	Pareto Optimum	GA-based Agents	Distance from PO
1	1.15	1.05	0.48
2	1.16	1.09	0.33
3	1.21	1.12	0.45
4	1.18	1.13	0.23
5	1.20	1.16	0.20
Mean	1.18	1.11	0.34

Table 12.2 depicts the comparison between the joint payoff achieved by the GA-based agents and the Pareto optimum for each negotiation case. According to our experiment, the GA-based agents can achieve near optimal outcomes under dynamic negotiation environment. Based on Euclidean distance [8], the average absolute distance from the solutions identified by the GA-based agents to the Pareto optimum is 0.34. As the maximal distance of our negotiation space is $d_{max} = \sqrt{4^2 + 4^2 + 4^2 + 4^2 + 4^2} = 8.944$, the relative distance of these

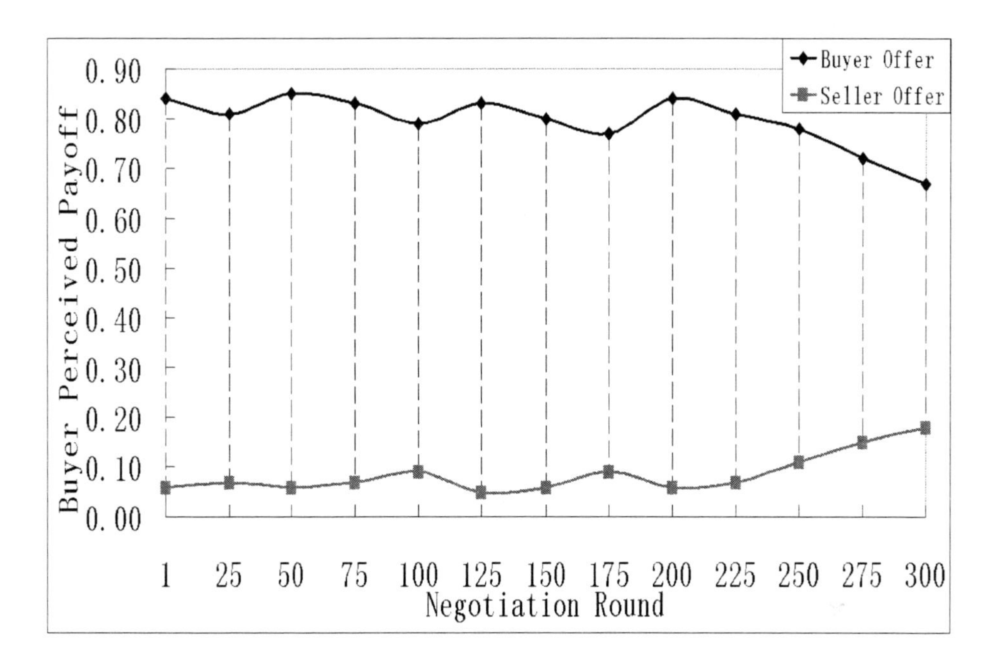

Fig. 12.6. Distance Between the Offers Under Dynamic Negotiation Environment

solutions with respect to the Pareto optimum is 3.7% only. On the other hand, the average joint payoff is only 5.7% less than the Pareto optimum. Figure 12.6 further shows the distances (from the perspective of the buyer agent) between the offers proposed by the buyer agent and the offers proposed by the seller agent. These distances were computed according to Eq.(12.4). The decreasing distances between the offers after 50th and 125th round respectively indicate that the GA-based agents can adapt to each other's preferential changes after some rounds of negotiations. In other words, the GA-based learning mechanism is effective. Similarly, the agents can learn and adapt to the preferential changes of their opponents after the third change point (the 200th round). This is evidenced by the smaller distance between the offers made by the buyer and the seller. From the buyer's point of view, the seller's counter-offers are becoming more attractive (a higher utility) because the seller has gradually learn the buyer's preferences. Similarly, the buyer also adapts to the seller's changing preferences by conceding more (a lower utility) after the change points.

12.7 Conclusions

Evolutionary negotiation agents are promising in supporting real-world negotiations in e-marketplaces because they can perform multi-party multi-issue negotiations based on a parallel and distributed decision making model. These agents can deliberate near optimal negotiation solutions with incomplete and uncertain information about dynamic negotiation spaces. Our experiments shows that the evolutionary negotiation agents outperform a classical negotiation mechanism which guarantees Pareto optimum under dynamic negotiation conditions such as the presence of time pressure. In addition, these agents can also achieve near optimal negotiation outcomes in dynamic negotiation environment. Our research work opens the door to the application of intelligent agents to streamline real-world e-marketplaces. Future work includes the enhancement of our existing genetic algorithm by exploring adaptive genetic operators. Moreover, the interaction effects among various agent attitudes and market conditions will be further examined based on real business negotiation scenarios.

References

1. Barbuceanu, M., Lo, W.K.: Multi-attribute utility theoretic negotiation for electronic commerce. In: Dignum, F.P.M., Cortés, U. (eds.) AMEC 2000. LNCS (LNAI), vol. 2003, pp. 15–30. Springer, Heidelberg (2001)
2. Bazerman, M., Magliozzi, T., Neale, M.: Integrative bargaining in a competitive market. Organization Behavior and Human Decision Processes 35, 294–313 (1985)
3. Bui, H.H., Venkatesh, S., Kieronska, D.: Learning other agents' preferences in multiagent negotiation using the bayesian classifier. International Journal of Cooperative Information Systems 8(4), 275–293 (1999)
4. Chavez, A., Maes, P.: Kasbah: An agent marketplace for buying and selling goods. In: First International Conference on the Practical Application of Intelligent Agents and Multi-Agent Technology, pp. 75–90 (1996)
5. Cheng, C.B., Chan, C.C., Lin, K.C.: Intelligent agents for e-marketplace: Negotiation with issue trade-offs by fuzzy inference systems. Decision Support Systems 42(2), 626–638 (2006) (in press)
6. Choi, S.P.M., Liu, J., Chan, S.-P.: A genetic agent-based negotiation system. Computer Networks 37(2), 195–204 (2001)
7. Cross, J.: The Economics of Bargaining. Basic, New York (1969)
8. Duda, R., Hart, P.: Pattern Classification and Scene Analysis. John Wiley & Sons, New York (1973)
9. Dumas, M., Governatori, G., ter Hofstede, A.H.M., Oaks, P.: A formal approach to negotiating agents development. Electronic Commerce Research and Applications 1(2), 29–50 (2002)
10. Ehtamo, H., Ketteunen, E., Hämäläinen, R.: Searching for joint gains in multiparty negotiations. European Journal of Operational Research 130, 54–69 (2001)
11. Faratin, P., Sierra, C., Jennings, N.R.: Negotiation decision functions for autonomous agents. Journal of Robotics and Autonomous Systems 24(3-4), 159–182 (1998)
12. Faratin, P., Sierra, C., Jennings, N.R.: Using similarity criteria to make issue trade-offs in automated negotiations. Artificial Intelligence 142(2), 205–237 (2002)
13. Fatima, S., Wooldridge, M., Jennings, N.R.: Comparing Equilibria for Game-Theoretic and Evolutionary Bargaining Models. In: Proceedings of the International Workshop on Agent-Mediated Electronic Commerce V, Melbourne, Australia, pp. 70–77 (2003)
14. Fatima, S., Wooldridge, M., Jennings, N.R.: An agenda based framework for multi-issues negotiation. Artificial Intelligence 152(1), 1–45 (2004)

15. Gerding, E., van Bragt, D., La Poutré, H.: Multi-issue negotiation processes by evolutionary simulation, validation and social extensions. Computational Economics 22, 39–63 (2003)
16. Goldberg, D.: Genetic Algorithms in Search, Optimisation and Machine Learning. Addison-Wesley, Reading (1989)
17. Guttman, R., Moukas, A., Maes, P.: Agent-mediated electronic commerce: A survey. Knowledge Engineering Review 13(2), 147–159 (1998)
18. Harinck, F., De Dreu, C.: Negotiating interests or values and reaching integrative agreements: the importance of time pressure and temporary impasses. European Journal of Social Psychology 34(5), 595–611 (2004)
19. He, M., Jennings, N.R., Leung, H.: On agent-mediated electronic commerce. IEEE Trans. on Knowledge and Data Engineering 15(4), 985–1003 (2003)
20. Heiskanen, P.: Decentralized method for computing pareto solutions in multiparty negotiations. European Journal of Operational Research 117, 578–590 (1999)
21. Jaiswal, A., Kim, Y., Gini, M.: Design and implementation of a secure multi-agent marketplace. Electronic Commerce Research and Applications 3(4), 355–368 (2004)
22. Jennings, N., Sycara, K., Wooldridge, M.: A roadmap of agent research and development. Journal of Autonomous Agents and Multi-Agent Systems 1(1), 7–38 (1998)
23. Jennings, N.R., Faratin, P., Lomuscio, A.R., Parsons, S., Sierra, C., Wooldridge, M.: Automated negotiation: prospects, methods and challenges. Journal of Group Decision and Negotiation 10(2), 199–215 (2001)
24. Kim, J., Segev, A.: A web services-enabled marketplace architecture for negotiation process management. Decision Support Systems 40(1), 71–87 (2005)
25. Kraus, S.: Negotiation and cooperation in multi-agent environments. Artificial Intelligence 94(1–2), 79–97 (1997)
26. Kraus, S., Sycara, K., Evenchik, A.: Reaching agreements through argumentation: A logical model and implementation. Artificial Intelligence 104(1–2), 1–69 (1998)
27. Krause, P., Ambler, S., Elvang-Goransson, M., Fox, J.: A logic of argumentation for reasoning under uncertainty. Computational Intelligence 11, 113–131 (1995)
28. Krovi, R., Graesser, A., Pracht, W.: Agent behaviors in virtual negotiation environments. IEEE Transactions on Systems, Man, and Cybernetics 29(1), 15–25 (1999)
29. Lau, R.Y.K.: The State of the Art in Adaptive Information Agents. International Journal on Artificial Intelligence Tools 11(1), 19–61 (2002)
30. Lau, R.Y.K.: Towards Genetically Optimised Multi-Agent Multi-Issue Negotiations. In: Sprague, R. (ed.) Proceedings of the the 38th Hawaii International Conference on System Sciences (HICSS 2005), Big Island, Hawaii, January 3–6. IEEE Press, Los Alamitos (2005)
31. Lau, R.Y.K., Chan, S.Y.: Towards Belief Revision Logic-Based Adaptive and Persuasive Negotiation Agents. In: Zhang, C., W. Guesgen, H., Yeap, W.-K. (eds.) PRICAI 2004. LNCS (LNAI), vol. 3157, pp. 605–614. Springer, Heidelberg (2004)
32. Lau, R.Y.K., Tang, M., Wong, O.: Towards Genetically Optimised Responsive Negotiation Agents. In: Proceedings of the 4th IEEE/WIC International Conference on Intelligent Agent Technology, Beijing, China, September 20–24, pp. 295–301. IEEE Computer Society, Los Alamitos (2004)
33. Lau, R.Y.K., Tang, M., Wong, O., Milliner, S., Chen, Y.: An Evolutionary Learning Approach for Adaptive Negotiation Agents. International Journal of Intelligent Systems 21(1), 41–72 (2006)

34. Lomuscio, A.R., Jennings, N.R.: A classification scheme for negotiation in electronic commerce. Journal of Group Decision and Negotiation 12(1), 31–56 (2003)
35. Matos, N., Sierra, C., Jennings, N.R.: Determining successful negotiation strategies: an evolutionary approach. In: Demazeau, Y. (ed.) Proceedings of the 3rd International Conference on Multi-Agent Systems (ICMAS 1998), Paris, France, pp. 182–189. IEEE Press, Los Alamitos (1998)
36. Matwin, S., Szapiro, T., Haigh, K.: Genetic algorithm approach to a negotiation support system. IEEE Transactions on Systems Man and Cybernetics 21(1), 102–114 (1991)
37. Miechmann, T.: Learning and behavioral stability: An economic interpretation of genetic algorithms. Journal of Evolutionary Economics 9(2), 225–242 (1999)
38. Montano, B.R., Malaga, R.A.: A Weighted Sum Genetic Algorithm to Support Multiple-Party Multi-Objective Negotiations. IEEE Transactions on Evolutionary Computation 6(4), 366–377 (2002)
39. Narayanan, V., Jennings, N.: An adaptive bilateral negotiation model for E-commerce settings. In: Proceedings of the Seventh IEEE International Conference on E-Commerce Technology (CEC 2005), pp. 34–41 (2005)
40. Nash, J.F.: The bargaining problem. Econometrica 18, 155–162 (1950)
41. Oliver, J.R.: A machine learning approach to automated negotiation and prospects for electronic commerce. Journal of Management Information Systems 13(3), 82–112 (1996)
42. Oliver, J.R.: Artificial agents learn policies for multi-issue negotiation. International Journal of Electronic Commerce 1(4), 49–88 (1997)
43. Oliver, J.R.: On learning negotiation strategies by artificial adaptive agents in environments of incomplete information. In: Kimbrough, S.O., Wu, D.J. (eds.) Formal Modelling in Electronic Commerce, pp. 445–462. Springer, Berlin (2005)
44. Pruitt, D.: Negotiation Behaviour. Academic Press, London (1981)
45. Raiffa, H.: The Art and Science of Negotiation. Harvard University Press (1982)
46. Rosenschein, J., Zlotkin, G.: Rules of Encounter: Designing Conventions for Automated Negotiation among Computers. MIT Press, Cambridge (1994)
47. Rosenschein, J., Zlotkin, G.: Task oriented domains. In: Rules of Encounter: Designing Conventions for Automated Negotiation among Computers, pp. 29–52. MIT Press, Cambridge (1994)
48. Roth, A., Murnighan, J., Schoumaker, F.: The deadline effect in bargaining: Some experimental evidence. American Economic Review 78(4), 806–823 (1988)
49. Rubinstein, A.: Perfect equilibrium in a bargaining model. Econometrica 50(1), 97–109 (1982)
50. Sierra, C., Faratin, P., Jennings, N.R.: Deliberative automated negotiators using fuzzy similarities. In: Proceedings of the EUSFLAT-ESTYLF Joint Conference on Fuzzy Logic, Palma de Mallorca, Spain, pp. 155–158 (1999)
51. Sim, K.M., Choi, C.Y.: Agents that react to changing market situations. IEEE Transactions on Systems, Man and Cybernetics, Part B: Cybernetics 33(2), 188–201 (2003)
52. Singh, R., Salam, A.F., Iyer, L.: Agents in e-supply chains. Communications of the ACM 48(6), 108–115 (2005)
53. Sycara, K.: Multi-agent compromise via negotiation. In: Gasser, L., Huhns, M. (eds.) Distributed Artificial Intelligence II, pp. 119–139. Morgan Kaufmann, San Francisco (1989)

54. van der Putten, S., Robu, V., Poutré, H.L., Jorritsma, A., Gal, M.: Automating supply chain negotiations using autonomous agents: a case study in transportation logistics. In: Nakashima, H., Wellman, M.P., Weiss, G., Stone, P. (eds.) Proceedings of the 5th International Joint Conference on Autonomous Agents and Multiagent Systems (AAMAS 2006), Hakodate, Japan, May 8–12, pp. 1506–1513. ACM, New York (2006)
55. von Neumann, J., Morgenstern, O.: The Theory of Games and Economic Behaviour. Princeton University Press, Princeton (1994)
56. Wilkenfeld, J., Kraus, S., Holley, K., Harr, M.: Genie: A decision support system for crisis negotiations. Decision Support Systems 14(4), 369–391 (1995)
57. Wooldridge, M., Jennings, N.: Intelligent Agents: Theory and Practice. Knowledge Engineering Review 10(2), 115–152 (1995)
58. Ye, Y., Liu, J., Moukas, A.: Agents in electronic commerce. Electronic Commerce Research 1(1-2), 9–14 (2001)
59. Zeng, D., Sycara, K.: Bayesian learning in negotiation. International Journal of Human-Computer Studies 48(1), 125–141 (1998)
60. Zeuthen, F.: Problem of Monopoly and Economic Warfare. Routledge and Kegan-Pail, London (1967)

13

Security of Intelligent Agents in the Web-Based Applications

Chung-Ming Ou[1] and C.R. Ou[2]

[1] Department of Information Management, Kainan University, #1 Kainan Rd., 338 Luchu, Taiwan
cou077@mail.knu.edu.tw

[2] Department of Electrical Engineering, Huisping Institute of Technology, 412 Taichung, Taiwan
crou@mail.hit.edu.tw

Abstract. The major goal of this chapter is to discuss the following tow topics: first is the security issues related to web-based applications with intelligent agents; the second is the adaptation of intelligent agents to existing information security mechanisms. Mobile agents are considered to be an alternative to client-server systems. Security issues are discussed for generic agent-based systems, i.e. intelligent agents migrate to agent platforms. Public key infrastructure (PKI) is a major cryptographic systems deployed for agent-based systems. Cryptographic techniques such as digital signatures, hash function, proxy certificate and attribute certificate, are utilized for protecting both intelligent agents and agent platforms. Countermeasures to agent protections and agent platform protection are given, which are based on information security mechanisms such as authentication, authorization, access control and confidentiality. Other major security concern such as the identity binding and delegation between intelligent agent and its host are discussed with solutions based on proxy certificates and attribute certificates. For application layer security mechanism, non-repudiation and Secure Electronic Transaction (SET) are developed for agent-based applications.

13.1 Introduction

Intelligent agents are considered to be an alternative to client-server based web applications. A mobile agent of the host is a set of code and data which can execute the code with the data as parameter in agent platforms which are trusted processing environment (TPE). Development of the Internet invokes the need for web applications across heterogeneous networks. Although the client/server architecture has followed such trend, it has also confronted with problem of maintainability [1], in particular for heavy network loads. The situation gets worse with mobile terminals and portable devices. Here, Agent-based web applications make the full capabilities of the existing network infrastructures and information available in the Internet.

The ubiquitous network, which combines mobile communication systems (GSM, GPRS, UMTS), wireless network systems (WiFi, WiMAX) and broadband network (xDSL, Cable, FTTx), is becoming a next generation network (NGN) which is promoted with much effort by telecommunication companies nowadays. In order to

N.T. Nguyen, L.C. Jain (Eds.): Intel. Agents in the Evol. of Web & Appl., SCI 167, pp. 303–329.
springerlink.com

promote the ubiquitous applications utilizing varied mobile devices such as mobile phone, PDA, laptop PC, etc., it is feasible to consider agent-based architectures and protocols in NGN.

There are many definitions of intelligent agents, see [2] for a good reference. There are several issues related to security and trust while considering mobile agent-based web applications [3] [4] [5]. For such applications, we will point out a few security concerns while adopting intelligent agents, such as protecting agents from malicious hosts, and protection of agent platform from malicious agents. Other major security concern is the identity binding and delegation between intelligent agent and its host. Solutions from public key infrastructure (PKI), such as proxy certificate and attribute certificate, will be introduced. In this chapter, we define a reasonable intelligent agent suitable for web applications to be a mobile agent with cryptographically computing ability. In this way, intelligent agents may perform necessary security actions to protect themselves. On the contrary, agent platforms can protect themselves from malicious code's attacks. Moreover, for full-scale security consideration of agent-based applications, (PKI) is a major cryptographic systems deployed for agent-based systems. PKI is an infrastructure to protect Internet transactions as a whole; it is the dual public-key cryptosystems, which means there are two public-key and private-key pairs for each PKI entity. One key pair is for encryption/decryption; the other one is for digital signature generation/verification. Other issue from security viewpoint is the intelligence of mobile agents in web-based applications, i.e., how cryptographic techniques improve agent's intelligence to secure web-applications. Here cryptographic techniques include digital signatures, hash function, digital envelope, proxy certificate and attribute certificate, etc.

The goal of this chapter is to summarize researches in designing and developing security frameworks for intelligent agents from web application viewpoints. The following two topics are discussed: first is the security issues related to web-based applications with intelligent agents; the second is the adaptation of intelligent agents to existing information security mechanisms. Intelligent agents are the next significant software abstraction; they will soon be as ubiquitous as graphical user interfaces are today. With the emergence of agent technologies, intelligent agents with mobility have become an attractive paradigm to implement objects that can migrate from one computer to another over heterogeneous network. In general, the term intelligent agent refers to autonomous software with one or more objectives. The agent owner is called the host; the execution environment is called the agent platform. An agent is first launched from its host platform (or some broker) to the first migration platform; then it migrates to another platform, and so on, finally returns to its host platform. The set of visiting platform is called the agent's itinerary. Therefore, agent has to choose a platform according to this initial itinerary; it may ask for recommendation from the current platform, or it contacts to its broker.

Designing of security mechanisms has been a major challenge for agent-based system and technology. An agent system with no security mechanism is just like opening a door and invites anyone in; traditional client-server security architecture is a necessary but not sufficient for agent-system. Mobile agents themselves have even more threats to applications servers due to their mobility and intelligence. This threat is not strange to us: malicious codes have becoming security issue these days. The open nature of agents and execution environment make them vulnerable and subject to attacks over the network. Agent platforms are exposed to attacks from malicious agents, and intelligent agents are subject to attacks from malicious platforms; agent's behavior and critical information may be detected in this way. We classify these security problems into two categories: agent protection and agent platform protection.

In this chapter, we will also discuss adaptations of intelligent agents to existing Internet security protocols. As we know, there are lots of security protocols executed while running web applications such as SSL and S-HTTP, etc. For example, dispute of a transaction is a common problem that could jeopardize the mobile commerce [6]. The purpose of non-repudiation is to collect, maintain, make available and validate irrefutable evidence concerning a claimed event or action in order to resolve disputes on the occurrence or non-occurrence of the event or action [7] [8]. The agent-based system is becoming a promising architecture for both e-Commerce and mobile commerce; for example, dual signature-based SET protocol can be adapted in agent-based applications, namely SET/A, SET/A+, LITESET/A, etc. Another issue in this area is the fair exchange for web-based applications; trustiness in agent-based applications and platform is becoming a feasible solutions once considering suitable middleware such as TPE.

The arrangement of this chapter is as follows. In section 13.2, we introduce the architecture of agent-based web applications. In section 13.3, security issues of both agent protection and platform protection are discussed. In section 13.4, we propose an agent-based non-repudiation protocol and an agent-based secure payment protocol, namely Agent-based Secure Electronic Transaction with Non-repudiation (SETNR/A).

13.2 Background of Secure Intelligent Agents Systems

In this paper, we simply concentrate on intelligent agents suitable for web-based applications, namely, mobile agent systems with cryptographic modules. Intelligent agents from security perspective requires mobile agent have the capability to fulfill the security mechanism, in particular, cryptographic mechanism such as encryption/decryption. Signature generation/verification and hash function.

13.2.1 Related Works

There are different approaches to secure agent-based applications. Issues of access control for agent-based systems can be found in [41][42]. Secure infrastructure for multi-agent applications such as agent-oriented PKI are proposed in [15][19]. The authors introduce concepts of protecting mobile agents from malicious servers [5][31]. Designs of security protocols for agent-based commerce system are discussed in [43] [46]. Security mechanism for multi-agent systems can be found in [32]. The authors concentrate on security protocols of mobile agent-based applications [1][39]. Intelligent agent-based security frameworks are proposed in [29][33]. For specific multi-agent systems, the authors discuss JADE security and FIPA agent platform security in [34][36] respectively.

Many researches are adopting cryptographic techniques to secure agent-based system. Lin et al. [35] adopts the blind signature mechanism to secure mobile agent environment. Ou provides the role-based access control mechanism for both agent and platform protection [45]. Ou also discusses agent delegation mechanisms using proxy certificate [13]. Multi-signature scheme for secure mobile agent system can be seen in [44]. Other important topic of web-based applications is the adaptation of existing Internet protocols to agent-based systems. Ramao and Silva improve the SET protocol and propose the agent-based SET/A protocol guided by SET rules for better performance and efficiency in e-commerce [28]. Agent-based Certificate Authority can be found in [38]. The authors consider utilizing non-repudiation protocols to secure itineraries of mobile agents [25] [40]. Applications of mobile agents to secure payment system can be seen in [39].

According to Jansen [9], threats to the intelligent agents may fall into four classes: disclosure of information; denial of service; corruption of information; and interference or nuisance. For efficient discussion of security, it is sufficient to use a model of two components, namely, mobile agent and agent platform. The agent platform, which may be a secure component of web servers, or being a secure middleware, provides TPE for mobile agent executions. Quardani et al. [1] propose some security protocol to protect mobile agents from malicious platform based on agent cooperation.

Zhang et al. proposed a two-layer architecture with a security layer on top of FIPA-OS [10] [11]. The FIPA-compliant agent platform (FIPA-OS) is responsible for agent management and communication (using agent communication language, ACL), and the security layer provides four security-related services, authentication, authorization, secure communications and a secure execution environment. There are two types of authentication processes implemented in the security framework, namely, platform authentication and agent authentication. Based on this architecture, we propose a generic structure which is suitable for many on-self agent platforms simply by adapting the two-layer structure for Agent system security; it is a security layer with PKI-enable middleware architecture. For example, for mobile agent platform suitable for mobile transactions, this security layer will be added more necessary components. For more details, see section 13.4.

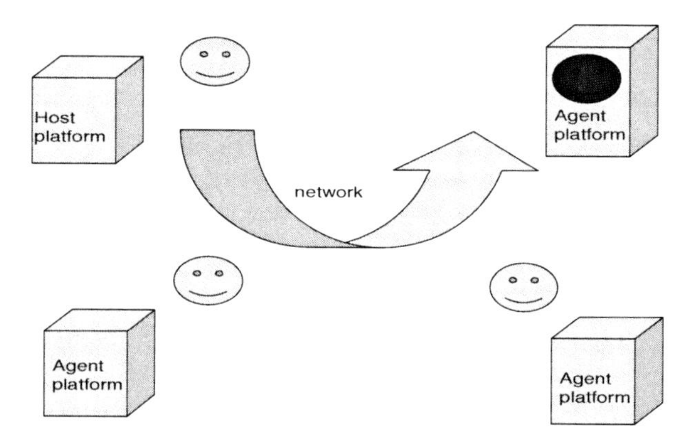

Fig. 13.1. Intelligent agent system model

13.2.2 Intelligent Agent System

Some literatures will adopt the terminology "software agent" to intelligent agents; we also see an agent to be a computer system. Autonomy is the central of this concept. For agent-based systems suitable for web applications, mobility is also crucial. A mobile agent system consists of two main components: mobile agents and their platform (Mobile agent platform). Therefore mobile agents are software agents which are goal-directed and can automatically suspend their execution on one platform and migrate to another platform, where they resume execution to accomplish their tasks within a network that can travel around.

Intelligent agents are also computational entities which physical exists in the form of the programs that run on computing devices. Generally speaking, an intelligent agent is a set of code of data generated by its host platform or some broker. The code is executable with reference to these carried data. An intelligent agent consists of the following components: agent owner, identifier, goal/result, life time and states. Intelligent indicates that the agent pursue their goals and execute their tasks such that they optimize some given performance measure. In particular, these agents can perform PKI operations for security measure, more details will come in the subsection 13.2.3.

There are several security threats related to intelligent agents such as repudiation, while an agent threatens another agent [12]. An agent may deny having exchanged message with other host. On the other hand, an agent may be modified by some malicious agents or hosts while transferring to some merchant server. The following steps are general guidelines for protecting these mobile agents by utilizing PKI.

1. The broker obtains the certified public key of the merchant server.
2. The broker encrypts this mobile agent by using merchant server's public key and sends it to the merchant server.

Each agent carries the item which is intended to be exchanged. For example, these items may include purchase orders and payment information (e.g. bank account number, credit card number, or micro payments account number, etc).

13.2.3 Countermeasure Based on Mobile Code Security

Mobile codes and mobile agents are similar in many aspects. Therefore, methodologies for securing mobile codes may be adapted to securing mobile agents. Jansen [9] has investigated these connections.

Countermeasures mean any action, device, procedure, technique or other measure that reduces the vulnerability or threat to a system. Many security mechanisms on client-server architectures can be used in agent-based applications. Several techniques have their special purposes for controlling mobile codes and executive content; for example, JAVA applets can be applicable to agent-based system [9].

Intelligent agent systems rely on some basic security assumptions. First is the trust of intelligent agent to its host platform. The second is that agent platforms are acting securely. The third is the well-establishment of PKI certificate management services, which is described as follows.

13.2.4 Public Key Infrastructure (PKI)

Agent platforms and agents are issued certificates by some certification authority (CA) within the PKI. An agent platform utilizes some secure module (hardware preferred) to store agent's host information such as its PKI components. These platforms are capable of verifying digital signatures to authenticate other platform, if necessary. We may deploy a middleware called the ***broker*** to help agents to authenticate agent platforms such that attackers cannot impersonate them. Broker can reduce lots of performance loading for agent's host.

PKI is the core cryptographic mechanism for digital signature-based protocol such as authentication and non-repudiation; it consists of two parts, one is the operation; the other is the entity. PKI entities must contain at least two public-private key pairs for encryption/decryption and signature generation/verification, respectively. These key pairs are generated by some CAs whose major task is to bind public key, private key and entity together. The public key will be stored in some certificate field; CA will issue (subscriber) certificates and server certificates to agent hosts and agent platforms, respectively. Agent hosts may issue proxy certificates to their mobile agents for transaction delegations [13]. The digital signature of a message is generated by using the private key of message owner and some hash function.

13.2.4.1 PKI Operations

Major PKI operation in security protocols is the digital signature generation and verification. Symmetric key and one-way hash function are also complementary. Let X be a certificate subscriber and Y the certificate issuer of X. Let K_X and K_X^{-1} be the public key and the corresponding private key of X, respectively. Let M be a message and $sS_X(M)$ the digital signature of M generated by the private key K_X^{-1}. We define signature and certificate verification as follows.

Signature verification of $sS_X(M)$:

$$sig\_ver(sS_X(M)) \text{ is successful if and only if } K_X(sS_X(M)) = H(M).$$

Verification of subscriber certificate $C\{X, K_X\}_{K_{Y^{-1}}}$:

$$Cert\_Ver(C\{X, K_x\}_{K_Y^{-1}}) \text{ is successful if and only if } sig\_ver(C\{X, K_X\}K_{Y^{-1}})$$

$$sig\_ver(S_X(M)) \quad \text{is successful, namely, } K_Y(C\{X, K_x\}_{K_Y^{-1}}) = H(C\{X, K_x\}_{K_Y^{-1}}).$$

13.2.4.2 PKI Entities

Certificate Authority *(CA)*

A CA issues (subscriber) certificates to subscribers and server certificates to servers, TTP, Home Revocation Authority (HoRA) and banks, etc. These entities can authenticate each other and transmit encrypted information. CA needs to provide certificate management service to ensure the validity of certificate; as mentioned in subsection 13.2.2, it is a basic assumption for agent-based web applications. These entities would continue PKI operations if and only if related certificates are valid.

Certificate Subject

We suggest that a certificate subject be token-based equipment for efficient signature generation and verification. For example, laptop, desktop PC, USIM-based 3G mobile handset, PDA, etc. Related public-key certificates are all issued by some CAs within this PKI. Private keys should be generated within secure tokens and contained in them afterwards.

13.2.5 Security Issues for Intelligent Agent Systems

13.2.5.1 Security Objectives

In general, a secure agent-based system, similar to a software system, should meet the following security objectives called the CI5AN, namely, confidentiality, integrity, authentication, authorization, accountability, assurance, availability and non-repudiation, also see Table 13.1. Reader may refer to [12] [14] for more details.

Table 13.1. CI5AN: Security objectives for agent-based systems

Security Objectives	Descriptions	Suggested Techniques
Confidentiality	Transmitted information is not revealed to unauthorized party.	encryption /decryption
Integrity	Data is not altered.	hash function, message authentication code (MAC)
Authentication (Au)	Verifying the identity of a user, program, process, or device.	hash, MAC, digital signature
Authorization (Ar)	Grant or deny access right to a user, program, process or device.	role-based access control (RBAC)
Accountability (Ac)	Traceability for each action to the unique entity	digital signature
Assurance (As)	Trustworthy mechanism for the objects of CI5AN	TPE
Availability (Av)	Service is provided without interruption according to its life cycle.	TPE
Non-repudiation	Transacting parties can not repudiate any involved actions.	digital signature

For intelligent agent systems composed by agents and platforms, there are three types of attacks: agent attacks platform, platform attacks agent and agent attacks agents. Since web applications can provide individual resource for each intelligent agent within its platform, we may consider the first two security issues. We also provide basic security measures for these two attacks.

13.2.5.2 Attacks to an Intelligent Agent

A mobile agent has to expose its own information to the platform it migrates for program execution. In particular, mobile agents are facing major security risks when it has to expose its code, data and state, which is called the strong migration. We categorize these security issues according to the degree of information expositions.

Code security

Intelligent agent's code has to be reachable by its migrating platform. This characteristic makes the attack by modifying mobile agent's codes possible. Therefore a malicious platform may interrupt the agent's program based on the modified codes. It is also possible that such malicious platform may explore complete agent's behavior.

A malicious platform may alter the agent's code temporarily to avoid detection of the modifying the code. For example, Java class files may be altered, deleted or replaced while transmitted; it is possible while information sharing among intelligent agents.

Data Security

Some agent data are sensitive such as private key, electronic cash, personal private information, etc. Leaking such data will cause great damage to agent hosts. In particular, if any malicious platform know the physical location of such sensitive information, this data may become a target of malicious behaviors.

Execution Security

A malicious platform can also take advantage of its ability to modify the agent's execution to execute agent's code in unintentional way. Another issue is the interruption of agent program execution, for example, by launching denial of service (DOS) attack to agent.

Table 13.2. Security categorization of attacks to intelligent agents

Security type	Malicious Actions	Security Objectives (CI5AN)
Code security	theft	C
Data Security	theft, eavesdropping, replay, impersonation,	C, I, Au
Execution Security	DOS, fraud	Av, N, Au, Ar

According to what we have discussed above, the security measure will be based on the capability of protecting a program from the processor which executes this program. Also the confidentiality of agent's execution is also crucial. Code hiding is another measure. In general, agent protection is a fairly new research topic which is relatively difficult from security perspective. Basically there are three directions of agent protection: agent program protection, confidentiality of agent's execution and agent code hiding. For more details see [14].

13.2.5.3 Attacks to an Intelligent Agent Platform

A malicious agent may exploit the security weakness of an intelligent agent platform and attack the platform. The possible attacks include masquerade, denial of service (DOS) and unauthorized access.

An intelligent agent platform is similar to a server in the traditional client-server architecture. Many security mechanisms for client-server environment could protect the mobile agents. According to [14], three measures can be considered, namely, safe environment, safety property of alien codes, and signed code & path history security.

Table 13.3. Security categorization of attacks to intelligent agent platform

Security type	Descriptions	Security Objectives (CI5AN)
Masquerade	A malicious mobile agent claims itself as another agent, also called faking.	Au
Denial of Service	Malicious agent consumes an excessive amount of the platform resources.	Av, Au, Ar
Unauthorized Access	Malicious agent accesses platform resource without proper authorization.	Ar

Table 13.4. Security measure for intelligent agent platform

Measure	Descriptions	Security Type	Major Techniques
Safe environment	Provided for secure execution of any alien program.	unauthorized access	Encryption
Safety property	Such property for alien code must be checked before executed in the platform.	unauthorized access, masquerade	Access control
signed code & path history security	Protect mobile agent platform from agent's tampering; code signing for authentication, of an object.	DOS, masquerade	PKI, Digital signature

13.3 Implementation of Secure Intelligent Agent Systems for Web Applications

In section 13.2, we introduce cryptography-based and conceptual security architecture for intelligent agent system. It will be better consider security problems from information security perspectives, since the viewpoints will be global and scalable for general web applications.

Each agent host delegates its intelligent agent to precede some web application. However, it raises a series problem of agent trust issue. Hu [15] defines the trust as a combination of authentication, authorization, and delegation verification problems involving digital certificates. Therefore, in order to mitigate the risk of security issues of agent-based systems, four security mechanisms are suggested, namely, trust delegation, mutual authentications, access control, and execution security.

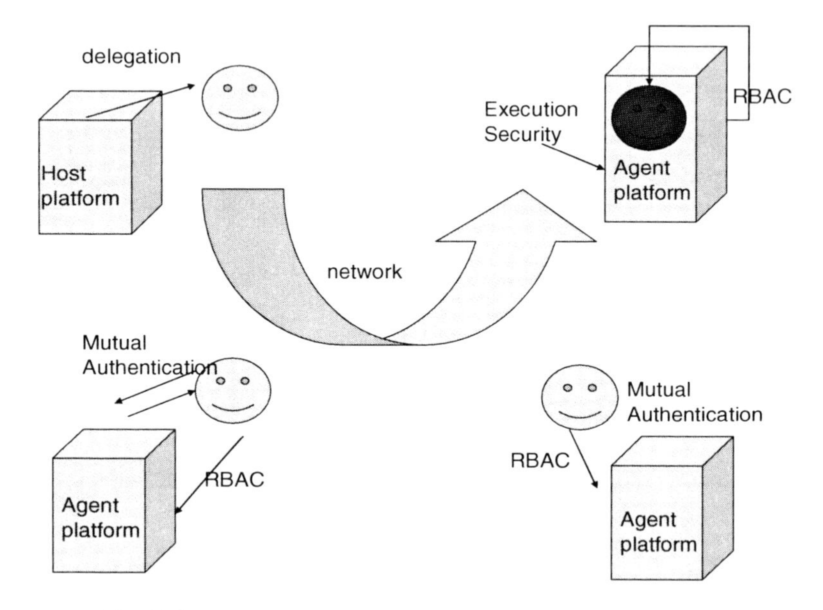

Fig. 13.2. Implementation of secure intelligent agent system

13.3.1 Agent Delegation Via Authorization

Agent delegation, which is an important concept for agents, needs to be addressed here. Utilizing the PKI, delegation can be achieved through several types of certificates, such as attribute certificates for access right delegation, and proxy certificate similar to traditional X.509 certificates for delegation of signing right. Some delegation logics with digital certificate have been discussed [15] [16]. Attribute certificate also has its major usage for access control; it will be discussed in subsection 13.3.2 and 13.3.3.

13.3.1.1 Proxy Certificates

The proxy certificate basically follows standard certificate format (such as ITU-X.509) with minor change. The major difference is the subject identifier (SID), which is the certificate field recorded the owner of this certificate; for a proxy certificate, its subject identifier is equal to the certificate issuer [17].

A proxy certificate of a mobile agent is issued and digitally signed by its owner. Beside standard certificate fields, this certificate contains a set of constraints which specifies valid operations that the agent is allowed to perform while using this certificate [17]. We use the notation $PC\{B, K_A, [D]\}_{K_B^{-1}}$ to represent the proxy certificate of a mobile agent A belonging to its owner B with additional data D. This proxy certificate is carried by this mobile agent along with its owner certificate such that a binding of mobile agent A with its owner B can be verified by applications through certificate validation process.

When migrating to a merchant's server, mobile agent will carry its proxy certificate. Verification of proxy certificate can be defined as follows.

$$cert\_ver(PC\{B, K_A, [D]\}_{K_B^{-1}}) \text{ is successful if and only if}$$

$$sig\_ver(PC\{B, K_A, [D]\}_{K_B^{-1}}) \text{ is successful, namely,}$$

$$K_B(PC\{B, K_A, [D]\}_{K_B^{-1}}) = H(PC\{B, K_A, [D]\}_{K_B^{-1}}) \cdot$$

For our agent-based web applications [13], an agent platform needs to verify mobile agents by checking its proxy certificate and its owner certificate according to standard certificate chain validation see Fig. 13.3.

13.3.2 Mutual Authentication Mechanism between Intelligent Agents and Agent Platforms

In order that the right intelligent agent will be executed at the right agent platform, both agent and agent platforms have to authenticate the identity of each other.

13.3.2.1 Authentication of Intelligent Agent by Migrated Platform

An agent platform authenticates the previous platform's identity provided by the migrating agent, and the identity of this incoming agent. For this purpose, an intelligent agent appends itself with the migrated platform identifier. The previous platform identifier should also be another helpful identifier; this so-called *identifiers pair* with the intelligent agent is sent to the migrated platform.

If this migrated platform cannot verify the digital signature appended by this intelligent agent correctly using the public key of the claimed host platform, it can conclude this intelligent agent does not really come from the claimed host platform. Next, the migrated platform identifies the agent itself. This authentication is done by checking the identifiers pair carried by this intelligent agent.

13.3.2.2 Authentication of Migrated Platform by Intelligent Agents

If the first authentication phase is passed successfully, the migrated platform has to prove its own identity to the agent. It sends this agent the digital signature of its identifier signed by its private key through internal communication channel. Intelligent agent verifies this signature using platform's public key. If the verification is valid, agent will execute its program through platform processor.

13.3.3 Access Control

The mobility of the agents brings the serious threats to the agent platforms; the latter could also pose some threats to mobile agents. These security concerns lead to the

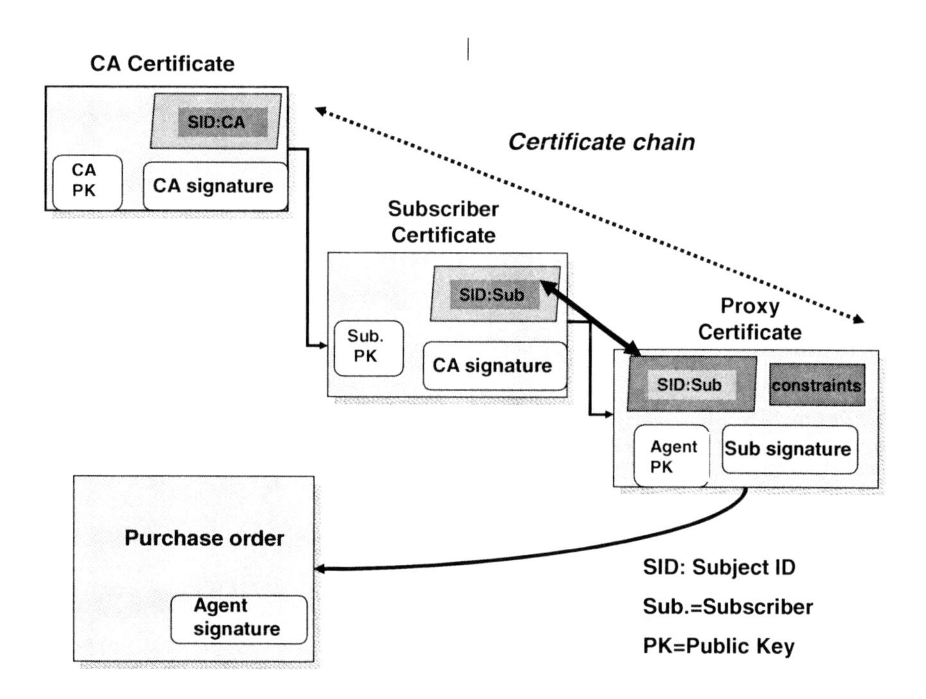

Fig. 13.3. Signature verification with proxy certificate

authorization issues of mobile agent migration to agent platforms. For example, SECMAP [18] is a mobile agent system which is especially focuses on security issues present in agent system. Hu and Tang [19] also suggests agent-oriented public-key infrastructure (APKI) for multi-agent e-services. Jansen [20] proposes a secure mobile agent platform based on attribute certificates. These structures are semantically similar but differ greatly in implementations. Compare to external data structure, agent systems using internal data structure to convey security policy cannot meet the dynamic policies change due to the constant threats from the Internet. On the other hand, each agent platform has its own internal data structure which hinders the interoperability of agent systems. The reason of adopting role-based access control using attribute certificates in [20] is to reach interoperability across varied agent platforms. It is very difficult to reach interoperability using internal data structure for security policy due to the variations in functionalities of trust relationships, policy expressions and the strength of policy protections, namely, assurance levels.

We propose a scheme that allocates, manages and enforces authorization policies in a flexible manner. This scheme, based on [20] and meaningful trusted levels by analyzing possible threats due to agent systems, adopts attribute certificate-based RBAC for both mobile agent and agent platforms.

13.3.3.1 Authorization Policies for Intelligent Agent Platform

The issue of security in the agent systems is related to unreliability. This means that mobile agents and agent platforms do not trust each other. We propose attribute certificate-based access control to mitigate the risks of trust problem.

Mobile Agent Systems with Authorization mechanism

A mobile agent system with authorization mechanism consists of three main components: mobile agent, agent platform and policy engine [20]. It cooperates with a PKI with at least two certification authorities (CAs); one is the Server CA (SCA) which issues server certificates; the other is the Agent CA (ACA) which issues agent certificates. The agent platform provides a computational environment in which agents operate. Policy engine verifies the platform policy by checking the policy certificate issued by SCA. On the other hand, mobile agents are issued attribute certificates by some ACA, see Fig. 13.4.

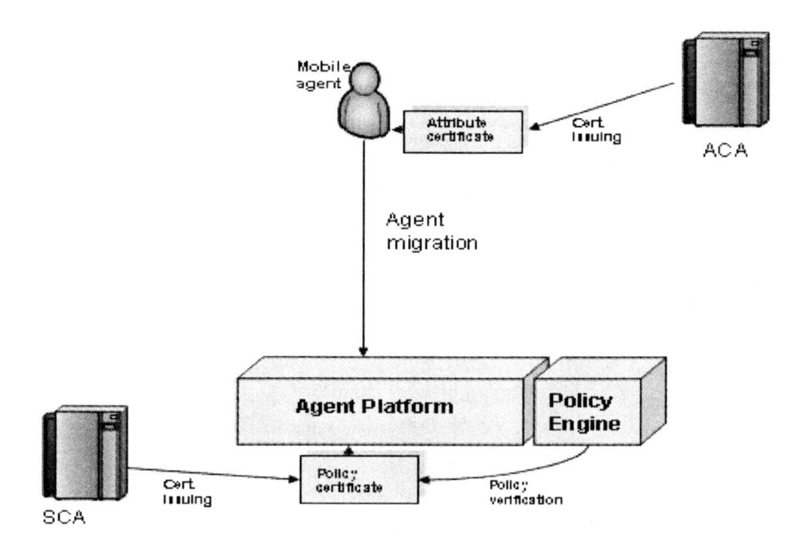

Fig. 13.4. Mobile agent system

13.3.3.2 Security Policy

A security policy of the agent system should consider all possible threats due to the use of computer systems and network connections. Both the mobile agents and agent platforms must perform only the activities that are permitted by verifying each action request against a set of policy rules. General guideline of security policies for agent systems is as follows.

1. An agent can be restricted to communicate with only certain agents on this agent platform.
2. An agent may be restricted to migrate to only certain platforms.
3. An agent platform may be restricted to not accepting any agent from certain other platforms.

Secure agent platforms should employ a policy-based authorization mechanism to permit or restrict agents to carry out certain classes of actions [20]. This policy can be conceptually described as follows.

1. Agent communication
2. Agent migration

3. Agent platform disk I/O
4. Access control to agent platform resources

All these policies can be relied on the attribute field of attribute certificates described in the following subsection.

13.3.3.3 Privilege Management Infrastructure (PMI) Based on Attribute Certificates

An attribute certificate is composed of a set of fields which is defined in [21], see Table 13.5. Issuers are in general CAs such as SCA or ACA in the agent system. The attribute field of an AC defines the RBAC policy for either an agent platform or a mobile agent. An attribute certificate (AC) must be digitally signed by the issuer to protect the certificate information from being altered. The platform authenticates the authorization status of every mobile agent. An AC holder and an AC verifier can be the Agent Platform-Agent or Agent-Agent Platform pair, respectively.

Table 13.5. Attribute certificate for agent systems

Certificate fields
Version
Holder:
Issuer:
Signature Algorithm
Validity Period
Attribute : role
Unique Identifier
Extension
CA signature

The attributes field gives information about the AC holder such as agent platforms and mobile agents. When an AC is used for authorization, this will often contain a set of privileges. Some of the attribute types make use of the *IetfAttrSyntax* type such as the **role** attribute [21]. It carries information about role allocations of the AC holders. The role can be classified as several trusted levels described in the next section.

13.3.3.4 RBAC and Information Flow

One major advantage for RBAC for intelligent agent systems is the capability of information flow control, namely, each intelligent agent can migrate only from agent platform with lower trusted level to the one with higher trusted-level. On the other hand, when two agents, say MA1 with lower trusted level and MA2 with higher trusted level, respectively, are aware of each other's existence in this agent platform and want to communicate with each other. According to their individual trusted levels, any information related to the contents of those agents can only flow from MA2 to MA1.

13.3.3.5 Access Control for Agent Platforms

The policy certificate, which is issued by some SCA, is installed to the access restricted area where only authorized entities can access to this certificate. According to

ISO 10181-3 [22], this certificate should be stored in some LDAP directory which is accessed according to AEF (access control enforcement function) and ADF (Access control Decision Function). AEF and ADF are two major functionalities of policy engine. This is somewhat different from the public-key certificate which is available to all legal PKI subscribers.

For illustration purpose, we consider 3 trusted-role levels for Agent platform authorizations, which are default, average and confidential from low to high levels, see Table 13.6. Policy engine will verify the role or the access privilege of Agent platform by checking the attribute field of this certificate.

Table 13.6. Trusted-role levels for agent platform

Level (from Low to high)
Default
Average
Confidential

13.3.3.7 Access privileges for Mobile Agents

The attribute certificates of mobile agents are issued by their corresponding ACA. The ACA generates this AC by checking the privilege of its agent host first. Basically, the trusted level of the mobile agent is related to its host behaviors. According to this privilege, the proper trusted level for this AC is defined by setting the role of the attribute field. For example, a host (such as a mobile phone subscriber) who has registered to some service provider for on-line ticketing service is requesting a mobile agent to purchase the cheapest airline ticket among all cooperated travel agencies. The ACA will issue a certain AC with some trusted level according to the request and the purchasing history of this host, for example, a customer-level AC for mobile agent. Also for illustration purpose, there are 4 trusted-role levels for AC-based authorizations for mobile agents, see Table.13.7.

Table 13.7. Trusted-role levels for mobile agents

Trusted Level (from low to high)
Inquirer
Customer
Agent
Super Agent

13.3.4 Execution Security

After an intelligent agent passes mutual authentication with agent platform and is authorized based on its trusted-role, it will be put into a queue of the platform and wait to be executed. Secure execution of an intelligent agent on an agent platform involved two aspects: first, executing agents should not affect functionalities of agent platform's and other programs running on the same device. Secondly, the intelligent agent should be executed correctly and securely by the platform. The former aspect is to protect a mobile agent platform from possible attacks of a malicious mobile agent. Problems in this category have been encountered in the traditional client-server environment and dealt by techniques used to protect servers in the traditional distributed system. While the latter concerns with protecting an intelligent agent from the malicious platforms. This aspect is seldom met before and remains a challenging area.

This stage of execution security can be categorized mainly as data security. Some of the agent data must be exposed to an agent platform for the normal execution; while some data which are not needed for this execution, data hiding can achieve such so-called data security.

Data hiding can be taken place in the so-called internal layer, which is not invisible outside. The knowledge base of an intelligent agent, which stores the data and other information, is declared as an internal state. The advantage of this operation is that agent platforms cannot access to agent's knowledge base. When an intelligent agent is executed, it exchanges data with this platform through some communication channels. This security can be achieved directly by computer program such as C++ and JAVA [14].

13.4 Adaptation of Intelligent Agents to Existing Internet Security Protocols

In section 13.2 and 13.3, we simply propose a secure architecture to solve general security problems of agent-based systems. However, as considering web applications, application layer security needs to be addressed. Among them, non-repudiation protocol and secure payment protocol are two crucial topics for web based e-commerce. In this section, we will focus on these two issues.

13.4.1 Agent-Based Non-repudiation Protocol

Non-repudiation services must ensure that when buyer B sends message to seller S over a network, neither B nor S can deny having participated in a part or the whole of this transaction. The basic idea is the following: an evidence of origin (EOO) is generated for buyer B and an evidence of receipt (EOR) is generated for seller S. In general, evidences are generated via PKI-based digital signatures. Disputes arise over the origin or the receipt of messages. For the case of origin dispute, B denies sending message while S claims having received it. As for the receipt dispute, S denies receiving any message while B claims having sent it. Buyers are at risks that sellers repudiate receiving purchase orders.

Many non-repudiation protocols have been proposed in a so-called "wrapper context", for example, the one proposed by Zhou and Gollman (we name it ZGP) [23]. In this situation, a party A wants to send a message M to B to enforce non-repudiation on M; namely, non-repudiation is enforced for one message (M). Lee and Yeh [24] proposed a delegation-based authentication protocol for mobile devices; it is achieved by utilizing proxy signature which basically delegates signing power to other end-entities. This concept is helpful for agent delegation.

ZGP is an efficient and fair non-repudiation protocol where TTP acts as a lightweight notary. It is also suitable for 3G communication by analyzing the capability of implementing cryptographic operations such as digital signature, symmetric key encryption/decryption, hash function and random number generations [25]. According to this investigation, we design a non-repudiation protocol adaptive to agent-based web applications.

13.4.1.1 Additional Servers for Non-repudiation Protocol

Broker

Intelligent agents are confronted with varied security threats both along their itineraries to agent platforms and within the agent platforms as mentioned in previous sections. The broker acts as a mediator between buyers and merchant servers in the Internet, see Fig. 13.5 (for mobile device as agent host). The broker must distinguish malicious servers from the honest ones according to Host Revocation List (HRL) to avoid sending agents to them. It is possible that honest server become malicious before HRL is updated. Esparza et al. [5] provides solutions to solve this mobile agent security. HoRA will issue an updated HRL to the broker if a merchant server is detected to be malicious.

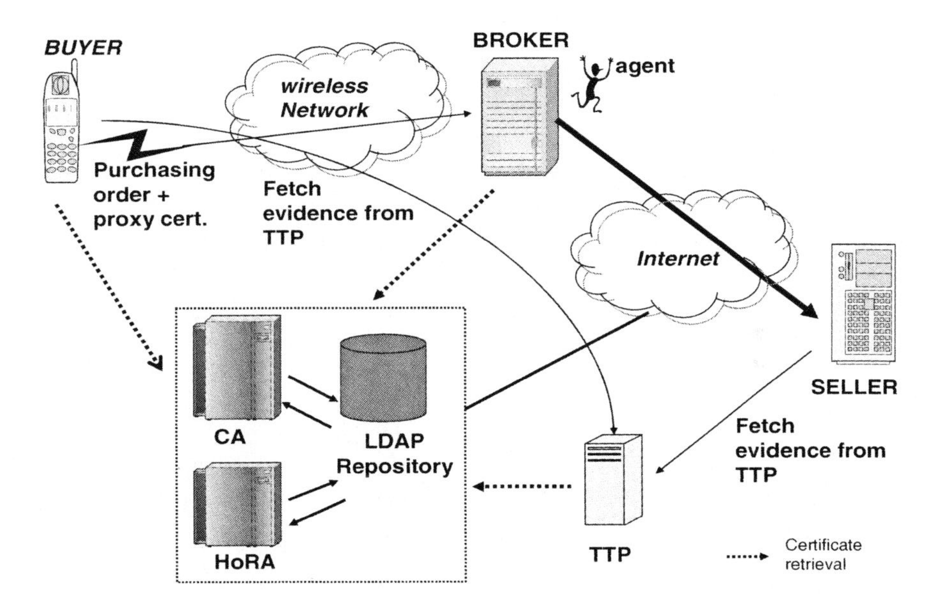

Fig. 13.5. Agent-based non-repudiation scheme: case of mobile transactions

Trusted Third Party (TTP)

The trusted third party here is a notary server which simply generates necessary evidences for buyers and sellers. TTP needs to perform PKI operations according to the non-repudiation protocol described in 13.4.1.2. Therefore TTP needs to access CA's repository to retrieve necessary certificates of buyers' (sellers') and verify digital signatures. TTP needs to store broker's public-key certificates and plays a role as the time stamp authority if necessary. For those generated evidences, TTP will store these information in its public directory from which buyers and sellers may fetch evidences.

TTP acts as a lightweight notary in this PKI-based non-repudiation protocol that only notarizes purchase orders by requests. TTP also provides directory services accessible to the public.

Host Revocation Authority (HoRA)

HoRA issues host certificates (HC) to merchant servers; these certificates bind mobile agent execution capability to the merchant host identity. When a merchant server acts maliciously, HoRA only needs to revoke this server's HC to prevent the broker from directing agents to it. The functionality of HoRA to detect the status of merchant servers can be referred to [26].

HoRA issues the HRL, which is a digital-signed list of revoked HCs. Before sending an agent of buyer to some merchant server, the broker must check the status of all servers on this agent's itinerary to see if any server is on the HRL. If the check is positive, broker will stop sending this agent to the merchant server.

13.4.1.2 Non-repudiation Protocol

A non-repudiation protocol is fair if it can ensure that at the end of a protocol execution, none or both of the two entities, the sender and the receiver, can retrieve all the evidences it expects. Fairness guarantees that neither sender nor receiver can gain advantage over the other.

Now we design a fair, agent-based non-repudiation protocol based on proxy certificates; this protocol also relies on the trust that broker will act according to the HRLs. We may assume (as one of the basic security assumptions described in 13.2.2) reasonably that Internet service providers provide brokers which are completely trusted by agent hosts. The purpose of this non-repudiation protocol is to transmit encrypted purchase order M and obtain non-repudiation evidences for buyer B and seller S. Purchase order M contains two parts, one is a commitment C, and the other is a key K. Notations are as follows.

M:	purchase order being sent from B to S.
K:	key generated by B.
A:	mobile agent generated by its owner B
$C=e_K(M)$:	commitment for purchase order with payment information M (e_K represents encryption by key K).
$sS_B(M)$:	signature of message M signed by B's private key.
L=H(M,K):	a label linking C and K (H represents a hash function).
f_i:	flag indicating the purpose of a signed message.
EOO_C:	evidence of origin of C, which is equal to $sS_B(f_{EOO}, S, L, C)$.
EOR_C:	evidence of receipt of C, which is equal to $sS_S(f_{EOR}, B, L, t_S, C)$.
*sub*_K:	authenticator of receipt of C, which is equal to $sS_B(f_{SUB}, S, L, t_B, K,$ EOO_C).
*con*_K:	evidence of confirmation of K issued by the TTP with time stamp T, which is equal to $sS_{TTP}(f_{CON}, B, S, L, T, t_B, t_S, K,$ EOO_C, EOR_C).

We include time information in this protocols; t_B is a time span defined by buyer B indicating that sub_K will be kept in TTP's private directory for t_B time units; t_S is a time span defined by seller S indicating that TTP will keep EOR_C in its private directory for t_S time units. T is the time stamp indicating the actual time TTP generate key confirmation con_K and make it public. This non-repudiation protocol is as follows.

Initiation phase

B generates a mobile agent A and its proxy certificate $PC\{B, K_A, [D]\}_{K_B^{-1}}$. B also generates EOO_C according to M, and f_{EOO}, S, L, C.

Evidence Exchange phase

Now B starts payment process with this specific seller S by delegating agent A.

1. $A \rightarrow_c S$: f_{EOO}, S, L, C, EOO_C, $PC\{B, K_A, [D]\}_{K_B^{-1}}$
2. $B \rightarrow TTP$: f_{SUB}, S, L, t_B, K, EOO_C, sub_K, $PC\{B, K_A, [D]\}_{K_B^{-1}}$
3. $S \rightarrow TTP$: f_{EOR}, B, L, t_S, EOO_C, EOR_C
4. $TTP \leftarrow B$: f_{CON}, B, S, L, T, t_B, t_S, K, EOR_C, con_K
5. $TTP \leftarrow S$: f_{CON}, B, S, L, T, t_B, t_S, K, EOR_C, con_K

"$A \rightarrow_c S$: M" means agent A carries information M to S; "$B \rightarrow S$: M" means B sends message M to S; "$TTP \leftarrow B$" means B fetches messages from TTP. The basic idea is that buyer B is able to send K, sub_K and t_B to TTP in exchange for con_K; on the other hand, seller S sends EOO_C, EOR_C and t_S to TTP in exchange for con_K. In step 1, S needs to verify EOO_C by retrieving B's (signature) public key from CA; EOO_C is saved as an evidence of origin for S. S also need to verify the proxy certificate of A. In step 2, after receiving sub_K, TTP keeps it in its private directory and delete it after t_B time units or until con_K is generated and published. In step 3, after receiving EOO_C, EOR_C and t_S from S, TTP needs to verify EOR_C using S's (signature) public key and compare EOO_C with the one sent by B in step 2. If either one is not true, TTP concludes that at least one party is cheating and it will not generate con_K. We call {f_{CON}, B, S, L, T, t_B, t_S, K, EOR_C, con_K } the evidence of this purchase order M. TTP also check if labels L from step 2 and 3 are coincident. If not, buyer B and seller S must be disagreed with this purchase order M. TTP will stop this protocol.

If steps 1-3 are shown positive results, TTP starts to generate con_K with time stamp T attached. In step 4, buyer B fetches K and con_K from TTP. In step 5, seller S fetches con_K from TTP to prove that K is available for S.

13.4.1.3 Security of Non-repudiation Protocol

The most important security issue of a non-repudiation protocol is the dispute resolution. We analyze the generated evidences of step 4-5 in the above non-repudiation protocol, dispute resolution mechanisms of buyer and seller to see whether non-repudiation can be reached. A trusted arbitrator will help solve the dispute according to submitted evidences.

The seller also checks if the constraint data D is bound by the buyer. Namely, it will first verify if

$$cert\_ver\{C\{B,K_B\}_{K_{CA}^{-1}}\} = H\{C\{B,K_B\}_{K_{CA}^{-1}}\};$$

Then this seller also verifies the proxy certificate of the mobile agent A by checking if

$$cert\_ver\{C\{B,K_A,[D]\}_{K_B^{-1}}\} = H\{C\{B,K_A,[D]\}_{K_B^{-1}}\}.$$

Validity of Evidence

Non-repudiation service will fail if bogus evidence is accepted or no evidence is received by either buyer or seller. Validity of non-repudiation evidence depends on the security of cryptographic keys used for generating evidences. These keys need to be revoked if they are compromised according to WPKI certificate policy practice.

According to PKI, buyer B, seller S and TTP could retrieve certificates of each other's from CA's repository to verify digital signatures. By the nature of hash functions, it is computationally hard to find two different key K and K' (with reasonable key length) with the same labels, namely L= H(M, K)= H(M, K')= L' and M= d_K(C)= $d_{K'}$(C)= M', where d_K represents decryption using key K. Therefore, TTP can investigate the validity of evidences by checking these labels.

Dispute of Origin

When buyer B denies having sent purchase order M to seller S, S may present EOO_C, EOR_C and con_K to some arbitrator in the following way:

S→ arbitrator: EOO_C, EOR_C, con_K, sS_S(EOO_C, EOR_C, con_K), L, K, M, C, $PC\{B,K_A,[D]\}_{K_B^{-1}}$

This arbitrator first verify the signature of S, sS_S(EOO_C, EOR_C, con_K); if the verification is positive, the arbitrator checks the following six steps:

step 1: if EOO_C is equal to sS_B(f_{EOO}, S, L, C).

step 2: if A is generated by B; namely, verification of proxy certificate, if $\text{sig\_ver}(PC\{B,K_A,[D]\}_{K_B^{-1}}) = H\{PC\{B,K_A,[D]\}_{K_B^{-1}}\}$

step 3: if EOR_C is equal to sS_S(f_{EOR}, B, L, t_S, C).

step 4: if con_K is equal to sS_{TTP}(f_{CON}, B, S, L, T, t_B, t_S, K, EOO_C, EOR_C).

step 5: if L is equal to H(M,K).

step 6: if M is equal to d_K(C).

If step 1 is checked positive, this arbitrator concludes that buyer B has sent seller S the encrypted purchase order C. If step 2 is checked positive, this arbitrator ensures that mobile agent A is generated by buyer B. If step 3 is checked positive, arbitrator concludes that S has sent all the correct payment information to TTP. For all 5 steps being checked positive, this arbitrator finally concludes that B has sent S the purchase order M, which is encrypted by K and presented to be C.

Dispute of Receipt

When seller S denies receiving the purchase order M from buyer B, buyer may present EOO_C, EOR_C, con_K to the arbitrator in the following way:

B→ arbitrator : EOO_C, EOR_C, con_K, sS_B(EOO_C, EOR_C, con_K), L, K, M, C, $PC\{B, K_A, [D]\}_{K_B^{-1}}$

The arbitrator first verifies the signature of buyer B, sS_B(EOO_C, EOR_C, con_K); if the verification is positive, the arbitrator checks all six steps same as those in the dispute of origin. For all six steps being checked positive, arbitrator concludes that seller S has received M, which is encrypted by K and presented to be C.

Private Key Security

Intelligent agent A must carry its own private key K_A^{-1} in order to sign purchase order. This private key, authorized by buyers to sign purchase order, is vulnerable to attacks and it is not under the direct control of buyer. Proxy certificates do not completely solve this private key compromise problem; however, Romao and Silva [17] suggested two methods to mitigate the risks of this private key.

1. Short certificate valid period for proxy certificate: This will reduce the risk that an attacker comprises the private key of mobile agent before the expiration of proxy certificate. Also the reverse engineering by attackers may also take too much time.
2. The constraints mechanism for mobile agent to allow limited number of tasks. This reduces the potential risk of private key discovered by attackers. That is, attackers can only sign specific documents within a very limited time.

13.4.2 Agent-Based Secure Electronic Transaction Protocol

The security of credit card-based payment system has been a concerned issue for a long time. In order to protect user's credit card information while transacting with payment systems, VISA and MasterCard, in association with major software and cryptography companies, developed SET (Secure Electronic Transaction) protocol [27]. One major advantage of SET is the separation of information disclosure, namely, ordering information and payment information. Merchant never knows credit card information; and financial institutes, which authorizes payment transaction, never knows ordering information. However, for mobile payment systems, SET may be too demanding for limited computational capacity, slower connection speeds such as mobile handsets. This investigation leads to the research of agent-based applications. Ramao and Silva [28] improved the SET protocol and proposed the agent-based SET/A protocol guided by SET rules for better performance and efficiency in e-commerce.

We improve SET/A by considering Broker as some trust verification center and adopting non-repudiation mechanism. This new protocol is called SETNR/A.

13.4.2.1 Agent-Based SET Protocol with Non-repudiation Mechanism (SETNR/A)

SET/A [28] is an implementation of the SET protocol on top of the mobile agent semantic. Yi et al. [29] suggested mobile agents sent to trusted verification center (for example, Broker) for signature verification to reduce the computational loading for agents.

SET/A can be improved for non-repudiation mechanism. Adding some trusted third party for dispute resolution in the improved protocol is one major consideration. The basic idea here is to embedding existing on-line TTP of ZGP for evidence generation before merchant sends payment gateway (PG) payment information (more specifically, authentication request).

We now design the SETNR/A protocol for credit card payment system with non-repudiation mechanism. The notations are as follows.

PRequest:	purchase request issued by C
PI:	payment information
OI:	order Information
PO:	purchase order being sent from C to M, which is equal to PI‖OI
DS:	dual signature for PO signed by C's signature private key
K:	symmetric key generated by C.
Cs(M):	signature certificate of M's
$C_k(PG)$:	key-exchange certificate of PG
$Env_{PG}(PI, K)$:	digital envelope of PI and K
CPO:	commitment for purchase order PO which is equal to $e_K(PO)$ (e_K represents encryption by key K).
$sS_C(PO)$:	signature of PO signed by C's private key.
L=H(PO,K):	a label linking PO and K (H represents a hash function).
f_i:	flag indicating the purpose of a signed message.
$e_{TTP}(.)$,$e_{BR}(.)$, $E_{PG}(.)$:	encryption by TTP's, Broker's and PG's public keys respectively
EOO_CPO:	evidence of origin of CPO, which is equal to $sS_C(f_{EOO}$, PO, L, CPO).
EOR_CPO:	evidence of receipt of CPO, which is equal to $sS_M(f_{EOR}$, C, L, t_M, CPO).
*sub*_K:	authenticator of receipt of CPO, which is equal to $sS_C(f_{SUB}$, M, L, t_C, K, EOO_CPO).
*con*_K:	evidence of confirmation of K issued by the TTP with time stamp T, which is equal to $sS_{TTP}(f_{CON}$, C, M, L, T, t_C, t_M, K, EOO_CPO, EOR_CPO).

t_C is a time span defined by agent A(C) indicating that sub_K will be kept in TTP's private directory for t_C time units; t_M is a time span defined by merchant M indicating that TTP will keep EOR_CPO in its private directory for t_M time units. T is the time stamp indicating the actual time TTP generate key confirmation con_K and make it public. SETNR/A is as follows.

1.	$C \rightarrow Broker \rightarrow_{A(C)} M$:	PRequest, e_{BR}(PO‖DS‖tc)
2.	$M \rightarrow A(C)$:	Cs(M), $C_k(PG)$
3.	$A(C) \rightarrow Broker$:	Cs(M), $C_k(PG)$
4.	$Broker \rightarrow A(C)$:	verification response
5.	$A(C) \rightarrow M$:	Cs(C), OI, $Env_{PG}(K, PI)$

6.	M→PG:	E_{PG}(K, PI)
	M→A(C):	Prequest_correct
7.	A(C) →M:	f_{EOO}, C, L, PO, EOO_CPO, t_C
8.	A(C) →TTP:	f_{SUB}, C, L, t_C, K, EOO_CPO, sub_K
9.	M→TTP:	f_{EOR}, C, L, t_M, EOO_CPO, EOR_CPO
10.	PG→ M, TTP :	authorization
11.	M→A(C):	response, Cs(M)
12.	TTP ← A(C), M:	f_{CON}, C, M, L, T, t_C, t_M, K, EOR_CPO, con_K

"C →Broker$\rightarrow_{A(C)}$ M: message" means C sends message to broker, then broker will generate an agent A(C) for C; message will be carried by this agent to M; "TTP ←M" means M fetches messages from TTP. The basic idea is that C is able to send K, sub_K to TTP in exchange for con_K; on the other hand, M sends EOO_C, EOR_C and t_S to TTP. We describe each step in details as follows.

- In step 1, cardholder generates a request (PRequest, the same elements as in the original SET) which is sent to Broker using Broker's public key; Broker decrypts PRequest, generates A(C) which is delegated to TPE of M with carried message.
- In step 2, M then sends certificates Cs(M), C_k(PG) to A(C) in TPE.
- In step 3, A(C) returns to Broker's TPE and Broker verifies these two certificates.
- In step 4, Broker returns verification results.
- In step 5, A(C) generates digital envelope of PI and K by using PG's (key-exchange) public key, then sends it along with OI and Cs(C) to M.
- In step 6, M verifies the certificate and the dual signature on OI; if it is correct, then M forwards this digital envelope and request verification to PG and A(C) respectively.
- In step 7, A(C) sends M the message "f_{EOO}, C, L, PO, EOO_CPO, t_C" within the TPE of M.
- In step 8, M needs to verify EOO_CPO by retrieving C's (signature) public key from the corresponding CA's repository. EOO_C is saved as an evidence of origin for M. On the other hand, A(C) sends TTP the message "f_{SUB}, C, L, t_C, K, EOO_CPO, sub_K " within TPE of the TTP's.
- In step 9, TTP keeps sub_K in its private directory after the verification of digital signature of C; TTP will delete it after t_C time units or until con_K is generated and published. M sends "f_{EOR}, C, L, t_M, EOO_CPO, EOR_CPO" to TTP. After receiving EOO_CPO, EOR_CPO and t_S from M, TTP needs to verify EOR_CPO using M's (signature) public key and compare EOO_CPO with the one sent by A(C) in step 7. If either one is not true, TTP concludes that at least one party is cheating and it will not generate con_K. We call {f_{CON}, C, M, L, T, t_C, t_M, K, EOR_C, con_K} the evidence of this purchase order PO.
- In step 10, TTP requests authorization from PG; then PG sends authorization to M and TTP
- In step 11, M sends authorization response to A(C) within its TPE, A(C) then verifies the certificate Cs(M) and the corresponding signature.

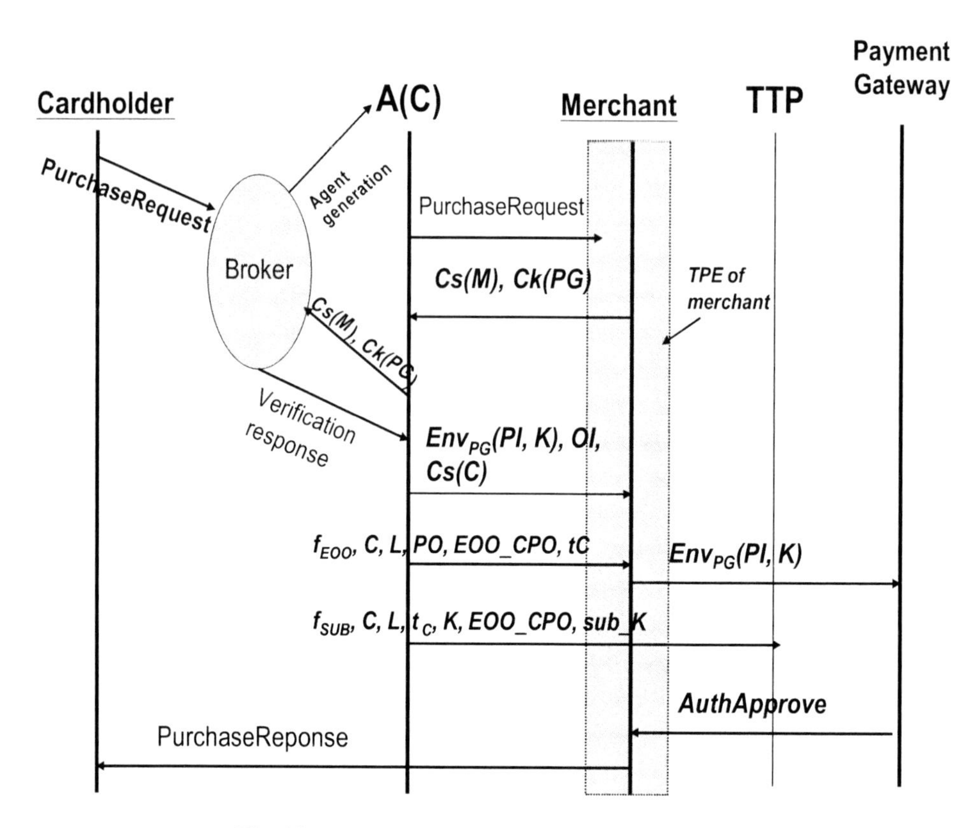

Fig. 13.6. Purchase request transaction in SETNR/A

- In step 12, A(C) can return to C's mobile device from TTP with carried evidence. M can also start evidence collection from TTP. C obtains K and con_K; M fetches con_K from TTP to prove that K is available for M.

13.5 Conclusion

We introduce cryptography-based and conceptual security architecture for intelligent agent system. Security problems from information security perspectives are considered such as security objectives and countermeasures. Before migrating agent to platform, each agent host delegates its intelligent agent to precede some web application with an agent trust mechanism. Agent trust is a combination of authentication, authorization, and delegation verification problems involving digital certificates. Therefore, four security mechanisms are suggested to mitigate the risk of security issues: trust delegation, mutual authentications, access control, and execution security. We also propose application layer security protocols such as a fair agent-based non-repudiation protocol based on PKI; an improved agent-based credit card payment system, namely, SETNR/A, is also discussed by "embedding" a non-repudiation mechanism to SET.

References

1. Ouardani, A., Pierre, S., Boucheneb, H.: A Security Protocol for Mobile Agents based Upon the Cooperation of Sedentary Agents. J. Network and Computer Applications 30, 1228–1243 (2007)
2. Weiss, G. (ed.): Multiagent Systems, A modern approach to distributed artificial intelligence. MIT Press, Cambridge (1999)
3. Pagnia, H., Vogt, H., Gartner, F., Wilhelm, U.: Solving Fair Exchange with Mobile Agents. In: Kotz, D., Mattern, F. (eds.) MA 2000, ASA/MA 2000, and ASA 2000. LNCS, vol. 1882, pp. 57–72. Springer, Heidelberg (2000)
4. Wilhelm, U., Staamann, S., Buttyan, L.: On the Problem of Trust in Mobile Agent Systems. In: Symposium on Network and Distributed System Security, Internet Society, pp. 114–124 (1998)
5. Esparza, O., Munoz, J., Soriano, M., Forne, J.: Host Revocation Authority: A Way of Protecting Mobile Agents from Malicious Hosts. In: Cueva Lovelle, J.M., Rodríguez, B.M.G., Gayo, J.E.L., del Ruiz, M.P.P., Aguilar, L.J. (eds.) ICWE 2003. LNCS, vol. 2722, pp. 289–292. Springer, Heidelberg (2003)
6. Zhou, J., Deng, R., Bao, F.: Evolution of Fair Non-repudiation with TTP. In: Pieprzyk, J.P., Safavi-Naini, R., Seberry, J. (eds.) ACISP 1999. LNCS, vol. 1587, pp. 258–269. Springer, Heidelberg (1999)
7. ITU-T, Recommendation, X.813: Information Technology-Open Systems Interconnection-Security Frameworks in Open Systems, Non-repudiation Framework (1996)
8. Li, B., Luo, J.: On Timeliness of a Fair Non-repudiation Protocol. In: InfoSecu 2004, pp. 99–106 (2004)
9. Jansen, W.: Countermeasures for Mobile Agent Security. Computer Communications 23, 1667–1676 (2000)
10. Zhang, M., Karmouch, A.: Adding Security Features to FIPA Agent Platform (2001), `http://www2.elec.qmul.ac.uk/~stefan/fipa-security/rfi-response/Karmouth-FIPA-Security-Journal.pdf`
11. Deng, X.: A Comparison of the Security Frameworks in Agent-Based Semantic Web. 20th Computer Science Seminar, SC2-T4-1
12. Jansen, W., Karygiannis, T.: NIST Special Publication 800-19: Mobile Agent Security (1999)
13. Ou, C.-M., Ou, C.R.: Adaptation of Proxy Certificates to Non-Repudiation Protocol of Agent-based Mobile Payment Systems. Applied Intelligence (Accepted and online first) DOI: 10.1007/s10489-007-0089-4
14. Ma, L., Tsai, J.: Security Modeling and Analysis of Mobile Agent Systems. Imperial College Press (2006)
15. Hu, Y.-J.: Trusted Agent-mediated E-commerce Transaction Services via Digital Certificate Management. Electronic Commerce Research 3, 221–243 (2003)
16. Lampson, B., Abadi, M., Burrows, M., Wobber, E.: Authentication in Distributed Systems; Theory and Practice. ACM Trans. Computer Systems 10(4), 265–310 (1992)
17. Romao, A., da Silva, M.: Secure Mobile Agent Digital Signatures with Proxy Certificates. In: Liu, J., Ye, Y. (eds.) E-Commerce Agents. LNCS (LNAI), vol. 2033, pp. 206–220. Springer, Heidelberg (2001)
18. Ugurlu, S., Erdogan, N.: Comparing Object Encodings. In: Ito, T., Abadi, M. (eds.) TACS 1997. LNCS, vol. 1281, pp. 415–438. Springer, Heidelberg (1997)

19. Hu, Y.-J., Tang, C.-W.: Agent-oriented Public Key Infrastructure for Multi-agent E-service. In: Palade, V., Howlett, R.J., Jain, L. (eds.) KES 2003. LNCS, vol. 2773, pp. 1215–1221. Springer, Heidelberg (2003)
20. Jansen, W.A.: A Privilege Management Scheme for Mobile Agent Systems. Electronic Notes in Theoretical Computer Science 63, 91–107 (2002)
21. IETF RFC 3281: An Internet Attribute Certificate Profile for Authorization
22. Information Technology-Open Systems Interconnection-Security Frameworks for Open Systems: Access Control Framework, ISO/IEC 10181-3:1996
23. Zhou, J., Gollmann, D.: A Fair Non-repudiation Protocol. In: Proc 1996 IEEE Symposium on Security and Privacy, pp. 55–61 (1996)
24. Lee, W.-B., Yeh, C.-K.: A New Delegation-based Authentication Protocol for Use in Portable Communication Systems. IEEE Trans. Wireless Communications 4(1), 57–64 (2005)
25. Ou, C.-M., Ou, C.R.: Non-Repudiation Mechanism of Agent-based Mobile Payment Systems: Perspectives on Wireless PKI. In: Nguyen, N.T., Grzech, A., Howlett, R.J., Jain, L.C. (eds.) KES-AMSTA 2007. LNCS (LNAI), vol. 4496, pp. 298–307. Springer, Heidelberg (2007)
26. Esparza, O., Munoz, J., Soriano, M., Forne, J.: Secure Brokerage Mechanisms for Mobile Electronic Commerce. Computer Communications 29, 2308–2321 (2006)
27. Stallings, W.: Cryptography and Network Security, 3rd edn. Prentice Hall, Englewood Cliffs (2003)
28. Romao, A., Silva, M.: An Agent-based Secure Internet Payment System for Mobile Computing. In: Lamersdorf, W., Merz, M. (eds.) TREC 1998. LNCS, vol. 1402, pp. 80–93. Springer, Heidelberg (1998)
29. Yi, X., Siew, C., Wang, X., Okamoto, E.: A Secure Agent-based Framework for Internet Trading in Mobile Computing Environment. Distributed and Parallel Database 8, 85–117 (2000)
30. Raghunathan, S., Mikler, A., Cozzolino, C.: Secure Agent Computation: X.509 Proxy Certificates in a Multi-lingual Agent Framework. J. Systems and Software 75, 125–137 (2005)
31. Benachenhou, L., Pierre, S.: Protection of a Mobile Agent with a Reference Clone. Computer Communications 29, 268–278 (2006)
32. Hutter, D., Mantel, H., Schaefer, I., Schairer, A.: Security of Multi-agent Systems: A Case Study on Comparison Shopping. J. Applied Logic 5(2), 303–332 (2007)
33. Kuo, M.H.: An Intelligent Agent-based Collaborative Information Security Framework. Expert Systems with Applications 32, 585–598 (2007)
34. Vila, X., Schuster, A., Riera, A.: Security for a Multi-agent System based on JADE. Computer & Security 26, 391–400 (2007)
35. Lin, M.-H., Chang, C.-C., Chen, Y.-R.: A Fair and Secure Mobile Agent Environment based on Blind Signature and Proxy Host. Computer & Security 23, 199–212 (2004)
36. Zhang, M., Karmouth, A., Impey, R.: Towards a Secure Agent Platform Based on FIPA. In: Pierre, S., Glitho, R.H. (eds.) MATA 2001. LNCS, vol. 2164, pp. 277–290. Springer, Heidelberg (2001)
37. Suna, A., Fallah-Seghrouchni, A.: A Mobile Agents Platform: Architecture, Mobility and Security Elements. In: Christianson, B., Crispo, B., Malcolm, J.A., Roe, M. (eds.) Security Protocols 2003. LNCS, vol. 3364, pp. 126–146. Springer, Heidelberg (2005)
38. Bsufka, K., Holst, S., Schnidt, T.: Ralization of an Agent-based Certificate Authority and Key Distribution Center. In: Albayrak, Ş. (ed.) IATA 1999. LNCS (LNAI), vol. 1699, pp. 113–123. Springer, Heidelberg (1999)

39. Das, A., Yao, G.: A Secure Payment Protocol Using Mobile Agents in an Untrusted Host Environment. In: Kou, W., Yesha, Y., Tan, C.J.K. (eds.) ISEC 2001. LNCS, vol. 2040, pp. 33–41. Springer, Heidelberg (2001)
40. Borrel, J., Robles, S., Serra, J., Riera, A.: Securing the Itinerary of Mobile Agents through a Non-Repudiation Protocol. In: Proceedings IEEE 33rd Annual, International Carnahan Conference, pp. 461–464 (1999)
41. Roth, V., Jalali-Sohi, M.: Access Control and Key Management for Mobile Agents. Comput. & Graphics 22(4), 457–461 (1998)
42. Chang, C.-C., Lin, I.-C.: A New Solution for Assigning Cryptographic Keys to Control Access in Mobile Agent Environments. Wireless Communications and Mobile Computing 6, 137–146 (2006)
43. Ou, C.-M., Ou, C.R.: SETNR/A: An Agent-based Secure Payment Protocol for Mobile Commerce. In: Nguyen, N.T., Jo, G.S., Howlett, R.J., Jain, L.C. (eds.) KES-AMSTA 2008. LNCS (LNAI), vol. 4953, pp. 527–536. Springer, Heidelberg (2008)
44. Seo, S.-H., Lee, S.-H.: A Secure Mobile Agent System Using Multi-signature Scheme in Electronic Commerce. In: Chung, C.-W., Kim, C.-k., Kim, W., Ling, T.-W., Song, K.-H. (eds.) HSI 2003. LNCS, vol. 2713, pp. 527–536. Springer, Heidelberg (2003)
45. Ou, C.-M., Ou, C.R.: Role-based Access Control (RBAC) Mechanism with Attribute Certificates for Mobile Agent. In: Proceedings of 10th Joint Conference of Information Sciences (JCIS 2007) (2007)
46. Liew, C.-C., Ng, W.-K., Lim, E.-P., Tan, B.-S., Ong, K.-L.: Non-Repudiation in an Agent-Based Electronic Commerce System. In: DEXA Workshop, pp. 864–868 (1999)

Author Index